Nematode Diseases of Fruit Crops and their Management

Nematode Diseases of Fruit Crops and their Management

Dr. P. PARVATHA REDDY

Former Director
Indian Institute of Horticultural Research
Bangalore

CRC Press
Taylor & Francis Group
Boca Raton London New York

CRC Press is an imprint of the
Taylor & Francis Group, an **informa** business

-EPH-

First published 2025
by CRC Press
4 Park Square, Milton Park, Abingdon, Oxon, OX14 4RN

and by CRC Press
2385 NW Executive Center Drive, Suite 320, Boca Raton FL 33431

CRC Press is an imprint of Informa UK Limited

British Library Cataloguing-in-Publication Data
A catalogue record for this book is available from the British Library

Print edition not for sale in South Asia (India, Sri Lanka, Nepal, Bangladesh, Pakistan or Bhutan).

ISBN: 9781041032670 (hbk)
ISBN: 9781041032687 (pbk)
ISBN: 9781003623106 (ebk)

DOI: 10.1201/9781003623106

Typeset in A Caslon Pro
by Elite Publishing House, Delhi

-EPH-

Contents

Preface

The destructive plant-parasitic nematodes are one of the major limiting factors in the production of crop plants throughout the world. For centuries, man has been plagued by these microscopic organisms feeding on the roots of crop plants essential to his survival and wellbeing. Roots damaged by the nematodes are not efficient in the utilization of available moisture and nutrients in the soil resulting in reduced functional metabolism. Visible symptoms of nematode attack often include reduced growth of individual plants, varying degrees of chlorosis, wilting of the foliage and sometimes death of plants. Furthermore, roots weakened and damaged by nematodes are easy prey to many types of fungi and bacteria which invade the roots and accelerate root decay. These deleterious effects on plant growth result in reduced yields and poor quality of crops. Estimated overall average annual yield loss of the world's major crops due to damage by plant parasitic nematodes is 12.3%. In India, plant parasitic nematodes on fruit crops cause 25.8% crop losses amounting to ₹ 31370.08 million annually. Citrus (₹ 9828.22 million) and banana (₹ 9710.46 million) among fruit crops suffered comparatively more losses. Nematode management is therefore, important for high yields and quality that are required by the high cost of modern crop production.

The information on various aspects of nematode diseases of fruit crops and their management is very much scattered and there is no book at present which comprehensively and exclusively deals with the above aspects. The present book on "Nematode Diseases of Fruit Crops and their Management" gives the detailed description of the causal organism, distribution, nature of damage and symptoms, crop losses, host range, biology and life cycle, interaction with other organisms, spread and survival, and management of nematode diseases of tropical, sub-tropical, temperate, and semiarid fruit crops.

The book is divided into six sections. The First Section describes the fruit crop scenario (both international and Indian); area, production and productivity; bridging yield gap; constraints; productivity enhancing technologies; food and nutritional security; and future thrusts and conclusions. This section also deals with different aspects of nematode diseases in fruit crops such as economic importance, emerging nematode problems, interaction with other pathogens (fungi, bacteria and viruses), and nematode management methods (regulatory, physical, cultural, chemical, biological, host resistance and integrated methods).

The nematode diseases of tropical fruit crops such as banana, citrus, papaya, pineapple, and mulberry and their management are discussed in detail in Section Two. Section Three describes nematode diseases of sub-tropical fruit crops such as guava, grapevine, passionfruit, litchi, avocado, and mango and their management. The nematode diseases of temperate fruit crops such as peach, strawberry, raspberry, cherry, walnut, kiwifruit, apple, and plum and their management are discussed in Section Four. Section Five deals in detail with nematode diseases of semi-arid fruit crops such as pomegranate and fig and their management.

Section Six deals with vision for future including current nematode management strategies, new paradigms for nematode management, transfer of technology, future approaches, and conclusions.

This book will be of immense value to scientific community involved in teaching, research and extension activities related to fruit crop nematodes and their management. The material can be used for teaching post-graduate courses. The book can also serve as a very useful reference to policy makers and practicing farmers. Suggestions to improve the contents of the book are most welcome (E-mail: reddypp42@gmail.com). The publisher, Elite Publishing House, New Delhi, deserves commendation for their professional contribution.

P. Parvatha Reddy

About the Author

Dr. P. Parvatha Reddy obtained his Ph. D. degree jointly from the University of Florida, USA, and the University of Agricultural Sciences, Bangalore.

Dr. Reddy served as the Director of the prestigious Indian Institute of Horticultural Research (IIHR) at Bangalore from 1999 to 2002 during which period the Institute was honored with "ICAR Best Institution Award". He also served as the Head, Division of Entomology and Nematology at IIHR and gave tremendous impetus and direction to research, extension and education in developing bio-intensive integrated pest management strategies in horticultural crops. These technologies are being practiced widely by the farmers across the country since they are effective, economical, eco-friendly and residue-free. Dr. Reddy has about 34 years of experience working with horticultural crops and involved in developing an F1 tomato hybrid "Arka Varadaan" resistant to root-knot nematodes.

Dr. Reddy has over 250 scientific publications to his credit, which also include 50 books. He has guided two Ph.D. students at the University of Agricultural Sciences, Bangalore.

Dr. Reddy is serving as Senior Scientific Advisor, Dr. Prem Nath Agricultural Science Foundation, Bangalore. He had also served as Chairman, Research Advisory Committee (RAC), Indian Institute of Vegetable Research, Varanasi; Member, RAC of National Centre for Integrated Pest Management, New Delhi; Member of the Expert Panel for monitoring the research program of National Initiative on Climate Resilient Agriculture (NICRA) in the theme of Horticulture including Pest Dynamics and Pollinators; Member of the RAC of the National Research Centre for Citrus, Nagpur; and the Project Directorate of Biological Control, Bangalore. He served as a Member, QRT to review the progress of the Central Tuber Crops Research Institute, Trivandrum; AICRP on Tuber Crops; AICRP on Nematodes; and AINRP on Betel vine. He is the Honorary Fellow of the Society for Plant Protection Sciences, New Delhi; Fellow of the Indian Phytopathological Society, New Delhi; and Founder President of the Association for Advancement of Pest Management in Horticultural Ecosystems (AAPMHE), Bangalore.

Dr. Reddy has been awarded with the prestigious "Association for Advancement of Pest Management in Horticultural Ecosystems Award", "Dr. G.I. D'souza Memorial Lecture Award", "Prof. H.M. Shah Memorial Award" and "Hexamar Agricultural Research and Development Foundation Award" for his unstinted efforts in developing sustainable, bio-intensive and eco-friendly integrated pest management strategies in horticultural crops.

Dr. Reddy has organized "Fourth International Workshop on Biological Control and Management of Chromolaena odorata", "National Seminar on Hi-Tech Horticulture", "First National Symposium on Pest Management in Horticultural Crops: Environmental Implications and Thrusts", and "Second National Symposium on Pest Management in Horticultural Crops: New Molecules and Biopesticides".

Chapter - 1

Fruit Crops: An Overview

1.1. INTRODUCTION

Horticulture sector is an integral element for food and nutritional security in the country. Horticulture is the main segment, while its various sub-segments are fruits, vegetables, aromatic and herbal plants, flowers, spices and plantation crops. All these are regarded as the essential ingredients of economic security. The wide range of agro-climatic conditions of India is conducive for growing a large variety of horticultural crops, including, root and tuber crops, mushrooms, ornamental crops, plantation crops like coconut, areca nut, cashew and cocoa. The Government of India has recognized horticulture crops as a means of diversification in agriculture in an eco-friendly manner through efficient use of land and optimum utilization of natural resources. Horticulture seeks to create ample opportunities for employment, particularly for unemployed youths and women folk. India has maintained leadership in the production of many commodities like mango, banana, acid lime, coconut, areca nut, cashew, ginger, turmeric and black pepper. Presently, it is the second largest producer of fruits and vegetables in the world next only to China. India has also made noticeable advancement in the production of flowers. Further, it is the largest producer, consumer and exporter of spices. India is home to a wide variety of spices like black pepper, cardamom (small and large), ginger, garlic, turmeric, chilli and a large variety of tree and seed spices. Almost all the States in the country grow one or more spices.

Horticultural crops form an integral part of wholesome food containing carbohydrates, proteins, minerals, dietary fibers and vitamins in adequate amount. They produce more edible energy and protein per unit area and time than many other food crops. There is an urgent need to increase horticultural productivity in order to provide nutritional security to the fast-growing population of the country. Among the horticultural crops, fruits recorded a three-fold increase in area and four-fold increase in production over the last five decades.

The UN General Assembly designated 2021 as the International Year of Fruits and Vegetables (IYFV). The official slogan of the campaign 'Fruits and vegetables, your dietary essentials' underlines the importance of fruits and vegetables in nutrition and overall health. The objectives of IYFV are to raise awareness of the health benefits of fruit and vegetable consumption; advocate for healthy diets through increased consumption of fruits and vegetables; promote international efforts to boost fruit and vegetable production and value chains in a sustainable and safe way; bring in focus on the need to reduce losses and waste in fruit and vegetable supply chains from production to consumption; and invite relevant stakeholders to strengthen the capacities of developing countries to adopt innovative approaches and technologies in combating loss and waste of fruits and vegetables. The IYFV 2021 is a step towards achieving UN Sustainable Development Goals (SDG).

1.2. FRUIT CROPS SCENARIO

1.2.1. Global scenario

Fruit crops are a major part of horticultural production. Based on production quantities, the most popular fresh fruits worldwide in order are bananas, apples and grapes. Bananas and apples are grown on trees, and watermelons are grown on thick vines.

According to the Food and Agricultural Organization (FAO), countries with a significant production amount include China, the United States and Turkey. Rated among the most popular apple varieties in the United States are Gala, Honeycrisp and Fuji.

Fruit and vegetables are seen as a strong part of a healthy and balanced diet. The Dietary Guidelines recommend increasing fruit and vegetable intake.

This statistics shows that the world produced 865.88 million tons of fruits in 65.24 million ha during 2016 (Table 1.1).

Table 1.1. Country-wise area, production, and productivity of fruit crops in 2016

Country	Area (million ha)	Production (million tons)	Productivity (tons/ha)
China	16.59	272.08	16.40
India	6.98	90.89	13.03
Brazil	2.33	39.69	17.02
USA	1.19	27.11	22.73
Mexico	1.42	21.43	15.13
Spain	1.63	19.05	11.66
Indonesia	0.77	18.52	24.17
Philippines	1.68	16.32	9.72
Italy	1.14	18.0	15.79
Turkey	1.38	21.74	15.79
Others	28.60	320.62	11.21
WORLD TOTAL	65.24	865.88	13.27

1.2.2. Indian scenario

During 2021-22, India produced 107.10 million metric tonnes of fruits from an area of 7.019 million hectares. In addition, India leads the world in the production of banana, papaya, mango and guava.

Area, production, and productivity of different fruit crops in India during 2021-22 (Second advance estimates) are presented in Table 1.2.

Table 1.2. Area, production, and productivity of fruit crops during 2021-22 in India (Second advance estimates).

Fruit crop	**Area ('000 ha)**	**Production ('000 million tons)**	**Productivity (tons/ha)**
Almond	10	11	1.1
Amla/Gooseberry	104	1272	12.230
Apple	313	2437	7.786
Banana	959	35131	36.633
Ber	53	586	11.056
Citrus (i) Lime/Lemon	322	3742	11.621
(ii) Mandarin	481	6399	13.303
(iii) Sweet orange (Mosambi)	231	4249	19.394
(iv) Other citrus fruits	72	460	6.389
Citrus total (i to iv)	1106	14850	13.426
Custard apple	47	402	9.533
Grapes	162	3445	21.265
Guava	315	4916	15.606
Jackfruit	188	1946	10.351
Kiwifruit	5	17	3.400
Litchi	99	728	7.353
Mango	2350	21011	8.940
Muskmelon	70	1544	22.057
Papaya	145	5885	40.586
Passion Fruit	12	58	4.833
Peach	18	110	6.111
Pear	43	296	6.883
Pecan nut	1	0	0.000
Pineapple	105	1741	16.580
Plum	23	84	3.652
Pomegranate	278	3188	11.467
Sapota	78	863	11.064
Strawberry	3	14	4.666
Walnut	109	282	2.587
Watermelon	123	3461	28.138
Other Fruits	299	2828	9.458
Total Fruits	7019	10710	1.525

State-wise area, production, and productivity of fruit crops during 2017-18 in India are presented in Table 1.3

Table 1.3. State-wise area, production, and productivity of fruit crops during 2017-18 in India.

State	Area ('000 ha)	Production ('000 tons)	Productivity (tons/ha)
Andhra Pradesh	650.52	15215. 85	23.390
Arunachal Pradesh	48.13	125.70	2.611
Assam	147.26	2123.62	14.420
Bihar	296.65	5117.12	17.249
Chhattisgarh	229.67	2666.20	11.608
Gujarat	422.41	8996.02	21.296
Haryana	64.02	793.35	12.392
Himachal Pradesh	230.85	565.31	2.448
Jammu and Kashmir	327.43	2355.19	7.193
Jharkhand	104.30	1081.69	10.370
Karnataka	431.64	7133.94	16.527
Kerala	309.45	2045.75	6.611
Madhya Pradesh	354.06	7416.91	20.948
Maharashtra	753.18	11728.66	15.572
Manipur	47.61	455.59	9.569
Meghalaya	32.81	316.51	9.646
Mizoram	63.19	340.51	5.388
Nagaland	39.50	380.52	9.633
Odisha	340.48	2402.30	7.055
Punjab	90.63	1908.85	21.062
Rajasthan	57.22	761.95	13.316
Sikkim	19.36	54.90	2.835
Tamil Nadu	291.37	5680.52	19.495
Telangana	167.13	1939.39	11.604
Tripura	53.75	547.52	10.186
Uttar Pradesh	476.64	10539.78	22.112
Uttarakhand	178.65	669.94	3.750
West Bengal	261.02	3850.56	14.752
Others	17.32	143.36	8.277
TOTAL	6506.23	97357.51	14.963

The biggest fruit-producing states are Andhra Pradesh, Maharashtra, Uttar Pradesh, Gujarat, Madhya Pradesh, Karnataka, and Tamil Nadu.

India is taking efforts to export the fruits and vegetables of economic importance to gain foreign exchange earnings. Production and export of Indian fruits have increased in recent years. During 2021-22, India exported fresh fruits worth US$ 750.7 million and processed fruits and juices worth Rs.4550.62 crores/ USD 610.69 million. The major fruits exported are grapes, pomegranates, mangoes, bananas and oranges. India mostly exports to its neighbors, namely the UAE, Bangladesh, Pakistan, Saudi Arabia, Sri Lanka and Nepal. India's global market share is still only about 1%, but its horticulture products are becoming increasingly popular due to advancements in cold chain infrastructure, research, contemporary post-harvest technologies, supportive governmental policies and quality control procedures.

The details of the exportable quantity of major fruits during the year 2020-21 was presented in Table. Banana found to be dominated in the export scene of fruits which has gained 19.37 per cent of growth rate annually and it is expected to reach the level of 1.88 lakh tons by the year 2020-21. The prediction of export quantity of grapes during the year 2020-21 was arrived at 2.38 lakh tons (Table 1.4).

Table 1.4. Quantity of fruits exported abroad from India in a decade's time ('000 tons)

Fruit crop	Quantity exported during 2006-07	Quantity exported during 2016-17	CAGR (%)	Expected export quantity during 2020-21
Apple	25.936	22.550	(-) 4.39	19.707
Banana	11.475	110.750	19.37	188.386
Grapes	85.562	198.471	6.31	238.512
Mango	79.060	52.760	(-) 3.02	47.200
Orange	30.110	48.114	5.52	56.538

1.3. BRIDGING YIELD GAP

Even though India is leading in productivity of some fruit crops like grapes, still there is scope to increase productivity in other fruit crops compared to several countries.

In India, prospects of increasing production of fruit crops by increasing land under cultivation is limited. Hence, it is essential to increase productivity of fruit crops in order to meet the future demand and fulfilment of country's commitment to the sustainable fruit development with the objective of nutritional security at individual level, poverty alleviation and employment generation through fruit sector (Table 1.5).

Table 1.5. Productivity of major fruit crops in India *vis-à-vis* other countries

Fruit crop	India (tons/ha)	World highest (tons/ha)	Country	Increase over India (%)
Banana	34.00	52.54	Costa Rica	54.50
Mango	6.75	12.50	Brazil	85.00
Citrus	8.40	25.00	Brazil	297.60
Guava	12.00	14.90	France	42.80
Grape	25.69	25.69	India	---
Papaya	39.60	22.30	Indonesia	69.50
Apple	7.68	84.01	Austria	993.88
Pear	5.59	283.19	Austria	4966.01

Pomegranate	6.90	24.70	Indonesia	85.80

1.3.1. Interventions needed for bridging yield gap

It has been observed that in most of the fruit crops, the productivity gap between national level and leading states is more than 100% as compared to demonstration plots or experimental station yields. Major focus on research should be devoted to productivity enhancement technologies like development of high yielding varieties/F1 hybrids, use of micro-irrigation, integrated nutrient management, fertigation, protected cultivation of strawberry, use of biotechnological approaches, use of bioregulators, post-harvest management, integrated weed, pest, disease and nematode management. This calls for efficient transfer of technology and training of the stake holders both private and public coupled with providing appropriate inputs, technical knowledge and infrastructure facilities.

Key strategies to reduce yield gaps, therefore, include:

- Increasing the efficiency of technology transfer.
- Use of recommended practices: sowing date, seed rate, fertilizer amount, rotation, use of proper farm machinery, and disease and pest management practices.
- Proper targeting of varieties to production zones.
- Timely availability of inputs: quality seed, water, and fertilizers.
- Government intervention and policies to strengthen input availability and crop marketing.

1.4. CONSTRAINTS

The Indian fruit sector is facing severe constraints such as low crop productivity, limited irrigation facilities and underdeveloped infrastructure support like cold storages, markets, roads, transportation facilities, *etc.* There are heavy post-harvest and handling losses, resulting in low productivity per unit area and high cost of production. Common problems encountered in production and productivity of horticultural crops include:

- Inadequate availability of disease free, high quality planting material.
- Micro-propagation techniques are under exploited.
- Slow dissemination and adaptability of improved high yielding cultivars/hybrids.
- Inadequate facility for identification of nutrient deficiency and disorders.
- Lack of diseases and pests' outbreak forecast service.
- Unavailability of refined intensive integrated production systems.
- Lack of quality standards.
- Lack of technologies in value addition.
- Lack of post-harvest management technology and infrastructure.
- Weak database and poor market intelligence.
- Poor marketing practices and infrastructure.
- Instability of prices, with no support price mechanism.
- Inadequate technical manpower/human resource in farming systems.
- Poor credit supply, high rate of interest coupled with inadequate crop insurance schemes.
- Late implementation of government policies and schemes.
- Absence of fruit crops suitability map of India based on agro-climatic conditions depicting

most suitable areas for optimum productivity of a particular crop.

However, on the other hand, India's long growing season, diverse soil and climatic conditions comprising several agro-ecological regions provide ample opportunity to grow a variety of fruit crops. Thus, efforts are needed in the direction to capitalize on our strengths and remove constrains to meet the goal of moving towards a formidable pomology growth in India. The foreign trade policy in 2004-11 emphasized the need to boost agricultural exports, growth and promotion of exports of horticultural products.

1.5. FRUIT CROPS DEVELOPMENT

Fruit crops development had not been a priority in India until recent years. It was in the post-1993 period that a focused attention was given to fruit crops development through an enhancement of plan allocation and knowledge-based technology. Despite of this decade being called a "golden revolution" in horticultural production, the productivity of fruit crops has increased only marginally from 1991-92 to 2004-05 (NHB, 2010). The National Horticulture Mission was launched in 2005-06 by the Government of India with a mandate to promote integrated development of horticulture, to help in coordinating, stimulating and sustaining the production and processing of fruits and vegetables and to establish a sound infrastructure in the field of production, processing and marketing with a focus on post-harvest management to reduce losses. The area and production trend shows that there has been a sharp rise in area and production of fruit crops with an area of 7.09 million ha and production reaching 107.1 million tons during the year 2021-22. As a result of this huge spurt in fruit crops produce, India has become the second largest producer of fruits in the world, next only to China.

1.5.1. Current scenario

Fruit crops, over the past few years, has made remarkable progress in terms of expansion in area under different crops, increase in productivity, crop diversification, technological interventions for production and post-harvest and forward linkages through value addition and marketing. Critical areas like pest and disease management have been given principal focus with intensification of efforts on containing decline problems especially in orange. Massive rejuvenation programs in orange has brought back some ray of hope in the direction of revival of this crop. A number of nurseries have been established for producing healthy and quality planting materials of orange. The tissue culture laboratories, in private sector, have been able to supplement the requirement of quality planting materials.

Infrastructure development to augment marketing and post-harvest management are being accorded top priority, Cold rooms have been set up in various production clusters to aid production and marketing. Technology Mission for Integrated Development of Horticulture, sponsored by the Government of India has played the most vital role in integrating all ongoing initiatives. In addition to this, other initiatives under RKVY, TSP/SCSP, BADP and State Plan/Non-Plan have made noteworthy contribution in supplementing the programs spearheaded through Technology Mission. The main success factors for successful program implementation are use of high-quality planting materials, adoption of clusters and constant monitoring.

The most significant development that happened in the last decade is that fruit crops production has moved from rural confines to commercial production and this changing scenario has encouraged private sector investment in production system management. The last decade has seen technological infusion like micro-irrigation, precision farming, greenhouse cultivation, and improved post-harvest management impacting the development, but during the process various issues have emerged.

1.6. TECHNOLOGICAL INTERVENTIONS TO ENHANCE PRODUCTIVITY

1.6.1. High yielding varieties

During the last seven decades (since 1950), several improved varieties of fruit crops have been developed. As a result, productivity of banana has gone up three times. Regular-bearing mango hybrids and export quality grapes, have been developed.

1.6.2. High density planting

Planting of fruit trees rather at a closer spacing than the recommended one using certain special techniques with the sole objective of obtaining maximum productivity per unit area without sacrificing quality is often referred as high-density planting. This technique was first established in apple in Europe during sixties and now majority of the apple orchards in Europe, America, Australia and New Zealand are grown under this system. In this system, four planting densities are recognized for apples viz., low HDP (< 250 trees/ha), moderate HDP (250-500 trees/ha), high HDP (500 to 1250 trees/ha) and ultra-high HDP (>1250 trees/ha). Recently, super high-density planting system has been also established in apple orchards with a plant population of 20,000 trees per ha. In some orchards, still closer, planting of apple trees is followed (say 70,000 trees/ha) which is often referred as 'meadow orchards.'

Accommodation of the maximum possible number of plants per unit area to get the maximum possible productivity and profit per unit of the tree volume without impairing the soil fertility status is called High Density Planting (HDP). Basically, the availability of a dwarf plant is the first and foremost pre-requisite for establishing any high-density orchard. HDP is one of the methods to enhance productivity per unit area both in short duration and perennial horticultural crops. In perennial, it is more useful since it permits efficient use of land and resources, realizing higher yields, net economic returns per unit area, easy canopy management suited for farm mechanization and cultural operations, efficient spray and weed control, improvement in fruit quality and easy and efficient harvest of high-quality produce, *etc.*

1.6.3. Micro-irrigation

In traditional surface irrigation methods, the losses in water conveyance and application are large (Table 1.6). These losses can be considerably reduced by adopting drip and sprinkler irrigation methods. Among all the irrigation methods, the drip irrigation is the most efficient and it can be practiced in a large variety of crops, especially in orchard crops.

Table 1.6. Water losses under various irrigation systems

Method of irrigation	Water losses (%)
Surface	30-45
Gate pipe	15-20
Sprinkler irrigation	5-10
Drip irrigation	3-6

Micro-irrigation system is an irrigation system with high frequency application of water in and around the root zone of plant system, which consists of a network of pipes along with a suitable emitting device. With micro-irrigation, the yields of crops are better in addition to the saving of water.

Micro irrigation can be classified as drip irrigation and sprinkler irrigation. In drip irrigation, water is applied near the plant root through emitters or drippers, on or below the soil surface, at a low rate varying from 2-20 liters per hour. The soil moisture is kept at an optimum level with frequent irrigations. Drip irrigation results in a very high water application efficiency of about 90-95 per cent.

Drip irrigation system is suitable for all orchard crops. The National Committee on Plasticulture Applications in Horticulture (NCPAH), Ministry of Agriculture, Government of India (GOI), has estimated a total of 27 million hectares area in the country that has the potential for drip irrigation application.

There is a need to promote increased irrigation water use efficiency. Participatory action research program initiated by the Ministry of Water Resources, Government of India, during 2007-09 in over 2000 villages all over the country to assess the impact of water saving technologies has shown that yield and income can be increased by 50 to 100% in most fruit crops by using water saving technologies.

The advantages of micro-irrigation are of two types – qualitative and quantitative. The qualitative advantages are production of quality horticultural produce and increase in water and fertilizer use efficiency. Excellent soil health could be maintained due to optimum soil moisture content, accelerated microbial activities and maximization of biomass decomposition converting into humus. The quantitative advantages are related to growth of micro-irrigation in terms of area coverage and overall increase in the productivity. Quantity of water or fertilizer saved due to imposition of drip system, the energy conservation and reduced labor cost are the other advantages. Data on increase in crop yields and percent increase over traditional surface irrigation methods by using micro-irrigation system are presented in Table 1.7. It shows that micro-irrigation could save water ranging from 23.58 % to 67.90 % with enhancement of yield from 23.1% and 514.0 % under varied agro-climatic and crop conditions. The higher yield under micro-irrigation system is attributed to the fact that during crop growth favorable soil moisture (generally at ⅓ bar) is maintained and the required volume of soil mass is wetted through irrigation water, which provides optimum environment for root growth of the plant.

Table 1.7. Water saving and yield increase with drip irrigation system in fruit crops

Fruit crop	Yield (tons/ha)		% water saving	% increase in yield
	Surface irrigation	Drip irrigation		
Banana	575.0	875.0	45.0	52.2
Grape	264.0	325.0	47.2	23.1
Papaya	130.0	230.0	67.9	76.9
Pomegranate	34.0	67.0	23.8	97.0
Water melon	82.1	504.0	65.3	513.9

1.6.4. Fertigation

Among the various factors that are responsible for increasing the productivity of horticultural crops, a regular and optimum supply of water and fertilizers at appropriate times is very crucial as it enables the crops to attain higher growth rate and give increased yields. As both water and fertilizers are costly inputs and are becoming scarce in recent years, there is an immediate need for adoption of techniques with which these inputs can be utilized more efficiently. Fertigation is one such technique of applying fertilizers to the crops along with water through micro-irrigation systems (drip/sprinkler) on a continual basis in controlled manner so as to allow steady uptake of nutrients by the plants and to effect savings in the costly inputs including water and fertilizers.

There are three beneficial effects of fertigation:

1.6.4.1. Maximizing water and fertilizer use efficiency

- » Delivers water and nutrients to the root zone.
- » Increases the availability and uptake of nutrients.
- » Decrease nutrient losses from leaching.
- » Accurate and uniform application of water and nutrients.
- » Maximum saving of water and nutrients.

1.6.4.2. Maximizing crop and soil productivity

- » Higher yields and high quality produce.
- » Steep-sloped and marginal lands can be made cultivable.

1.6.4.3. Minimizing production cost

- » Economy in labor, water and nutrients.

1.6.5. Integrated nutrient management

Integrated nutrient management refers to maintaining the soil fertility and plant nutrient supply to an optimum level for sustaining the desired crop productivity through optimization of the benefits from all possible sources of plant nutrients in an integrated manner. It is a holistic approach, to know what exactly is required by the plant for an optimum level of production, in what different forms these nutrients should be applied in soil and at what different timings in the best possible method, and how best these forms should be integrated to obtain the highest productive efficiency on the economically acceptable limits in an environment friendly manner.

The encouraging results obtained by conjunctive use of fertilizers, organic manure, green manure, biofertilizers and fertigation can provide leads to the future strategies for the rational use of fertilizers and manures for enhancing the productivity of horticultural crops without detrimental effect to the environment.

The crops do not efficiently utilize fertilizers applied under traditional methods of irrigation. The efficiency of different nutrients and methods of application is presented in Table 1.8.

Table 1.8. Efficiency of different nutrients and methods of application

Element	Method of application	Efficiency (%)
N	Fertigation	90-95
	Surface flow	50-70
	Broadcasting (Solid fertilizers)	30-50
P	Fertigation	40-45

	Surface flow	10-20
K	Fertigation	75-80
	Surface flow	< 50

1.6.5.1. Optimum leaf nutrient status: The concentrations of the nutrient in the leaf at a specific growth stage are related to the performance of the crop. The leaf nutrient levels required for optimum production in some important fruit crops have been presented in Table 1.9. Efforts in the nutritional management program should be directed towards the maintenance of these levels in tree leaves before flowering.

Table 1.9. Optimum leaf nutrient status of some important fruit crops

Fruit crop	Nutrients											Reference
	%						ppm					
	N	P	K	Ca	Mg	S	B	Fe	Mn	Zn	Cu	
Mango	1.23	0.66	0.54	1.71	0.91	0.12	---	171	66	25	12	Samra *et al.*, 1978
Banana	3.29	0.44	3.11	2.12	0.24	---	---	---	---	---	---	Ramaswamy, 1971
Citrus	2.50	0.14	0.90	4.20	0.43	0.25	---	90	112.5	62.5	10.5	Reuther *et al.*, 1962
Guava	1.70	0.50	1.30	2.10	0.46	---	10	---	---	77.5	75	Singh, 1981
Grape	1.75	0.56	1.66	0.94	---	0.20	---	---	125	92	---	Bhargava & Chadha, 1993
Litchi	1.47	1.03	---	---	---	---	---	---	---	---	---	Ghosh *et al.*, 1986
Sapota	1.66	0.08	0.80	0.83	0.48	0.07	---	100	39.3	15.7	6.7	Annapurna *et al.*, 1988
Papaya	1.66	0.50	5.21	1.81	0.67	0.38	---	---	---	---	---	Bhargava *et al.*, 1990
Pineapple	1.40	0.16	3.70	---	---	---	---	---	---	---	---	Chadha & Reddy, 1993
Pomegranate	2.50	0.19	1.47	---	---	---	96	199	196	55	---	Shende, 1977 Bhambal, 1987
Ber	2.50	0.25	1.82	1.21	0.45	---	---	---	---	---	---	Khanduja & Garg, 1984

1.6.6. Plant bioregulators

Ever since the discovery that the plant hormones have the ability to influence several physiological processes in plants, attempts have been continuously made to manipulate the growth and development process in different horticultural crops to streamline the production. As the years passed, besides the natural compounds, a number of synthetic compounds were also found to have growth regulatory roles, which are now collectively known as bio-regulators. Many of the research accomplishments have led to the commercial applications.

The use of bioregulators to increase the vitality and therewith the productivity of fruit crops is a distinct possibility. In several experiments, the positive effect of different bioregulators used (Humate, Lactate, *Bacillus subtilis*) could be discovered if one of the growth factors was at a critical level. Humates are effective in nutrient and salt control as well as subtropical temperature. Lactates have a good effect if the pH-value is unstable. Microorganisms like *Bacillus subtilis* are effective under different stress conditions. They have an unspecific positive effect in particular if the nutritional status is suboptimal. However, fruit crops with long vegetative or fruiting phase are often confronted with different stress factors. A combined application of bioregulators could be beneficial in such complex stress situations in order to improve the productivity of the fruit crops.

Bioregulators have been used for many years to alter the behavior of fruit or fruit trees for the economic benefit of the fruit grower. Control of vegetative vigor, stimulation of flowering, regulation of crop load, reduction of fruit drop, and delay or stimulation of fruit maturity and ripening are important examples of processes in fruit and fruit trees that can be regulated with exogenous applications of bioregulators.

Bioregulators help in improving the fruit size of banana and grapes. Spraying of 100 ppm GA_3 twice at 25 and 50 days after the shooting resulted in an increase of 32.5, 23.9 and 16.98% of length, girth and weight of the fingers, respectively in banana cv. Rajapuri (syn. Munavalli) which is popular in Belagavi district of Karnataka (Rao, 1997).

Application of GA_3 has caused multi-fold increase in berry size and consequently productivity has been raised from 6 to 8 tons/ha to 15 to 18 tons/ha in Thompson Seedless cv. of grapes.

Bioregulators have been used in improvement of color and ripening in grapes. Rao and Shankarnarayanan (1982) could overcome the problem of uneven ripening in Gulabi cv. of grape by dipping the clusters twice at 5 days interval during the lag phase in 250 ppm ethrel containing 2% urea. Use of 50 ppm 2,4,5-TP was also found effective in this respect (Hanumashetti *et al.*, 1981).

Bioregulators help in controlling the berry drop in Anab-e-Shahi cv. of grape. Spraying of 100 ppm NAA or PCPA a week before harvest reduced the berry drop (Rao *et al.*, 1968). Mixing of 100 ppm BA along with 100 ppm NAA was much more effective. NAA helps in thickening of the pedicel attachment.

1.6.7. Canopy management

Canopy management is the manipulation of tree canopies to optimize the production of quality fruits. The canopy management, particularly its components like tree training and pruning, affects the quantity of sunlight intercepted by trees, as tree shape determines the presentation of leaf area to incoming radiation. An ideal training strategy centers around the arrangement of plant parts, especially, to develop a better plant architecture that optimizes the utilization of sunlight and promotes productivity. Light is critical to growth and development of trees and their fruits. Strong bearing branches tend to produce larger fruits. The problem of a fruit grower is initially to build up a strong and balanced framework of the trees, then equip them with the appropriate fruiting. Obviously, pruning in the early years has to be of a training type to provide strong and stocky framework with well-spaced limbs or any other desired shape.

1.6.8. Quality planting material production

Availability of planting material of good quality is one of the most important elements of successful horticultural production. Planting material available to small-scale farmers in different areas is often of insufficient quality, which undermines potential yield and performance of crop production. There has been a significant increase in the production of horticultural crops in recent years in India. At present in India, more than 4,409 fruit nurseries including 1,575 under government sector and 2,834 under private sector are functioning which have an annual target of producing 1,387 million fruit plants. This accounts for 30-40% of the demand of planting materials of fruit sector.

In order to produce true to type planting materials, most of the fruit crops are propagated by asexual or vegetatively except the crops like papaya, phalsa and mangosteen.

1.6.9. Genetically modified crops

GM crops offer vast potential for improving the efficiency of crop production, thereby lowering the cost and increasing the quality of food. The tools of biotechnology can provide scientists with new approaches to develop higher yielding and more nutritious crop varieties, to improve resistance to pests including weeds and adverse conditions, or to reduce the need for fertilizers and other expensive agricultural chemicals.

In the first 20 years (1996 to 2015), biotech crops were planted by up to 18 million farmers (up to 90% were small/poor farmers) in 28 countries annually. With an increase of 100-fold from 1.7 million hectares in 1996 to 179.7 million hectares in 2015, this makes biotech crops the fastest adopted crop technology in recent times – the reason – biotech crops have the trust of millions of farmers because they deliver significant and multiple benefits. Accordingly, the number of biotech countries has more than quadrupled from 6 in 1996, to 16 in 2002 and 28 in 2015.

1.6.10. Precision horticulture

Precision horticulture can be defined as a comprehensive system designed to optimize agricultural production through the application of crop information, advanced technology, and management practices. To be a truly comprehensive system, it must begin during the planning stages of the crop or commodity and continue through the post-harvest processing phase of production. Information, technology, and management are the keys to success in this production system.

An information and technology-based farm management system identifies, analyses and manages variability in fields by conducting crop production practices at the right place and time and in the right way, for optimum profitability, sustainability and protection of the land resource. Precision horticulture technologies system approach helps to re-organize the total system of agriculture towards low-input, high-efficiency and sustainable agriculture. Precision farming is an approach where inputs are utilized in precise amounts to get increased average yields, compared to traditional cultivation techniques.

Sustainable precision horticulture is this century's most valuable innovation in farm management that is based on using Information and Communication Technologies (ICTs). This is the most recent innovation technology based on sustainable agriculture and healthy food production and it consists of profitability and increasing production, economic efficiency and the reduction of side effects on the environment.

Precision horticulture is a comprehensive system designed to optimize production. The precision agriculture can be used to increase production efficiency, improve product quality, improve the efficiency of crop chemical use, conserve energy, and protect the environment. Technology and management practices such as field scouting, field mapping, variable rate control, yield mapping, and post-harvest processing can be readily adapted to fruit crop production. Much of this technology is still in its infancy. More research will be necessary to allow the systems to reach maturity. While technically feasible, further research is also needed to clarify the economic and environmental benefits of many elements of precision agriculture.

1.6.11. Nanotechnology

In the last decade, nanotechnology has emerged as a recent evolving sector in the field of science and technology. Nanoparticles due to their small size show exceptional physical, chemical, optical and magnetic properties. These unique properties of nanoparticles make them useful in horticulture for obtaining sustainability in production through their application in production of quality planting material, nano-fertilizers for improving plant nutrition, nano-sensors for monitoring plant health and quality of soil and water, nano-pesticides for disease and pest management and improved packaging for enhancing shelf-life. Nanoparticles can be used for improving germination percentage of seeds and controlling contamination rate in *in vitro* propagation. They can also improve stress tolerance in plants against various biotic and abiotic stresses. Nanotechnology provides the possibilities of exploring nanoscale materials in the form of nano-fertilizers or nano-pesticides which can improve the nutrient use efficiency, enhance the efficacy and specificity of pesticides and at the same time minimize the risk of pollution to ecosystem. Smart packaging developed through the use of nanomaterials can reduce the occurrence of food-borne diseases and improve food safety by ensuring fresh and tasty purchase to the consumers. Nano biosensors can not only be used for monitoring soil and plant health but also can help in predicting the outbreak of pest or diseases, thus reducing the economic loss. Better understanding about the interaction of nanomaterials in the horticultural field, however, is required for eliminating the adverse effects of nanoparticles in the horticultural crops.

Various kinds of nanomaterials have shown high potential for enhancing shelf-life, reducing post-harvest damage and improving the quality of horticultural crops. Antimicrobial nanomaterials as nanofilm on harvested products and/or on packaging materials are suitable for the storage and transportation of fruits. The nanotechnology could be used to generate cutting-edge techniques towards promoting productivity and quality of fruit crops to ensure food and nutritional security of ever-increasing population of the world.

1.6.12. Post-harvest management

There is a considerable gap between the production and net availability of the horticultural produce due to post-harvest losses due to lack of elaborate harvesting equipment, collection centers in major producing areas, suitable containers, commercial storage plants and lack of cold chain in entire post-harvest operation.

The total post-harvest losses in fruits are estimated to be 17 to 35% amounting to nearly Rs. 2000 crores annually. In a systematic study on post-harvest losses in fruits have been estimated between 17.1 to 36.7% in mango, 12 to 19% in banana, 8.3 to 30.7% in orange, 10 to 25% in apple, and 23 to 90% in grapes. Considerable losses occur during transport. Minimizing these losses has, therefore, been recognized as an effective means of increasing fruit availability without bringing additional areas under these crops. In fact, the reduction in post-harvest losses is a complementary means of increasing the productivity.

1.6.12.1. Pre-harvest treatments

» Pre-harvest application of GA delays ripening at market stage and improves storage life in mango and guava and color development in citrus.

- Application of thiophanate methyl (0.05%) effectively controls post-harvest losses in mango by controlling maturity and delaying ripening.
- Post-harvest decay of Nagpur mandarin, anthracnose and stem-end rot of mango can be controlled by 3 pre-harvest sprays of Benlate (0.1%) or Topsin-M (0.1%) at 15 days interval before harvest.
- Pre-harvest sprays of 0.6% calcium chloride 10 to 12 days prior to harvest improved shelf-life and reduces physiological loss in weight in Thompson seedless grapes.
- Application of 2.5% calcium chloride at 50 ml/sucker in the core of pineapple plant about 300 days after planting showed better ripening and improved overall quality.

1.6.12.2. Post-harvest treatments

i) Use of chemicals: Post-harvest losses can be reduced by use of chemicals, fungicides, waxing etc. Protective skin coating with wax increases storage life of fruits at ambient temperatures. This treatment protects fruits against excessive moisture loss and higher rate of respiration. To protect against microbial spoilage, fungicides can be added to the wax emulsion.

ii) Disinfection: Vapor heat treatment of tropical fruits at 48°C for 1 hr eliminates all stages of fruit flies.

iii) Cold storage: Storage at low temperatures immediately after harvest reduces the rate of respiration resulting in reduction in build-up of field heat, thermal decomposition, microbial spoilage and also helps in retention of color and quality and freshness for a longer period. At present, there is a cold storage capacity of 30 million cubic meters in India. Fruits and vegetables together utilize only about 8.1% of total cold storage capacity.

Different methods employed in cold storage include controlled atmospheric (CA) storage and zero energy cool chamber.

1.6.12.3. Value addition: Processing of fruits for value addition is an indispensable part of horticulture industry. In countries like USA and Brazil, nearly 60 to 70% of fruits and vegetables are used for processing as against 1 to 2% in our country. Steps should be initiated for achieving at least 15 % level of processing by 2025 which would generate an investment of Rs. 1 lakh crores and create almost half a million direct employment and another 2 million indirect employment. Hence, attempts to promote processing industries in this country is very essential. In mango and banana, our share is nearly 50 and 15 % of the total world production and hence a great scope exists for development of newer products from these commodities.

1.6.13. Integrated pest management

1.6.13.1. Weeds: Integrated weed management system (IWMS) involves co-ordinated efforts towards development of a long lasting and environment friendly practice of weed management. IWMS includes all preventive, crop husbandry and direct control measures aimed at reducing and maintaining weed population at sub-economic level. In this, all weed control methods like preventive, cultural, mechanical and biological are included in different permutation and combinations are tried to achieve maximum weed control which is not possible in any one single method. IWMS is well suited for adoption in different cropping systems of fruit crops.

1.6.13.2. Insect pests, diseases pathogens and nematodes: Integrated Pest Management (IPM) is an important principle on which sustainable crop protection can be based. IPM allows farmers to manage pests in a cost effective, environmentally sound and socially acceptable way. According to FAO, IPM is defined as "A pest management system that in the context of the associated environment and the population dynamics of the pest species, utilizes all suitable techniques and methods, in a compatible manner as possible and maintains the pest populations at levels below those causing economic injury". IPM strategies are based on 3 main components/tools like prevention, monitoring/observation/regulation, and intervention.

Biointensive integrated pest management is a systems approach to pest management based on an understanding of pest ecology. It begins with steps to accurately diagnose the nature and source of pest problems, and then relies on a range of preventive tactics and biological controls to keep pest populations within acceptable limits. Reduced-risk pesticides are used if other tactics have not been adequately effective, as a last resort, and with care to minimize risks.

i) Proactive options: Proactive options, such as crop rotations and creation of habitat for beneficial organisms, permanently lower the carrying capacity of the farm for the pest. The carrying capacity is determined by the factors like food, shelter, natural enemy complex and weather, which affect the reproduction and survival of a pest species. Cultural control practices are generally considered to be proactive strategies. Proactive practices include crop rotation, resistant crop cultivars including transgenic plants, disease-free seed and plants, crop sanitation, spacing of plants, altering planting dates, mulches, *etc.*

ii) Reactive options: The reactive options mean that the grower responds to a situation, such as an economically damaging population of pests, with some type of short-term suppressive action. Reactive methods generally include inundative releases of biological control agents, mechanical and physical controls, botanical pesticides and chemical controls.

1.7. POTENTIAL OF HORTICULTURAL CROPS FOR ENSURING FOOD AND NUTRITIONAL SECURITY

The fruit sector of horticulture is most potent to fulfil food requirements of the country and elevate the nutritional and economic status of people (Rana, 2010). In general, antioxidants, vitamins, minerals, and fibers are considered as main nutrients contributed by fruit crops to a balanced diet, and thus special attention should be addressed to this group of nutrients (Siddiqui *et al.*, 2015).

Fruits provide substantial amount of nutrients, important for human health and can play significant role for improving the nutritional intake especially of predominantly vegetarian population. They are important source of micronutrients, vitamins, minerals and folic acid (Siddiqui and Dhua, 2010). Fruits have proved to be essential for a balanced diet as a rich source of vitamins, protein, minerals, organic acids, carbohydrates, and antioxidants (Siddiqui *et al.*, 2013, 2015). Man needs a wide range of nutrients to perform different metabolic functions and to lead a healthy life and their deficiency leads to a number of diseases in children and adults. Fruits contain nutraceutical substances that provides medical and health benefits and their daily consumption have been strongly associated with reduced risk for some forms of cancer, heart disease, stroke, and other chronic diseases (Siddiqui and Dhua, 2010; Siddiqui *et al.*, 2013, 2015). Some components of fruits (phytochemicals) are strong antioxidants and function to modify the metabolic activation and detoxification/disposition of carcinogens, or even influence processes that alter the course of the tumor cell. It is better to consume a variety of commodities rather than limiting consumption to a few with the highest antioxidant capacity as it varies greatly among fruits (Kalt, 2002). Nutritive constituents of fruits that have a positive impact on human health and their sources are present in Table 1.10.

Table 1.10. Bioactive substances present in some fruit crops

Crop	Bioactive substances
Amla	Vit. C, Polyphenols
Apple	Flavonoids, polyphenols
Citrus	Vit. C, bioflavonoids, Chalcone
Grapes	Flavonoid, phenols
Strawberry	Flavonoid, phenols

Source: Rana, 2010; Ayala-Zavala *et al.*, 2011; Siddiqui *et al.*, 2013b.

1.8. FUTURE THRUSTS AND CONCLUSION

The fruit crops sector in India possesses unique advantage because of availability of varied climatic conditions for production of various fruit crops and several ICAR institutes and SAUs are working for the benefit of farmers and other stakeholders. Besides, there is a huge varietal diversity in most of the cultivated fruits, which increase the harvesting season and more revenues to the farmers. The production and supply of quality planting materials, implementation of INM, IDM and IPM, multi-cropping system, integrated farming system (IFS), pollinators, pre- and post-harvest management, forward and backward linkages in marketing are very important for successful cultivation of fruit crops. In spite of higher production, the quantity of processed products from the raw materials are poor and very low in quantity produce, which is due to the lack of GAP for most of the fruit crops and poor post-harvest handling. Several prediction models have suggested that the area under horticultural crops is going to increase and cereal crops will be decreasing, hence, it is time to breed the climate resilient varieties to withstand the climatic aberrations. Hence, a comprehensive plan covering R and D aspects for all fruit crops is a must and convergence of scientists, policy makers and public is the need of the hour for marching ahead and to take the fruit crops production to newer heights.

Development of fruit crops in India needs some critical management inputs particularly that of supply chain management - collaboration among various stakeholders along with efficient vertical and horizontal integration. The fruit crop sector in particular has to prioritize development of research in the issues of genetics, biotechnology, integrated and sustainable production systems, post-harvest handling, storage, marketing and consumer education. Diversification offers an attractive option and a major source of pushing up growth of fruit crops sector. While technological up-gradation and associated institutional changes are identified as thrust areas for future development of the fruit sector, exports are considered to be most important for the growth of the sector. India can look forward to emerge as a major producer of fruit products and thus secure reasonable market access for its agro exports, which are largely dependent on the competitive technologies that will help in enhancing export potential. This development will also help in overall growth of the economy through generation of extra foreign exchange, creating employment opportunities and also upliftment of the small and marginal farmers, with definite positive implications on income and employment. The government should create a positive environment that will ensure a mutually beneficial relationship between farmers and organized sector.

Fruit crop diversification should be encouraged by intercropping fruit crops with other crops. This will yield more food, more income and better soil health. The use of vegetables in intercropping also helps in increasing the incomes of farmers during the period when the fruit orchard has not become commercially viable. To increase the production and productivity of fruits, adoption of recommended package of practices is very important. The diversification plan of the fruit crops sector needs to identify potential crops area-wise and the area under low yielding fruits should be shifted to more productive and profitable ones. There is a strong need to strengthen the research on fruit crops to develop demand-driven technology by improved variety, pest management, *etc.* The fruit crops development requires a minimum set of basic production factors, an optimal crop management infrastructure, post-harvest infrastructure, entrepreneurial management and horticultural expertise, logistical infrastructure and supporting financial infrastructure.

Development of fruit crops sector should be accompanied by the growth of the agro processing industry. The opportunity exists to promote the industry by intensifying production of a required, appropriate variety of mangoes, pineapples, lemons, *etc.*, for the products like ketchup, sauce, juice and pickles. Thus, the production strategy should target not only meeting the domestic and export demand of fresh products but also of the processed products. There is the need to improve post-harvest operations related to handling, storage and marketing of fresh and processed produce. Volumes saved in post-harvest losses are actually the surpluses generated, without additional cost.

The fruit crops sector has an immense potential for generating additional employment by developing the horticulture based agro-processing units. Empirical analytical evidence has shown that the fruit crops export has increased the opportunity for higher earnings for smallholders and that the much higher land sizes owned by smallholders are indeed a cause or consequence for their participation in the sector. This sector needs to be developed as an organized industry and has to be run collectively by all the stakeholders with farmers as entrepreneurs and private sectors.

Our future line of action should focus on strengthening of integrated cold chains at farmgate pack houses; expanding the harvesting window by mainstreaming of fruit crops even during offseason to increase availability; developing nutrient dense biofortified varieties, and last but not the least, considering perennial fruit crops as a source of promoting clean environment. It goes without saying that our last mile, more than 700 Krishi Vigyan Kendras (KVKs), need to be aggressively plugged into farmgate supply and value chains to advise on increasing productivity, yield and market reach of fruit crops. India can achieve the goal of bringing the plate of fruit crops to the poorest of the poor, not as a special food but as a daily necessity.

Chapter - 2

Nematode Problems in Fruit Crops: An Overview

2.1. INTRODUCTION

Plant parasitic nematodes are one of the major pests of fruit crops. Many biotic and abiotic factors play an important role in distribution of these nematodes. The first report of nematode damage to fruit crops was reported on citrus by *Meloidogyne* sp. (Thirumala Rao, 1956). The infestation of *Radopholus similis* on banana was reported from Kerela (Nair *et al.*, 1966). These were some earlier record of plant parasitic nematode infestation on fruit crops in India.

Plant parasitic nematodes cause considerable economic losses in fruit crops. The main loss is the destruction of roots which hinders the movement of nutrients and water through the vascular system, so, there is drastic reduction in fruit or bunch weights, the quality of fruits is deteriorated and there is a drastic reduction in plant numbers. The root-knot nematodes, *Meloidogyne* spp.; the burrowing nematode, *Radopholus similis* and citrus nematode, *Tylenchulus semipenetrans* are the major nematode pests that infect these fruit crops. Around 30–40% loss in yield is due to these nematodes. Furthermore, roots damaged by nematodes are easy prey to fungi and bacteria which invade the infected roots and feed on them and thus roots decay rapidly. The nematode infestation in fruit crops not only aggravate disease complexes but also breaks down disease resistance in certain varieties of fruit crops. These nematodes mainly spread through infested planting material to other uninfested sites. For an instance, in banana, paring or trimming of suckers is carried out before planting which is usually not adequate to eliminate deep infections in the suckers. The residual nematode population builds up and disseminates when they enter the irrigation system. Other routes of their dispersal include soil adhering to tractor tyres, shoes of labor and tillage implements.

Most of the fruit crops are grown as monocrop which provides a congenial environment to the nematodes. If the nematode population is unchecked then it increases to an alarming level and may finally wipe out the whole crop. Nematode management is therefore important for high yields and quality that are required by the high cost of modern crop production. The major fruit growing states are facing remarkable loss due to nematode infestation. The major fruit crops which suffer severe nematode infestation, are discussed here.

2.2. ECONOMIC IMPORTANCE

2.2.1. International scenario

On a worldwide basis, the 10 most important genera of plant parasitic nematodes were reported to be as follows (Sasser and Freckman, 1987):

» Meloidogyne
» Pratylenchus
» Heterodera
» Ditylenchus
» Globodera
» Tylenchulus
» Xiphinema
» Radopholus
» Rotylenchulus
» Helicotylenchus

Crop losses due to nematodes are difficult to calculate accurately, with global estimates varying considerably from US$ 80 billion (101) to US$ 157 billion per year (Abad *et al.*, 2008). On a global scale, annual economic losses due to nematode infection of crops have been estimated at $173 billion (Elling, 2013).

2.2.2. Indian scenario

Based on data generated through AICRP on Nematodes over the years, a critical analysis has been made on losses in different fruit crops. Overall, plant parasitic nematodes on fruit crops cause 25.8% crop losses amounting to ₹ 31370.08 million annually. Citrus (₹ 9828.22 million) and banana (₹ 9710.46 million) among fruit crops suffered comparatively more losses. The details of crop losses in different fruit crops are provided in Table 2.1 (Walia and Chakraborty, 2018).

Table 2.1. Crop losses due to nematodes in different fruit crops

Fruit crop	Nematode	Production (million tons)	Yield loss (%)	Price per metric ton (₹)	Monetary loss (₹ in million)
Banana	*Meloidogyne incognita*	02.92 (29.22)	15	22170	9710.46
Citrus	*Tylenchulus semipenetrans*	01.16 (11.65)	27	31380	9828.22
Grapes	*M. incognita*	00.28 (02.82)	30	46910	3940.44
Guava	*Meloidogyne spp.*	00.40 (03.99)	28	20990	2350.88
Papaya	*M. incognita + Rotylenchulus reniformis*	00.49 (04.91)	30	17120	2516.64
Pomegranate	*Meloidogyne* spp.	00.18 (01.78)	23	73030	3023.44
Mean yield loss in fruit crops-25.5%		**Total monetary loss in fruit crops - ₹ 31370.08 million**			

2.3. EMERGING NEMATODE PROBLEMS

The changing cropping patterns, introduction of new crops, crop diversification, and agronomic practices *etc.* change the spectrum of plant parasitic nematodes. For example, drip irrigation in horticultural crops; diversification towards horticultural crops; and widespread and unchecked movement of planting materials from fruit crop nurseries, have led to the emergence of new nematode problems in newer areas and intensification of existing nematode problems. A number of new nematode problems have emerged in the intensive cropping systems.

2.3.1. Root-knot nematode on acid lime

The root-knot nematode, *Meloidogyne indica* is devastating the acid lime in Gujarat. Acid lime orchards in North Gujarat, particularly in Banaskantha district were observed to be heavily infected by *M. indica*. The citrus root-knot nematode infection was observed in Banaskantha, Mahesana, and Anand districts of Gujarat (Patel *et al.*, 1999).

2.3.2. Root-knot nematode on pomegranate

There has been an outbreak of serious root-knot nematode problems in the recent years associated with pomegranate (*Meloidogyne incognita* Race 2) disseminated through infected planting material and capable of killing the tress outright within three months after the onset of symptoms (Khan *et al.*, 2005). Besides causing direct damage, the root-knot nematode species are responsible for causing wilt disease complexes in association with soil-borne fungal plant pathogens. The root-knot nematode complex has similarly created serious situation with *Ceratocystis fimbriata* on pomegranate in Maharashtra, Karnataka, and North Gujarat states.

2.3.3. Root-knot nematode on guava

Till recently, *Meloidogyne incognita* and *M. javanica* have been reported associated with guava from various parts of the country (Ansari and Khan, 2012); however, these were not considered highly pathogenic to guava. The situation turned alarming with the interception of an exotic species of root-knot nematode, *Meloidogyne enterolobii* (syn. *M. mayaguensis*) on guava in Ayyakudi area (District Dindigal) of Tamil Nadu (Poornima *et al.*, 2016). Young guava trees witnessed heavy mortality (30-50%) within three months of first appearance of the symptoms (Poornima *et al.*, 2016; Ashokkumar and Poornima, 2019). Subsequent surveys revealed its occurrence in many districts of Tamil Nadu (Ashokkumar *et al.,* 2019) besides 10 other states (Andhra Pradesh, Telangana, Karnataka, Gujarat, Kerala, Haryana, Uttar Pradesh, Rajasthan, Uttarakhand, West Bengal of India) (AICRP-Nematodes Centres' Reports).

2.3.4. Root-knot nematode on mulberry

The root-knot nematode, *Meloidogyne incognita* is a major constraint in mulberry cultivation and plays an important role in reducing herbage yield and quality of leaves besides the life span of mulberry plants. It is responsible for 10-12% herbage yield loss and adversely affects the silk industry in Karnataka (Govindaiah *et al.*, 1991).

2.4. MAJOR NEMATODES OF FRUIT CROPS

The major nematode problems on fruit crops include root-knot, reniform, root lesion, burrowing, reniform, and citrus nematodes.

2.4.1. Root-knot nematodes, *Meloidogyne* spp.

Meloidogyne spp. are most damaging with wide host range. *Meloidogyne incognita, M. arenaria, M. javanica,* and *M. hapla* are the major species of this genera. Root-knot nematodes infect citrus, banana, papaya, pineapple, grapevine, passion fruit, fig, olive, kiwi fruit, and other fruit crops. In the northern part of country, *M. enterolobii* also becoming the emerging problem in guava orchards. Second- stage juvenile (J2) is the only infective stage. This stage survives in soil for several days in normal condition and in adverse condition it converts into anhydrobiotic phase. The nematode takes 25-30 days to complete its life cycle at 25- 30 ^{0}C.

2.4.2. Citrus nematode, *Tylenchulus semipenetrans*

This nematode cause 'slow decline of citrus'/ 'citrus dieback' which is characterized by a general reduction in tree growth, lack of vigor, yellowing of foliage and small size of fruits resulting in heavy loss to citrus crops (Family: Rutaceae). Some biotype attacks on olive, grapes, persimmon, loquat *etc.* Second-stage juveniles (J2) cause infection to fine feeder roots of plants. Generally, nematode completes its life-cycle in 6-8 weeks. The nematode is often unnoticed in the seedlings in nursery which causes widespread distribution.

2.4.3. Burrowing nematode, *Radopholus* spp.

Radopholus similis was first time recorded in 1966 on banana in India. *Radopholus citrophilus*, which is a serious pest of citrus in Florida, USA, is responsible for 'spreading decline' disease. This nematode has wide host range; however, it causes serious problem on banana, avocado, and other fruit crops. It is the causal organism of 'blackhead toppling disease'/ 'banana rot disease' of banana. All life stages of this nematode are capable of infection, except males. The life-cycle is completed in 4-5 weeks at 25-30 ^{0}C and all stages are found inside the roots.

2.4.4. Root lesion nematode, *Pratylenchus* spp.

This nematode is migratory endoparasitic in nature and having all life stages infective. *Pratylenchus* spp. are known worldwide, and they have very wide host range (>400 host plant species). The vernacular name 'root lesion nematode' is derived from the necrotic discolored patches (lesion) which develop on roots. Lesion nematodes infect citrus, banana, avocado, olive, kiwi fruit, and other fruit crops. *P. coffeae* infects banana and *P. penetrans* and *P. vulnus* in temperate fruits are serious problems. Life-cycle is completed in about 27 days at 26-32 ^{0}C in the banana crop.

2.4.5. Reniform nematode, *Rotylenchulus reniformis*

R. reniformis is the most prevalent species and attacks over 150 plant species of 50 families (Khan, 2005). It can injure several tropical fruit crops like papaya, avocado, and pineapple. The adult female of reniform nematode is kidney shaped, obligate and sedentary semi-endoparasite. It is associated with old-age banana decline. The reniform nematode infects pineapple, papaya, passionfruit, and other fruit crops.

2.4.6. Dagger nematodes, *Xiphinema* spp.

Dagger nematodes are migratory ectoparasites, that attack roots from the outside, feeding on epidermal cells with their long stylets. *Xiphinema index* is associated with significant decline of grapevine (*Vitis* sp.) and associated with virus transmission. Dagger nematode is an economic concern in orchards because it can vector viruses including tomato ringspot virus (TRSV), the cause of apple union necrosis and decline, the Cherry rasp leaf virus (CRLV) which causes flat apple, and Peach rosette mosaic virus (PRMV).

2.4.7. Ring nematodes, *Hemicriconemoides* spp., *Mesocriconema* spp.

Hemicriconemoides mangiferae is the main nematode damaging mango (*Mangifera indica*). Ring nematode is common to soft fruit orchard soils and in concert with the bacterial canker pathogen, *Pseudomonas syringae* and freeze damage. It has been associated with the phenomenon termed peach tree short life (Cao *et al.*, 2006; Ritchie and Clayton, 1981). Root feeding in Pome and Stone fruit can result in loss of up to 85% of small roots causing significant water and nutrient stress (Westerdahl and Duncan 2015). As higher populations of *Mesocriconema* are known to occur in sandy soils with large pore spaces.

2.5. RECENT ADVANCES IN NEMATODE MANAGEMENT

Presently practiced nematode management measures mainly orient towards chemical control. The excessive dependence on chemical nematicides leads to the development of resistance in nematode pathogens, outbreaks of biotypes and occurrence of residues in food chain. The chemical control also has other limitations such as high cost, low cost benefit ratio; poor availability; selectivity; temporary effect; efficacy affected by physico-chemical and biological factors; health hazards; toxicity towards plants, animals and natural enemies; environmental pollution, *etc.* Because of highly intensive horticultural practices and chemicalization, the age-old environment friendly pest management practices like sanitation, crop rotation, mixed cropping, adjustment of date of planting, fallowing, summer plowing, green manuring, composting, *etc*, to combat plant nematodes are not being practiced. The pace of development and durability of resistant varieties had been slow and unreliable.

Considering these limitations, there has been a growing awareness and increasing demand for novel and improved nematode management approaches to guarantee effective and sustainable fruit production which offers new opportunities for crop protection research. The most recent advances include precision nematode management, nano-formulations of nematicides, biofumigation, biotechnological approaches, bio-priming of seeds, biointensive integrated nematode management and others.

2.5.1. Biofumigation

As originally defined, the term "biofumigation" demonstrates the suppressive effects of Brassicaceae plant family on noxious soil-borne pathogens and is specifically attributed to the release of biocidal isothiocyanates (ITCs) due to the hydrolysis of glucosinolates (GSLs, thioglucosides) present in crop residues, catalysed by myrosinase (MYR, β-thioglucoside glucohydrolase) isoenzymes (Matthiessen and Kirkegaard, 2006). Stapleton *et al.* (1998) defined biofumigation as an agronomic practice that release volatile biotoxic compounds into the soil atmosphere during the decomposition of organic amendments. Cultivation and incorporation of cover/rotation crops, especially Brassicaceae plants, occasionally suppresses soil-borne diseases, including nematodes. Utilization of volatile toxic compounds, such as isothiocyanates generated from the glucosinolates in such crops, for soil-borne disease control is generally termed biofumigation (Kirkegaard *et al.*, 1993).

Biofumigation is a sustainable agronomic practice using naturally produced plant compounds to manage soil pests, including plant-parasitic nematodes (PPNs). This practice primarily relies on volatile organic compounds (VOCs) when they or their by-products are incorporated into soil. Many plants produce VOCs; however, the biofumigation practice is dominated by the use of glucosinolates (GSLs) that are hydrolyzed into isothiocyanates (ITCs) capable of killing or driving away PPNs. Biofumigation efforts concentrate on the Brassicaceae family, with mustards and radish at the forefront of success (Daneel *et al.*, 2018). Plant material is grown as a cover crop and tilled into the soil at maturity, where it breaks down to release ITCs. Biofumigation plants or by-products may also be incorporated as seed meal or applied as extractions added to water (Estupiñan-López *et al.*, 2017).

The use of certain crops as biological fumigants ahead of crop production to manage soil-borne pests is receiving considerable interest in recent times. The crops that have shown the potential to serve as biological fumigants include plants in the mustard family (such as mustards, radishes, turnips, and rapeseed) and Sorghum species (Sudan grass, sorghum-Sudan grass hybrids). The crops from the mustard family show some promise to reduce soil-borne pests by releasing naturally occurring compounds called glucosinolates in plant tissue (roots and foliage). When chopped plant tissues are incorporated in soil, they are further broken down by enzymes (myrosinase) to form chemicals (glucosinolates) that behave like fumigants. Isothiocyanates are the breakdown products of glucosinolates, which are the same chemicals that are released from metam-sodium (Vapam) and metam-potassium (K-Pam), commonly used as chemical fumigants.

Biofumigation research focuses on the most economically damaging PPNs, including root-knot nematodes (RKNs) (*Meloidogyne* spp.) and lesion nematodes (LNs) (*Pratylenchus* spp.) (Jones *et al.*, 2013).

2.5.2. Seed bio-priming or biological seed treatment

Biological seed treatments for control of seed and seedling diseases offer the grower an alternative to chemical fungicides. While biological seed treatments can be highly effective, it must be recognized that they differ from chemical seed treatments by their utilization of living microorganisms. Storage and application conditions are more critical than with chemical seed protectants and differential reaction to hosts and environmental conditions may cause biological seed treatments to have a narrower spectrum of use than chemicals. Conversely, some biocontrol agents applied as seed dressers are capable of colonizing the rhizosphere, potentially providing benefits to the plant beyond the seedling emergence stage (Nancy *et al.*, 1997).

Seed treatment with bio-control agents along with priming agents may serve as an important means of managing many of the soil and seed-borne diseases, the process often known as 'bio-priming'. The bio-priming seed treatment combines microbial inoculation with pre-plant seed hydration. Bio-priming involves coating seed with bacterial and fungal biocontrol agents such as *Pseudomonas aureofaciens* AB254, *Trichoderma harzianum, Purpureocillium lilacinum, etc.* and hydrating for 20 hr under warm (23 °C) conditions in moist vermiculite or on moist germination blotters in a self-sealing plastic bag. The seeds are removed before radical emergence. The biocontrol agents may multiply substantially on seed during bio-priming (Callan *et al.*, 1990).

PGPR application through seed bio-priming involves soaking the seeds for pre-measured time in liquid bacterial suspension, which starts the physiological processes inside the seed while radicle and plumule emergence is prevented (Anitha *et al.*, 2013) until the seed is sown. The start of physiological process inside the seed enhances the abundance of PGPR in the spermosphere (Taylor and Harman, 1990). This proliferation of antagonist PGPR inside the seeds is 10-fold than attacking pathogens which enables the plant to survive those pathogens (Callan *et al.*, 1990) increasing the use of biopriming for biocontrol too.

Bio-priming process had potential advantages over simply coating seed with *P aureofaciens* AB254. Seed priming often results in more rapid and uniform seedling emergence and may be useful under adverse soil conditions.

Induction of defense-related enzymes by biocontrol agents and chemicals, osmotic priming of seeds using poly ethylene glycol (PEG) solution is known to improve the rate and uniformity of seed germination in several crops (Smith and Cobb, 1991).

A successful antagonist should colonize rhizosphere during seed germination (Weller, 1983). Priming with PGPR increase germination and improve seedling establishment. It initiates the physiological process of germination, but prevents the emergence of plumule and radical. Initiation of physiological process helps in the establishment and proliferation of PGPR on the spermosphere (Taylor and Harman, 1990). Bio-priming of seeds with bacterial antagonists increase the population load of antagonist to a tune of 10-fold on the seeds thus protected rhizosphere from the ingress of nematode pathogens (Callan *et al.*, 1990).

2.5.3. Bioprotectants

The expanding hazardous effects of nematicides have led to check for eco-friendly biological control of PPNs, and hence biological nematicides are becoming more popular and adding as major component of biomanagement practices for PPNs. Bio-products containing fungal and bacterial antagonists rank high among the bio-nematicides used against PPNs (Askary, 2015a, b; Eissa and Abd-Elgawad, 2015). Their effect on root-knot and cyst nematode, has been well demonstrated in previous studies (Stirling, 1991; Meyer, 2003).

Biopesticides are the substances extracted from the natural materials such as living organisms (natural enemies) or their products (microbial products, phytochemicals) or they may be their by-products (semi-chemical), which are used in crop protection and management (Dybas, 1989). These biopesticides include broad area of microbial pesticides and biochemicals derived from the natural sources and microorganisms (bacteria, fungi, and nematodes). Biopesticides have a significant role in control and management of nematode population under the economic thresholds. They are released in masses to control nematode pathogens. In the integrated nematode management programs, biopesticides are an important component and are considered as effective and eco-friendly compared to synthetic materials, and also biopesticides do not persist in high number in the crop environment.

Microbial pesticides include microbes that are bacterium, fungus, virus, and protozoans, which act as biological control for plant pathogens. Many microorganisms colonize in the rhizoplane and rhizosphere of plants. Plant growth promoting microbes (PGPM) produce plant growth promoting substance as well as antibiotics, which are having capability to protect the plants from nematode disease (Siddiqui and Mahmood, 1996). Most of the research has been carried out among the nematode-antagonistic organisms including nematophagous fungi and bacteria (Askary and Martinelli, 2015). Among the nematophagous fungi, *Purpureocillium lilacinum* has antagonistic effects on root-knot nematodes (Sharon *et al.*, 2001). Abamectin, which is a mixture of macrocyclic lactone metabolites of fungus *S. avermitilis*, is used to control plant-parasitic nematode. Abamectin is active against root-knot and reniform nematodes (*M. incognita* and *R. reniformis*) and lesion nematodes (*Pratylenchus* spp.) (Faske and Starr, 2007). Use of *Trichoderma* spp. for management of plant-parasitic nematode has been confirmed (Haseeb and Khan, 2012).

Another parasitic bacterium, *Pasteuria* spp., is found to be effective on 323 nematode species (both plant-parasitic nematodes and free-living), which adds a plus point for its use as biopesticide (Chen and Dickson, 1998). Bacteria belonging to *Bacillus* spp. have shown greater potential in nematode management. *Bacillus cereus* strain S2 treatment in *M. incognita* resulted in a mortality of 90.96% (Gao *et al.*, 2016).

2.5.4. Precision nematode management

Uniform application of nematicides across the field and on an area-wide basis brings in habitat homogeneity which is congenial for nematode outbreaks. Similarly, eradication of all non-crop plants in the field and its vicinity by herbicide applications may make it easier for nematodes to locate crop plants and deprives beneficial predators and parasitoids of important food sources (Shelton and Edwards, 1983). Hence, there is a need to improve the efficacy of nematode-control strategies and to minimize adverse environmental impact by responding to the spatial variability in nematode occurrence and habitats. Crop diversity through the use of multi-line cultivars, varietal mixtures and intercropping enhances natural enemy populations that reduce the density of nematode populations through predation, parasitism and competitive interactions. The nematode population outbreaks can be retarded by fine-tuning of nematicide applications that are made only at 'hot-spots' where nematode densities reach action thresholds, which would create a mosaic of communities that differ in species composition and inter-specific interactions. The release of toxic chemicals into the environment would be greatly reduced by agro-ecologically based nematode management strategy i.e., precision nematode management (Weisz *et al.*, 1996; Brenner *et al.*, 1998).

Application of broad-spectrum nematicides is responsible for killing of natural enemies of the target nematode which can also lead to the resurgence of nematode populations (Hardin *et al.*, 1995). Site-specific applications through precision farming would minimize the exposure of predators and parasitoids to nematicides, and favor biological control which subsequently reduces nematode damage. Site-specific nematode management is responsible for the buffering effect of inter-patch dispersal and would make nematode population outbreaks less likely because it links sink and source populations. Sampling procedures are considerably influenced by the alteration of nematode distribution patterns in the field (approaches random or even distribution) requiring less sampling effort.

The environmental nematicide load is significantly reduced by precision farming leading to the reduction of material costs, as the necessary nematicide amount is 8-10% lower (calculated in active ingredient) than in case of traditional treatment. Takács-György *et al.* (2014) estimated that the amount of pesticides saved on the level of EU-25 countries is 31.7-84.5 thousand tons in case 15% of farms apply precision farming, 63.4-169.1 thousand tons in case 25% of them introduce it, while in the most favorable case (40%) it is 126.8-338.1 thousand tons.

The information needed to make soil and crop management decisions that fit the specific conditions found within each field is provided by precision nematode management that combines the best available technologies. It enables to take more informed management decisions by using GPS, GIS, and RS to revolutionize the way data are collected (at resolutions of 1 to 5 m) and analyzed. Precision nematode management has the potential to have detailed records covering every phase of the crop production process, thus enhancing sound nematode management decisions.

The agricultural crop production costs and crop and environmental damage can be potentially reduced by following precision nematode management.

2.5.5. Nanotechnology

Nowadays, the rapid development of nanotechnology presents a new way to improve the performance of conventional pesticide formulations through the construction of nanotechnology-based agricultural systems such as drug carriers and a controllable drug targeting and releasing system (Nakamura *et al.*, 2017). Many nano-products or nanomaterials were used against plant-parasitic nematodes such as nano-silver and nano-sulfur (Hardman 2006).

Nano nematicides are formulations of active ingredient of a nematicide in nanoform that have slow degradation, targeted delivery, and controlled release of active ingredient for longer period that make them environmentally safe and less toxic in comparison with conventional chemical nematicides. Several studies have reported an enhancement in the efficacy of certain biological substances on nematodes and a reduction of losses due to physical degradation through encapsulation of these substances in nanoparticulate systems.

Use of nanoparticles in nematode management is a novel and fancy approach that may prove very effective in future with the progress of application aspect of nanotechnology (Ladner *et al.*, 2008). The effect of applications of nanoparticles in nematode management can be divided into two perspectives; direct effect of nanoparticles on nematodes and use of nanoparticles in formulating nematicides. Due to their ultra-sub microscopic size, nanoparticles gain the high degree of reactivity and sensitivity and thus have potential to prove very useful in control of plant parasitic nematodes (PPNs), in addition, nematicidal residue analysis. As chemical and physical properties of nanoparticles vary greatly as compared to larger form, thus it has become imperative to evaluate the effect of nanoparticles on nematodes to harness the beneficial effects of this technology in plant protection, especially against PPNs. Ultra-small size and very high reactivity will affect the activity of PPNs (Gatoo *et al.*, 2014).

An indigenous bacterial strain isolated from indigenous gold mines identified as *Bacillus* sp. GPI-2 was found to convert $AuCl_4$ into AuNP of size 20 nm with the spherical shape. These bio gold AuNPs were evaluated for their ability to kill nematodes (Siddiqi and Husen, 2017). Gold and silver NPs possess nematicidal activity that may provide an alternative to high-risk synthetic nematicides or inconsistent biological control agents. High frequency (biweekly) and high application doses (90.4 mg/m^2) of gold and silver NPs may be required to achieve effective field efficacy for root-knot nematode (Kalishwaralal *et al.*, 2008). Combining AgNP with an irrigation system such as fertigation or tank-mixture with compatible chemicals that supplement the AgNP nematicidal effect may increase applicability of AgNP. Further understanding of the mechanism in the nematicidal action of Silver NP also warrants improvement of gold and silver NPs efficacy (Ganesh Babu and Gunasekaran, 2009).

Silver nanoparticle (AgNP) has shown evidence of being a potentially effective nematicide (Roh *et al.*, 2009), and its toxicity is associated with induction of oxidative stress in the cells of targeted nematodes (Lim *et al.*, 2012). Ag-nanoparticles of *Urtica urens* extracts concomitant with rugby were effective in the management of *M. incognita*, since it increased nematicidal activity 11-fold more than the least toxic extract against eggs (Nassar, 2016). The toxicity of three nanoparticles, silver, silicon oxide and titanium oxide, to the root-knot nematode, *M. incognita*, was recorded in laboratory and pot experiments (Ardakani, 2013).

2.5.6. New molecules of chemicals

2.5.6.1. Nimitz (Fluensulfone) (Adama): Nimitz is a systemic fluoroalkenyl compound that recently received U.S. Environmental Protection Agency (EPA) registration for use in crop plants. In crop production systems, Nimitz has received much interest as a methyl bromide alternative in part because it causes mortality of target nematodes within 24 to 48 hours of product application. This nematicide controls several types of plant-parasitic nematodes (e.g., root-knot, stubby root, sting, lesion, and needle nematodes) resulting in improved fruit quality and yield of several crops. Nimitz can be broadcast or applied through drip irrigation seven days before seeding or planting crops. Nimitz has shown the greatest efficacy in the control of plant-parasitic nematodes when used as pre-plant application. When applied on plant foliage, the active ingredient of Nimitz moves from the application point downwards into the roots, where it might affect parasitic nematodes. To prevent phytotoxicity and subsequent crop losses, foliar application of Nimitz is not recommended.

2.5.6.2. Velum Prime (Fluopyram) (Bayer CropScience): Velum Prime is manufactured as a liquid formulation and is one of the few available non-fumigant nematicides with systemic properties labelled for control of multiple species of plant-parasitic nematodes. This nematicide can be broadcast or applied through soil drench and drip irrigation at planting. Velum Prime is registered for vegetable crops including potatoes and sweet potatoes; cucurbit vegetables such as cucumbers, pickling cucumbers, squash, watermelons, and cantaloupes; fruiting vegetables such as tomatoes, eggplants, okra, and peppers; and brassica vegetables such as cabbage and broccoli.

2.5.7. Avermectins

Excitement is being generated by the results of a new class of insecticide - nematicide which has demonstrated a broad spectrum of anthelminthic activity. Promising results for the control of root-knot nematodes have been reported.

Avermectins (abamectin and emamectin) can be utilized to control plant parasitic nematodes because of their chemical and biological properties, as well as relative safety. Avermectins have short half-lives and their residues can be eliminated easily through different food processing methods. Both abamectin and emamectin are very effective nematicides which proved capability of reducing PPNs significantly in various crops.

Abamectin was applied as soil treatment for the management of citrus nematode in oranges (Table 2.2).

Table 2.2. Effect of abamectin on citrus nematodes in oranges

Compounds/Application methods	Crop/Target nematode	Effect on nematode reduction parameters (%)	Reference
Tervigo (2% SC) Soil application	Orange trees, *Tylenchulus semipenetrans*	Population-78-87	El-Tanany *et al.*, 2018

2.5.8. Biotechnological approaches

Due to an increasing awareness of potential adverse effects of pesticides on the safety of foods and on the environment, the development of new and safer pest control methods, including alternatives to widely used chemical pesticides, has become a serious priority for the entire agri-food industry.

2.5.8.1. Recombinant DNA technology: The genetic engineering leading to transgenic plants harboring nematode resistance genes has demonstrated its significance in the field of plant nematology. The strategies include natural resistance genes, cloning of proteinase inhibitor coding genes, anti-nematodal proteins and use of RNA interference to suppress nematode effectors. Furthermore, the manipulation of expression levels of genes induced and suppressed by nematodes has also been suggested as an innovative approach for inducing nematode resistance in plants.

2.5.8.2. Genome editing: The emergence of genome manipulation methods promises a real revolution in biotechnology and genetic engineering. Targeted editing of the genomes of living organisms not only permits investigations into the understanding of the fundamental basis of biological systems but also allows addressing a wide range of goals towards enhancing productivity including conferring resistance to various biotic stresses. A major area of application of genome editing approaches in plant breeding is to create varieties resistant to various pathogens and/or pests. These methods have been used for the modification of the key plant immunity stages at different levels in several crops.

2.5.9. Biointensive integrated nematode management

Integrated nematode management (INM) is an economically viable and socially acceptable approach to crop protection. INM is a system that, in the context of the associated environment and the population dynamics of the nematode species, utilizes all suitable techniques and methods in as compatible manner as possible and maintains the nematode populations below those causing economically unacceptable damage or loss. The criteria for judging the effectiveness of INM are productivity, stability, sustainability and equity. Sound knowledge of nematodes, cropping systems and the environment are important for the implementation and success of INM.

Biointensive integrated nematode management (BINM) is a systems approach to pest management based on an understanding of nematode ecology. It begins with steps to accurately diagnose the nature and source of nematode problems, and then relies on a range of preventive tactics and biological controls to keep nematode populations within acceptable limits. Reduced-risk nematicides are used if other tactics have not been adequately effective, as a last resort, and with care to minimize risks.

An important difference between conventional and biointensive INM is that the emphasis of the latter is on proactive measures to redesign the agricultural ecosystem to the disadvantage of a nematode and to the advantage of its parasite and predator complex

2.5.9.1. Proactive options: Proactive options, such as crop rotations and creation of habitat for beneficial organisms, permanently lower the carrying capacity of the farm for the nematode. The carrying capacity is determined by the factors like food, shelter, natural enemies complex and weather, which affect the reproduction and survival of a nematode species. Cultural control practices are generally considered to be proactive strategies. Proactive practices include crop rotation, resistant crop cultivars including transgenic plants, disease-free seeds and plants, crop sanitation, spacing of plants, altering planting dates, mulches, *etc.*

The proactive strategies (cultural controls) include:

» Healthy, biologically active soils (increasing below-ground diversity).
» Habitat for beneficial organisms (increasing above-ground biodiversity).
» Appropriate plant cultivars.

2.5.9.2. Reactive options: The reactive options mean that the grower responds to a situation, such as an economically damaging population of nematodes, with some type of short-term suppressive action. Reactive methods generally include inundative releases of biological control agents, mechanical and physical controls, botanical pesticides, and chemical controls.

2.5.10. Mitigation of climate change

Climate change will cause alterations in the spatial and temporal distribution of nematodes and consequently the control methods will have to be altered to suit these new situations. Assessments of the impact of climate change on nematode infestations and in crops provide a basis for revising management practices to minimize crop losses as climate conditions change (Ghini *et al.*, 2008).

Global warming resulting in elevated carbon dioxide (CO_2) and temperature in the atmosphere may influence plant pathogenic nematodes directly by interfering with their developmental rate and survival strategies and indirectly by altering host plant physiology. Available information on effect of global warming on plant pathogenic nematodes though limited, indicate that nematodes show a neutral or positive response to CO_2 enrichment effects with some species showing the potential to build up rapidly and interfere with plant's response to global warming. These findings underline the importance of understanding the impact of climate change on soil nematodes and its implications to crop production while developing mitigation and adaptation strategies to address impact of climate change on agriculture.

Prediction of changes in geographical distribution and population dynamics of nematodes will be useful to adapt the management strategies to mitigate the adverse effects of climate change on crop production. Nematode outbreaks might occur more frequently, particularly during extended periods of drought, followed by heavy rainfall. Some of the components of nematode management such as host plant resistance, bio-pesticides, natural enemies, and synthetic chemicals will be rendered less effective as a result of increase in temperatures and UV radiation, and decrease in relative humidity. Climate change will also alter the interactions between the nematodes and their host plants. As a result, some of the cultivars that are resistant to nematodes, may exhibit susceptible reaction under global warming. Adverse effects of climate change on the activity and effectiveness of natural enemies will be a major concern in future pest management programs. Rate of nematode multiplication might increase with an increase in CO_2 and temperature. Therefore, there is a need to have a concerted look at the likely effects of climate change on crop protection, and devise appropriate measures to mitigate the effects of climate change on food security.

Crop protection professionals routinely develop and deploy strategies and tools based on well-established principles to manage plant nematodes and many may also be applicable under climate change when projected changes, processes and interactions are factored in. Therefore research to improve adaptive capacity of crops by increasing their resilience to nematodes may not involve a totally new approach, although managing plant nematodes may have the added advantage of mitigating rising CO_2 levels.

Recent observations suggest that nematode pressure on plants may increase with climate change (Ghini *et al.*, 2008). As a result, there may be substantial rise in the use of nematicides in both temperate and tropical regions to control them. Non-chemical nematode management methods (green manuring, crop rotation, mulching, application of organic manures, *etc.*) assume greater significance under changing climate scenario.

A major shift in nematode-management strategies is occurring, from almost exclusive reliance on soil-applied nematicides, to the use of combinations of alternative strategies such as crop rotation, host plant resistance, cultural manipulations and biological control (Ferris *et al.*, 1992). Greater understanding of nematode genetic response and adaptation to abiotic factors will be important in optimizing the design of cultural management tactics such as manipulations of planting and harvest times, wet or dry fallow, and soil solarization. Wet or dry fallowing starve nematodes while active (wet fallow) or die from extreme moisture stress (dry fallow). Soil solarization involves natural heating of soil under plastic cover to attain the thermal death point of nematodes. Avoidance may include changes in planting and harvest dates, such as delaying planting in the fall to avoid infection activity, and early planting or late harvest of crops to avoid additional nematode generations. Soil amendments such as green manures and various bio-protectants show promise for nematode suppression in some systems. Some potential biological control agents are known to give effective management of nematodes. Current research is addressing many areas of nematode management, including: development and deployment of nematode-resistant plants; crop rotation to reduce population densities of these pathogens; cover crops, trap crops and soil amendments including green manures to reduce population densities; the role of weed hosts in bringing about phenotypic changes in nematode populations; characterization of resistance genes and resistance responses; the development of molecular diagnostic protocols for nematode identification and the reference databases necessary for their implementation. Efforts to develop improved techniques for nematode diagnostics, understanding of nematode diversity and genetic variability, processes of nematode fitness and adaptation, and the incorporation of this knowledge into the design and analysis of improved nematode management strategies needs to be intensified.

2.6. NEW PARADIGMS FOR NEMATODE MANAGEMENT

Biological control amplifies biotic imbalances, because the economic pressures compel drastic methods based on the physical elimination of antagonists. High grade results achieved by crop protection researches carried out on crop cultural practices, plant resistance to nematodes or biocontrol, are very diverse. But, all these control practices, included or not in integrated pest management strategies, seem not to be sufficient: they all target some nematode species (population approach), and then involve changes in nematode communities but do not necessarily modify their overall pathogenicity. So, binary researches focused on plant-nematode or nematode-parasite relationships should be extended to ecological investigations on nematode communities.

These reasons support eco-epidemiological approaches, based on the ecology of communities (interspecific competitions, biological, and edaphic constraints) and focused on the management of the natural enemy biodiversity, appearing as a new research alternative for crop protection.

Specific and functional evolution and rapid multiplication of plant-parasitic nematodes are enhanced by agriculture intensification and by environment adaptation. In fact, plant-parasitic nematodes are only considered as predators, and not as one of the soil components. Up-to-date, all control strategies developed in agriculture are focused on the eradication of target species. They induce biotic gaps, community rearrangements, insurgence of virulent races, increased aggressiveness of minor species, *etc.* Then, the "soil cleaning" strategy appears to be not sustainable.

The development of sustainable management strategies must move from the "therapeutic approach", mostly adopted in research programs carried out in the world, to the "ecological approach". This approach will seek for information and knowledge about biotic trade-offs in ecosystems, in order to introduce them in agro-systems (resilience). That questioning can be declined as follows:

» Why seek with plant-parasitic nematode eradication?

» Can agronomic problems be solved by agronomic strategies only?

» Can nematode diversity in communities be considered as an auxiliary for nematode management?

Effector biogenesis has not been studied for 30 years because the system was intractable. With recent development in functional genetics, the E-biogenesis is now well placed to dissect this strategically important knowledge gap. E-biogenesis aims to investigate methods of blocking crop parasitism by attacking the biogenesis machinery that produces the molecules that cause disease. The vision of the E-biogenesis is centred on the ability of nematodes to produce molecules that cause disease; so called "effectors". Blocking effectors blocks parasitism. However, blocking individual effectors is insufficient. Rather, there is a need to attack the features that unite most effectors; their biogenesis machinery.

This research will highlight a multiplicity of attractive targets for the biotechnological control of globally important nematode pests. To achieve this, there is a need to open a whole new area of effector study, contribute new fundamental knowledge of strategic relevance, and provide a platform for a paradigm shift from individual effectors to a holistic view of effector biogenesis. Surely, effector biogenesis (*E-biogenesis*) is the Achilles heel of the nematode. Recently (2020), a project on E-biogenesis has been granted by European Research Council to Dr. Sebastian Eves-van den Akker, Crop Science Centre, University of Cambridge.

2.7. FUTURE OUTLOOK

Current research for nematode management is focusing on the development of effective cultural practices such as those traditionally practiced before the advent of broad-spectrum nematicides. In looking to the future, what are some of the things that we need to do from a realistic and practical stand point of view?

Through surveys and identifications, determine more accurately the numbers and kinds of nematodes which actually exist and pinpoint their distribution. Success of crop rotations and resistant cultivars depend upon knowing what nematodes occur in which areas and fields.

With regard to viruses transmitted by soil-borne nematodes, it is advisable to explore the possibility of finding dorylaimid nematodes from the rhizosphere and their role in transmitting soil-borne viruses mainly in crops like fruits, vegetables and ornamentals.

We should continue research on nematode management to include efforts to obtain (i) nematicides which are more effective, can be used at lower dosage, and cause less pollution; (ii) resistant cultivars for major crops and at least for the most important nematode species; and iii) knowledge necessary to utilize the benefits and capabilities offered by integrated nematode management systems.

There is a need to develop a better understanding of the biological aspects of all our more important plant nematodes. This would include the molecular genetics of closely related species and races; host-parasite relationships and histochemical details which may provide clues to the nature of plant resistance to nematodes; ecological factors, such as temperature, moisture and soil types which may influence nematode development and survival which in turn affect control measures; and make full use of current biotechnologies as appropriate in this and other nematology research.

There is a need for development of cropping systems based nematode management technologies. Use of green manuring plants such as *Crotalaria spectabilis* prevents the nematode from reproducing. Coastal Bermuda grass *(Cynodon dactylon)* incorporated before planting tobacco or vegetable transplants protects against root-knot nematodes (*Meloidogyne* spp.) (Burton and Johnson, 1987). These plants could be even more effective if they could be genetically engineered to produce nematode attractants or pheromones.

Another area of concern is to advance biological control to its fullest potential, utilizing biological control agents plus organic amendments, sewage, sludge, *etc*. This would include cultural methods for biocontrol parasites, understand their nature, make them practical, and genetically modify the biocontrol agents to improve their effectiveness.

The present research thrust on biological control of nematode pests need to be intensified. More basic studies in biological control and interactions in the rhizosphere are required. Improved assay techniques for assessing the biological antagonism coefficients of various soils must be developed.

Many fungal and bacterial bioagents have been found effective under *in vitro* conditions. Efforts should be made to test them under field conditions. Further, our efforts should be directed towards developing mass production technologies for effective bioagents and easy field application techniques.

In recent years, it has been observed that the combination of fungal bioagents perform better in reducing nematode population and increasing the plant health. Thus the combination of highly toxic fungus, *Aspergillus niger* (killed most of the infective second stage juveniles) and an egg parasite, *Cladosporium oxysporum* (invaded and killed the eggs in egg sac) both at half the doses reduced significantly more *M. incognita* population and exhibited better plant growth than when either of the fungal bioagents in brinjal (Goswami and Singh, 2002).

Another promising area of research in nematode management will be to identify and develop new generation chemicals which may not kill the nematodes but either change their behavior or delay the development and force them into the inactive stages or alert the plant to defend the nematode attack.

The selection and development of varieties for resistance and tolerance to nematode stress will continue. This may involve incorporation of appropriate genetic material into varieties already selected for production, economic, and marketing qualities. It is still important to develop biological and chemical nematicides that are systemic, easily associated with the root system, target specific organism, or a combination of these factors. These pesticides will allow flexibility in management decisions and compensation for management errors that have promoted or amplified nematode stress problems in a particular production system.

Phytoalexins effectively induce plant resistance mechanisms to nematodes, particularly to sedentary ones. However, the mechanism by which phytoalexins are accumulated in plants is not known. Further research in accurately elucidating the effect of potential phytoalexins on the nematodes is urgently needed.

Although valuable improvement has been obtained in development of nematode resistant horticultural crop cultivars, there is a need for breeding varieties with combined resistance to different nematode species and to other organisms such as soil-borne fungi and bacteria which cause wilt and root rot disease complexes.

The development of transgenic plants with durable nematode resistance keeping pace with the evolution of resistance breaking populations should be considered. Biotechnological applications including RNAi or miRNA represent a potential breakthrough in the application of functional genomics for plant nematode control. Here, a comparison is made between some old and modern nematode management practices but recent data shows that application of RNAi or miRNA has a better option of nematode control in some crop plants.

Once the INM technologies are developed, it is necessary that they should be put to practice under field conditions. The implementation of these programs will require scientists/trained technicians and efficient extension media. In this direction, National Institutes and SAU's can play an important role by holding training courses and carrying out problem oriented researches. Since integrated approach may involve more than one discipline, it is therefore, essential that professional scientists must collaborate in its planning and execution.

Our future research programs need to be more based towards development of sustainable and non-chemical methods for the management of nematode pests. We have made much progress in nematology; but we still have a long way to go in the battle against these destructive and insidious plant nematodes. Through our continued dedication and strong research efforts, greater knowledge and better understanding of nematodes will be achieved. This will result in further reductions in losses caused by nematodes and a significant increase of food in our countries.

2.8. CONCLUSION

Considering the importance of economic losses caused by nematodes and the fact that the restrictions governing the use of chemical nematicides are elevating, it is clear that there is need for the development of new environmentally benign strategies. It is also crucial to continue improving the current green methods in search of making them more efficient. For example, some questions to be addressed for antagonistic fungi/bacteria which include what are the optimum rate, timing, frequency, and method of application for biocontrol agents especially under field conditions?

It is critical to consider the development and improvement of multidisciplinary management strategies for nematodes such as combining microbial strategies using both bacterial and fungal agents with other cultural control practices or host resistance. Both biocontrol and application of soil amendments have been studied to some extent against PPN, but with the recent advances in technology there is room for deeper studies on how these two strategies can be synergistically used.

It is thus necessary to support a cognitive approach to study ecology of nematode communities which will aim to include/understand assemblage rules, their impacts on populations in communities (life trait evolution, adaptation), and the consequences on system functions and management.

Crop practices favor monocultures, which reduce soil biodiversity. That supports proliferation of nematodes. Imbalances induced by these situations strengthened by large inputs, would solve economic questions only. Such strategies delay the effective control of nematodes. On the other hand, taking into account the whole nematode diversity, according to an ecological approach, makes it possible to consider more suitable strategies for sustainable management of these parasites in agriculture, based on an auto-regulation of the global pathogenic effect of the nematode communities. It is consequently critical to focus control practices on agronomic methods because they involve a great impoverishment of the inputs generated by the agro-industrial research (pesticides, biopesticides, and resistant seeds), conferring an acute brittleness in terms of sustainability at the same time.

Integrated pest management should hence flank the development of nematodes more than killing them. We must be convinced that both biological balances in soil and environmental factors are fundamental and non-avoiding helps for plant-parasitic nematode management. Every strategy processed in order to eradicate plant parasites would hence invariably prove to be unsustainable.

Biotechnology-, molecular biology-, precision agriculture, and nanotechnology-based approaches have added a new dimension to nematode disease diagnosis and management. Identification of genes that reduce nematode's ability to reproduce has allowed the breeding of nematode-resistant plants. Marker-assisted selection, genetic engineering, and RNA interference to confer resistance in crop plants, nematode suppression using host plant proteinase inhibitors, and genome-editing technologies have helped tremendously in developing management strategies for plant-parasitic nematodes.

In conclusion, future studies should focus on environmentally benign approaches which are based on multidisciplinary strategies that can fill the gaps of single sided management methods. Such approaches will also reduce the chance of resistance as the complexity of different nematicidal components would make resistance highly improbable. Whatever strategy is devised, future attempts should focus on important factors such as synergism between nematode antagonists, environmental conditions, sustainability, studying the effect of new treatments on non-target organisms, and association of individual plants with nematode antagonists of interest. In summing up, a sustainable management of plant-parasitic nematodes is feasible when two or more compatible tactics are applied concurrently while appraising environmental protection.

Chapter - 3

Banana, *Musa* spp.

3.1. INTRODUCTION

3.1.1. Origin

Banana originated in Asia.

3.1.2. Area, production and productivity

Country-wise area, production and productivity of banana is presented in Table 3.1.

Table 3.1. Country-wise area, production and yield of banana (2016)

Country	Area (million ha)	Production (million tons)	Productivity (tons/ha)
India	0.85	29.12	34.43
China	0.43	13.32	30.98
Philippines	0.46	5.83	12.77
Brazil	0.47	6.76	14.40
Indonesia	0.14	7.01	50.06
Ecuador	0.18	7.01	38.86
Guatemala	0.08	3.78	48.27
Angola	0.13	3.86	29.35
Baurundi	0.20	0.91	4.67
Others	2.0	113.28	16.89
WORLD	5.49	113.28	20.62

State-wise area, production and productivity of banana is presented in Table 3.2.

Table 3.2. State-wise area, production and yield of banana in India (2017-18)

State	Area ('000 ha)	Production ('000 tons)	Productivity (tons/ha)
Andhra Pradesh	88.96	5003.07	56.239
Arunachal Pradesh	2.21	14.08	6.371
Assam	53.08	913.27	17.205
Bihar	31.07	1396.39	44.943
Chhattisgarh	26.57	745.78	28.068
Gujarat	68.15	4472.32	65.624
Himachal Pradesh	0.08	0.35	4.375
Jharkhand	9.17	32.06	3.496
Karnataka	110.55	2328.90	21.066
Kerala	109.26	1119.16	10.243
Madhya Pradesh	26.38	1834.03	69.523
Maharashtra	80.88	4209.27	52.043
Manipur	6.93	93.48	13.489
Meghalaya	7.37	96.90	13.147
Mizoram	11.21	143.84	12.831
Nagaland	8.34	117.04	14.033
Odisha	24.20	449.82	18.587
Punjab	0.09	5.27	58.555
Rajasthan	0.03	0.68	22.666
Sikkim	1.28	3.71	2.898
Tamil Nadu	82.63	3205.04	38.787
Telangana	2.29	90.02	39.310
Tripura	10.29	109.40	10.631
Uttar Pradesh	69.38	3172.33	45.724
West Bengal	49.30	1200.00	24.340
Others	4.07	51.29	12.601
TOTAL	883.77	30807.50	34.859

The major banana producing districts in different states in India include:

» Tamil Nadu (Theni, Tiruchirappalli, Coimbatore, Tuticorin, Pudukkottai *etc.*)

» Gujarat (Bharuch, Anand, Narmada, Surat, Vadodara *etc.*)

» Andhra Pradesh (Kadapa, Anantapur, East Godavari, West Godavari, Vizianagaram *etc.*)

» Uttar Pradesh (Gorakhpur, Kushi Nagar, Fatehpur, Kaushambi, Maharajganj *etc.*)

» Maharashtra (Jalgaon, Nanded, Solapur, Akola, Pune *etc.*)

3.1.3. Commercial varieties

Selected commercially cultivated varieties in major banana growing states are listed in Table 3.3.

Table 3.3. State-wise commercially cultivated banana varieties

State	Varieties grown
Tamil Nadu	Virupakshi, Robusta, Rad Banana, Poovan, Rasthali, Nendran, Monthan, Karpuravalli, Sakkai, Peyan, Matti
Gujarat	Dwarf Cavendish, Lacatan, Harichal (Lokhandi), Gandevi Selection, Basrai, Robusta, G-9, Harichal, Shrimati
Andhra Pradesh	Dwarf Cavendish, Robusta, Rasthali, Amritpant, Thellachakrakeli, Karpoora Poovan, Chakrakeli, Monthan and Yenagu Bontha
Uttar Pradesh	G-9 (Tissue Culture), Dwarf Cavendish, Alpon, Chinia, Chini Champa *etc.*
Maharashtra	Dwarf Cavendish, Basrai, Robusta, Lal Velchi, Safed Velchi, Rajeli Nendran, Grand Naine, Shreemanti, Red Banana

3.1.4. Advance technology

- Planting of tissue cultured plants.
- High density planting

3.1.5. Pre-cooling

Forced air pre-cooling or hydro-cooling at 10 to 13 ^{0}C and RH at 90-95%, to remove field heat.

3.1.6. Storage conditions

- Holding room: 13° to 14°C
- Ripening room: 15° to 18°C
- Relative humidity: 90-95%
- Storage period: Mature green bananas can be stored up to 3 weeks in ethylene free air or up to 6 weeks in a controlled atmosphere at 14 ^{0}C

3.1.8. Export from India (2017-18)

- Quantity : 101.31 Thousand MT
- Value : 34,877.39 Lakh Rupees

Banana found to be dominated in the export scene of fruits which has gained 19.37 per cent of growth rate annually and it is expected to reach the level of 1.88 lakh tons by the year 2020-21.

3.1.8. Nematode problems

Among these major biotic stresses, plant parasitic nematodes are causing severe losses in banana production not only individually but also with the interaction of wilt fungi, *Fusarium oxysporum* f. sp. *cubense* (Gowen and Quénéhervé, 2005). The major nematode pests of banana is the burrowing nematode, *Radopholus similis*. *Pratylenchus coffeae, Meloidogyne incognita, Helicotylenchus multicinctus, Heterodera oryzicola* and *Rotylenchulus reniformis* are the other nematode parasites of banana that are of some economic importance.

3.2. BURROWING NEMATODE, *RADOPHOLUS SIMILIS*

Cobb (1893) first described the burrowing nematode from necrotic root lesions of banana growing in Fiji. The burrowing nematode is responsible for decline, blackhead, *Radopholus* root rot, and blackhead toppling disease of bananas. Nair *et al.* (1966) first reported this nematode on banana from Kerala state, India.

3.2.1. Distribution

The nematode occurs in all the tropical and sub-tropical areas of the globe. In India, it is distributed in all the banana growing states viz., Tamil Nadu, Kerala, Karnataka Goa, Maharashtra, Gujarat, Madhya Pradesh, Assam and Tripura.

3.2.2. Crop losses

The burrowing nematode is responsible for 30.76 to 41.00 per cent yield loss in banana (Rajagopalan and Naganathan, 1977; Nair, 1979; Parvatha Reddy *et al.*, 1996c; Vadivelu *et al.*, 1987). Root population of *R. similis* is indirectly correlated with the yield (Charles *et al.*, 1985).

3.2.3. Symptoms

The following are the symptoms:

» Dwarfing of plants

» The size and number of leaves are reduced and bunch weight is also reduced.

» Dark red lesions appear on roots which is the identification of fungal infection after nematodes.

» The whole plant topples down in windstorms or heavy rains due to less vigorous roots after infection of plants.

» Heavily infected plants have thin pseudo stems, while the leaves are yellowish or discolored, greenish yellow bands appeared along the leaf blades (Milne and Kuhne, 1968).

» Infected plants are also prone to wilting during hot days because the plants could not take water due to root damage.

These are migratory endoparasitic nematodes which live in inter cellular spaces in cortical parenchyma from where they penetrate the nearby cells, suck their cell sap and destroy them by forming cavities. After destroying one site, the nematode moves to another site for feeding. These cavities coalesce with one another and produce characteristic reddish brown lesions throughout the cortex.

Wounding of banana roots by the burrowing nematode usually induces reddish-brown cortical lesions which are diagnostic of the disease. The cortical lesions can be clearly observed when the infected root is examined immediately after it is longitudinally split open. Root damage due to *R. similis* is responsible for various above-ground symptoms like stunting of plants, yellowing of leaves, and uprooting of mature plants.

With the increase in nematode population, feeding roots are invaded and destroyed as fast as they are formed. The resulting setback in the uptake of plant nutrients leads to debility and smaller fruit. The lesioning of the primary roots together with the girdling and death of those roots which anchor the plant to the ground makes the plant prone to tip over under wind pressure.

3.2.4. Host range

Poucher *et al.* (1967) list 244 hosts for the burrowing nematode. It is known to infect *Musa* spp., *Ipomoea batatas, Pueraria phaseoloides javanica,* pepper, areca nut, coconut, ginger, turmeric, sugarcane, coffee, maize, avocado, tea, vegetables, ornamentals, trees grasses and weeds.

3.2.5. Life cycle

The burrowing nematode has a migratory endoparasitic habit and although the stages remain vermiform throughout, sexual dimorphism is apparent with adult males being somewhat degenerate and probably non-parasitic. Eggs are normally laid in infested tissue over 7-8 days at the rate of about 4 eggs per day. Eggs take 8 to 10 days to hatch, while larvae take 10 to 13 days to mature. The nematode completes its life cycle by about 20-25 days.

3.2.6. Spread and survival

The most important means whereby the burrowing nematode is spread from one locality to another is through planting material (corms). Trimming sets (paring), which is often done before they are planted, is usually not sufficiently rigorous to remove infection that extend deeply into a set.

R. similis may be disseminated by water that drains from infested areas and reenters either irrigation channels or streams from which water is being taken to irrigate bananas. Soil that adheres to implements, tires of motor vehicles and shoes of plantation workers may also spread *R. similis.*

After a plantation is destroyed, the corms may remain succulent for several months and nematodes in these corms probably survive for as long as the corms remain succulent. The nematode cannot survive for more than 12 weeks without its host.

3.2.7. Physiological races

Venkitesan (1976) reported four host types of *R similis*. The population of *R similis* isolated from pepper roots did not infect *Citrus sinensis, C. reticulata, C aurantifolia, Coffea arabica,* coconut and areca nut; but infected banana only. Banana population infected pepper, coconut and areca nut; coconut population infected pepper and banana but not areca nut; while areca nut population infected pepper and banana but not coconut.

3.2.8. Management

3.2.8.1. Regulatory methods: The unintentional planting of banana suckers already infected with nematodes is the principal means of nematode dispersal to uninfested sites. In order to restrict the nematode spread, the first line of defense is to obtain planting stock free from nematode infestation. Certified nematode-free sets can be obtained by either paring/heat treatment/selection/tissue cultured plants. Micro-propagation from nematode-free banana materials using sterile techniques offers a good way to obtain nematode-free planting materials.

3.2.8.2. Physical methods: Corms of suckers should be trimmed carefully to remove the nematode affected necrotic tissue (Paring) before planting.

The pared sets can be disinfected by dipping them in a hot water bath maintained at 55 °C for 25 minutes without causing significant set losses. Fungicides applied after treatment would probably decrease any post-treatment losses caused by the fungi.

Sun drying the rhizomes of banana suckers wrapped in plastic sheet for four hours before planting has been shown to be effective in the control of nematodes, which is being commonly practiced in Kerala.

Storage of large corms in the sun for two weeks prior to planting to kill the nematodes and their juveniles.

3.2.8.3. Cultural methods

i) Cover cropping: After destroying an infested plantation, planting of cover crops like *Panicum maximum* var. *trichoglume*, marigold, and perennial legume (*Phaseolus atropurpureus*) did not support *R. similis* population. Planting of sugarcane or Pangola grass (*Digitaria decumbens*) following the destruction of bananas eliminated *R. similis* after 10 weeks (Loos, 1961).

Marigold is suitable allelopathic cover crop due to its low growing habit which can be incorporated into soil at flowering for nematode control. Only living marigold root systems exhibit nematicidal properties. Farmers should aim to plant marigold at 4 to 5 months after banana planting to maximize marigold growth and minimize competition from banana growth.

Planting of cover crops like *T. patula* and sun hemp (*Crotalaria juncea*) prior to banana crop planting in field soil infested with plant-parasitic nematodes significantly reduced the nematode population densities after 5 months.

Use of cover crop calapogonium is very useful to avoid the burrowing nematode population.

ii) Fallowing: Fallowing for a period of 3 months after banana effectively suppressed the burrowing nematode population. Rajendran *et al.* (1979) reported combined management of the burrowing nematode and Panama wilt by way of five months flood fallowing.

iii) Crop rotation: Rotation of banana with paddy, sugarcane, green gram, coriander, cotton, Pangola grass, sorghum, tobacco or cassava suppressed the nematode population and increased the yield (Rajendran *et al.*, 1979). The nematode population can be reduced by rotation of banana with cereals, cotton and vegetables in garden lands and with rice in wet lands. Rice or green gram grown after banana suppressed the population of *R. similis, P. coffeae* and *H. multicinctus* (Rajendran *et al.*, 1979).

iv) Intercropping: The nematode population both in soil and roots were considerably reduced when banana was intercropped with nematode antagonistic plants such as marigold. *Crotalaria juncea* was most effective in reduction of soil nematodes and *Coriandrum sativum* was more effective in reduction of root nematodes. *C. juncea* gave maximum bunch weight (16.7 kg) recording 51% more than control (11.03 kg) followed by *Tagetes erecta* (16.2 kg), *Sesamum indica* (15.87 kg), *Acorus calamus* (15.67 kg) and Carbofuran (15.23 kg) (Charles and Venkitesan, 1993).

Intercropping with sun hemp, coriander, marigold, radish or lucerne reduced nematode population and increased yield (Vadivelu *et al.*, 1987). Intercropping sun hemp in banana field and plowing the green manure 45 days after sowing was found effective in reducing *R. similis, P. coffeae* and *H. multicinctus* population by 38.4%, followed by marigold and cowpea which recorded 29.0 and 22.3% reduction, respectively (Shanthi, 2003). Increase in plant growth and yield and reduction in nematode population is achieved by intercropping banana with *Crotalaria* (Charles *et al.*, 1985). Lakshmana Murthy (1983) reported that when banana was intercropped with papaya and marigold, *R. similis* was not encountered. Significant reduction in the burrowing nematode population was reported by intercropping with radish, *Crotalaria* and marigold.

v) Mulching: Mulching with black polythene at 20 per cent moisture depletion recorded highest yield with low population of *R. similis* (Vadivelu *et al.*, 1987). Bhattacharya and Rao (1984) found that reduction in the population of the burrowing and lesion nematodes in banana roots by mulching soil with banana and sugarcane trash, and black polyethylene sheet.

vi) Organic amendments: Castor cake at 100 g/plant is effective against *R. similis* in banana. Application of Neem cake at 800 g per plant (two split applications) or Neem coated urea at 110 g per plant reduced the population of the burrowing nematode and also increased the yield of banana (Vadivelu *et al.*, 1987). Maximum reduction in the burrowing nematode population and increase in plant growth and yield were observed by soil application with neem products such as Nimbicidine and Econeem.

Application of press mud obtained from sugarcane waste treated plants has shown significant reduction in nematode populations (Sundararaju *et al.*, 2004).

3.2.8.4. Chemical methods: Paring and pralinage with Carbofuran at 1.2 g a.i. per sucker recorded low incidence of R. similis and highest yield (Rajendran et al., 1979). Dipping of banana suckers in 750 ppm of Carbofuran for 10 min, 750 ppm of Fensulfothion for 10 min, 0.03% solution of Parathion for 20 min, 50 ppm of Phorate for 15-30 min increased plant height, bunch weight, fingers per bunch, induced early flowering and effectively reduced the burrowing nematode population.

Soil application of Fenamiphos at 3 kg a.i. per ha was effective in increasing plant height, plant girth, number of leaves and yield and in reducing the nematode population (Parvatha Reddy et al., 1991). Various other chemicals (Carbofuran, Ethoprofos, and Oxamyl,) have also been found to be effective for the management of R. similis on banana under field conditions.

3.2.8.5. Biological methods

i) Antagonistic bacteria: Soil application of *Pseudomonas fluorescens* at 20 g/plant (2.5 x 10^8 cfu/g) was effective in increasing the plant growth parameters (pseudo stem height and girth, number of leaves, root length and weight) and in reducing the nematode population in soil (62.34, 67.12 and 58.87% of *R. similis, P. coffeae* and *H. multicinctus*, respectively) and roots (60.35, 62.00 and 56.54% of *R. similis, P. coffeae* and *H. multicinctus*, respectively). Rhizosphere treatment with *P. fluorescens* at 20 g/250 ml water along with soil treatment with 100 g powder/rhizome gave 56.5% reduction in nematode population. During the second month the nematode reduction was 63.8% (Nirmal Johnson and Devarajan, 2004).

In a glasshouse experiment, maximum increase in plant growth of banana and reduction in nematode population in soil and roots and root lesion index was observed in pots treated with talc based formulation of *P. fluorescens* (isolate PfB 13) at 10 g/plant which was on par with *P. fluorescens* (isolate Pf 1) and carbofuran (Senthil Kumar *et al.*, 2008a). Field application of *P. fluorescens* (isolate PfB 13) at 10 g/plant significantly reduced nematode population in soil and roots and root lesion index; and increased plant height, pseudo stem girth, leaf area, number of leaves and fruit yield (Senthil Kumar *et al.*, 2008b) (Table 3.4).

Table 3.4. Effect of different isolates of *Pseudomonas fluorescens* on plant growth, yield and management of *Radopholus similis* infecting banana

Treatment	Plant height (cm) (270 DAP)	Pseudo stem girth (cm) (270 DAP)	Bunch weight (kg)	Root lesion index
P. fluorescens - PfB 13	328.2	57.4	15.6	14
P. fluorescens - PfB 19	287.3	46.4	13.4	32

P. fluorescens - PfB 24	270.3	47.3	11.4	28
P. fluorescens - PfB 32	266.8	41.2	11.2	42
P. fluorescens - PfB 35	287.8	45.2	13.1	46
P. fluorescens - PfB 39	278.6	46.1	10.1	36
P. fluorescens - Pf 1	275.4	45.3	14.3	18
Carbofuran	270.9	43.1	13.2	19
Control	210.9	39.7	10.9	54
CD (P = 0.05)	26.59	6.99	0.96	1.37

Streptomyces costaricus and *Bacillus thuringiensis* were effective for the management of *R. similis* and *Helicotylenchus multicinctus* on banana plants (Esnard *et al.*, 1998).

Soil application of two isolates of *Bacillus subtilis* (EPB 5 and EPB 31) gave significant increase in plant growth parameters, pseudo stem girth, number of leaves coupled with reduction in nematode population (*M. incognita, P. coffeae, R. similis* and *H. multicinctus*) on tissue cultured banana (Jonathan and Umamaheswari, 2006).

Soil application of biofertilizer, *Azospirillum* at 20 g/plant had significant effect in enhancing the plant growth and reduced the population of *R. similis* in soil to an extent of 50.1% over control (Shanthi *et al.*, 2004).

ii) Antagonistic fungi: Application of fungal antagonists like *Purpureocillium lilacinum, Trichoderma viride, Pochonia chlamydosporia* and bacterial antagonists like *Pasteuria penetrans* were found effective in reducing the nematode population in soil and roots of banana.

Soil application of bio-control agents like *Purpureocillium lilacinum* and *Trichoderma harzianum* to the plants along with neem cake and farm yard manure (FYM). One kg of each of these bio-agents can be used to enrich 1 ton of farm yard manure. Farm yard manure added with these bio-agents should be left under shade for a period of 15 days and in between it is advisable to mix FYM thoroughly at an interval of 3 days to uniformly enrich FYM with the bio-agents. This enriched FYM can be added at 1 kg per pit at the time of planting and thrice after planting with an interval of 2 months at the same dosage.

For effective control of nematodes, banana plantlets should be inoculated with *Purpureocillium lilacinum* strain 251 and re-inoculated into the soil at transplanting (Mendoza *et al.*, 2004).

Devarajan and Rajendran (2001) reported that *P. lilacinum* application at 30 g/kg soil (1 g sorghum grains contained 10^7 conidia) 60 DAP could substitute carbofuran in reducing *R. similis* population in banana.

Paecilomyces marguandii was effective for the management of *R. similis* and *Helicotylenchus multicinctus* on banana plants (Esnard *et al.*, 1998).

iii) Arbuscular mycorrhizal fungi: In an interaction study of arbuscular mycorrhizal fungus (AMF), *Glomus fasciculatum* and *R. similis*, the multiplication of nematodes in the root was suppressed and soil population was significantly lower. The banana plants inoculated with mycorrhizae also recorded higher N, P, K, Ca, Mg, sugars, phenols and amino acids in their root system suggesting AMF as a potential biocontrol agent in the management of *R. similis* (Umesh *et al.*, 1988). Inoculation of *G.*

mosseae during raising of banana nursery was found effective in reducing the nematode populations of *R. similis* and *M. incognita* besides enhancing plant growth and bunch weight under field conditions. There was 35.7% increase in bunch weight, 30% increase in number of hands and 39.6% increase in number of fingers in the mycorrhizae colonized plants (Sosamma *et al.*, 1998).

The study conducted to determine whether the AMF *Glomus intraradices* was able to incite systemic resistance in banana plants towards two plant parasitic nematodes *Radopholus similis* and *Pratylenchus coffeae* using a split-root compartmental set-up revealed that the AMF reduced both nematode species by more than 50%, even when the AMF and the plant parasitic nematodes were spatially separated. The results obtained demonstrate that AMF have the ability to induce systemic resistance against plant parasitic nematodes in a root system.

iv) Biofertilizers: Soil application of biofertilizer, *Azospirillum* at 20 g/plant had significant effect in enhancing the plant growth and reducing the population of *R. similis* in soil to an extent of 50.1% over control (Shanthi *et al.*, 2004).

3.2.8.6. Host resistance: Selections (tetraploids) from crosses between wild bananas (diploid male parent) and 'Gros Michel' (triploid female parent) show considerable resistance to R. similis as well as to Panama wilt and Sigatoka (Mycosphaerella musicola) diseases. A variety Bodles Altaport, a cross between Gros Michel and Pisang Lilin, has been released for growing on the hill sides of Jamaica. This variety has an extensive root system, withstands wind and is more resistant to R. similis than Lacatan.

In American Samoa, the banana hybrid FHIA-01 has been found relatively resistant to the burrowing nematode, while the hybrid FHIA-25 is relatively resistant to all the nematodes feeding on banana.

3.2.8.7. Integrated methods

i) Cultural and biological methods: Integration of organic amendments (FYM/ Neem cake/ Karanj cake) with bioagents (*Trichoderma harzianum/ T. viride/ Pochonia chlamydosporia/ Pastueria penetrans/ Pseudomonas fluorescens*) or endomycorrhiza (*Glomus mosseae*) was found effective against the burrowing nematode. Soil application of 2 kg FYM with *P. fluorescens* (with 1 x 10^9 spores/g) and *P. chlamydosporia* (with 1 x 10^6 spores/g) per plant at the time of planting and at an interval of 4 months significantly reduced the burrowing nematode by 64% compared to control.

Increase in plant growth and reduction in *R. similis* population by 50% was obtained by integration of soil application of *Pastueria penetrans* at 100 g soil (300 spores/g)/*T. harzianum/ T. viride* at 250 g/plant while planting and repeated 4 months after planting. Increase in banana fruit yield by 45 kg/plant and early harvest of bunches by 65-75 days was achieved by combined application of FYM enriched with *T. harzianum* at 2 kg/plant and Neem cake at 500 g/plant.

Combined application of *G. mosseae* (containing 25-30 chlamydospores/g of inoculum) at 100 g/plant with Karanj cake increased banana fruit yield (16.61 kg/plant as compared to 5.45 kg/plant in control), while application of *G. mosseae* along with Neem cake reduced the burrowing nematode population. Mycorrhizal root colonization and number of chlamydospores of *G. mosseae* were highest in Neem cake amended soil (Parvatha Reddy *et al.*, 2002).

Application of FYM enriched with bio-control agents like *Purpureocillium lilacinum* and *T. harzianum* to the banana plants along with Neem cake at the time of planting and repeating the same treatment at two months interval proved beneficial.

Soil application of FYM enriched with *P. lilacinum* and *P. fluorescens* at 2 kg/plant at the time of planting and subsequent application for four times at an interval of six months reduced the root population of *R. similis* and *M. incognita* by 64.5 and 75%, respectively and increased the fruit yield by 21% over control (Rao *et al.*, 2016).

ii) Physical and chemical methods: The removal of infected tissue along with some of the surrounding healthy tissue will normally disinfect sets (Loos and Loos, 1960). The pared sets are dipped in mud slurry and sprinkled with 40 g Carbofuran 3G per set.

iii) Physical, cultural and chemical methods: Paring of suckers followed by hot water treatment (55° C for 20 minutes) along with application of Carbofuran @ 0.5 g a.i./plant and Neem cake @ 1 kg/plant, the population of *R. similis* decreased by 72-82% at 4 and 8 week interval after treatment. Maximum yield (2.7 t/ha) was recorded in the above treatment which was 97% increase over control.

Integration of paring of banana suckers, dipping in 0.5% Monocrotophos solution for 45 min along with intercropping with marigold or sun hemp gave higher fruit yield (62.838 and 61.816 t/ha, respectively as compared to 57.000 t/ha in control) and benefit cost ratio (1.70 and 1.28, respectively) (Patil *et al.*, 1999).

iv) Physical and biological methods: Paring and pralinage of suckers and application of FYM/ Vermicompost enriched with *T. harzianum* + *P. lilacinum* at 2 kg/plant at planting and thrice afterwards at 3 months interval was found effective.

v) Cultural and chemical methods: Application of Neem cake @ 1 kg/plant (two split doses, first at planting and second 3 months after planting) and Carbofuran (3G) @ 16.6 g/plant in the pit at planting was found effective.

vi) Cultural, chemical and biological methods: Integration of neem cake at 400 g/plant, carbofuran at 20 g/plant, *Glomus fasciculatum* at 50 g/plant and *P. penetrans* at 100 g soil/plant was most effective in reducing the population of *R. similis* in banana (Channabasappa *et al.*, 1995).

Reduction in the burrowing nematode population both in soil and roots and increase in plant growth, yield, and the benefit: cost ratio (2.65) was obtained by combined application of Neemark, Carbofuran, *P. penetrans* and *G. fasciculatum* (Vidya and Reddy, 1998).

3.3. BURROWING NEMATODE AND PANAMA WILT DISEASE COMPLEX

3.3.1. Interaction

Incidence and losses due to Panama wilt caused by *Fusarium oxysporum* f. sp. *cubense* is enhanced in association with the burrowing nematode, *Radopholus similis* under high nematode population. This clearly indicates the existence of synergistic interaction between the burrowing nematode and Fusarium wilt pathogen in banana.

Newhall (1958) showed that the incidence of Panama wilt in banana caused by *F. oxysporum* f. sp. *cubense* was doubled in the presence of *R. similis* during the experimental period of three months. When Gros Michel bananas infected with *R. similis* were inoculated with *F. oxysporum* f. sp. *cubense*, the period between inoculation and the onset of wilt was also considerably shortened (Loos, 1959).

Rishbeth (1960) suggested that nematodes breakdown resistance to Panama wilt in Lacatan bananas.

Lesions formed after inoculation with both *R. similis* and *F. oxysporum* f. sp. *cubense* were more extensively necrotic and increased in size more rapidly than when *R. similis* alone was used (Blake, 1966). *F. oxysporum* f. sp. *cubense* readily establishes itself in the feeder roots of banana when they are invaded by the nematode *R. similis*, but the fungus has seldom been recovered from nematode-free roots (Blake, 1966).

3.3.2. Management

3.3.2.1. Cultural methods: Application of compost and Neem cake was most effective in reducing the disease intensity by *Meloidogyne* sp. and *F. oxysporum* f. sp. *cubense* on banana (Roy *et al.*, 1998).

3.3.2.2. Integrated methods: Soil application of Neem cake + *T. viride* + Carbendazim was found effective in reducing the burrowing nematode (*R. similis*) and wilt (*F. o.* f. sp. *cubense*) disease complex, lesion index (1.0 as compared to 4.0 in control), increasing the banana fruit yield by 53.55% (15.147 t/ha as compared to 9.887 t/ha in control) giving the benefit: cost ratio of 2.72 (Ravi *et al.*, 2001).

3.4. LESION NEMATODE, *PRATYLENCHUS COFFEAE*

The decline in banana plantations occurred due to the infestation the lesion nematode, *P. coffeae*.

3.4.1. Distribution

Widely distributed in most of the tropical regions of Asia (India, Japan, Thailand), Africa, Australia and America.

3.4.2. Crop losses

In India, crop loss caused by the root-lesion nematode in banana cv. Nendran was reported to be 45%.

3.4.3. Symptoms

The following are the symptoms:

» Extensive black or purple necrosis on roots, are found

» Stunting of plants as a characteristic symptom

» Reduction in size and number of leaves and in bunch weight.

The symptoms caused by *P. coffeae* are very similar to those caused by *R. similis*. Both *R. similis* and *P. coffeae* are associated with the blackhead toppling diseases of bananas. *P. coffeae* enter the roots and feeds on cortical cells resulting in the production of dark lesions in the cortex. The corm is similarly infected resulting in necrosis of the outer tissues.

3.4.4. Host range

The grapevine and citrus veld are the well-known hosts of *Pratylenchus coffeae* although it has a wide host range including many broad leaf weed species (Gowen and Quénéhervé, 2005; Kleynhans *et al.*, 1996)

3.4.5. Life cycle

The life cycle of *P. coffeae* is similar to that of *R. similis*. The egg to egg cycle is completed in about 27 days at 26-32 °C. The optimum temperature for reproduction is 29.5°C.

3.4.6. Survival and spread

The lesion nematode can survive for 6 months in the soil, in the absence of its host. It has been observed infesting the corm, and hence spread can occur through planting material.

3.4.7. Interaction with other pathogens

The lesion nematode causes most damage to banana in association with *Fusarium oxysporum.* f. sp. *cubense* (Vadivelu *et al.*, 1987).

3.4.8. Management

3.4.8.1. Cultural methods

i. **Fallowing:** A fallow period of 6 months after destruction of all banana plants eliminated *P. coffeae*.

ii. **Intercropping:** A significant reduction in *Pratylenchus* sp. population (85%) was observed in the banana field where *Tagetes erecta* was grown as an intercrop and significantly increased the fruit yield (12 kg/plant) when intercropped compared with the untreated control (7 kg/plant) (Sundararaju *et al.*, 2002). Reduction in the lesion nematode population by 38.4% was obtained with intercropping of banana with sun hemp, followed by marigold and cowpea which recorded a 29.0 and 22.3% reduction, respectively.

iii. **Crop rotation:** The lesion nematode population was also reduced by rotation of banana with green gram/rice.

iv. **Organic amendments:** A significant reduction in *P. coffeae* population and yield increase was recorded in plants that have received 50% N applied through Neem cake (Sundararaju and Kumar, 2003). A significant reduction in *P. coffeae* population and increased yield in plants treated with Press mud (15 t/ha) was recorded. Sundararaju *et al.* (2002) reported that eco-friendly and economical way of controlling nematodes is through soil application of press mud. In banana cv. Nendran, maximum reduction in lesion nematode population and significant increase in plant growth was obtained by soil application of Neem cake at 200 g/plant.

3.4.8.2. Biological methods

i) Antagonistic fungi: Application of *Trichoderma viride* at 20 g/plant at the time of planting and repeated 3 months after planting were found effective in controlling *P. coffeae* as well as reducing the incidence of Panama wilt in banana.

Application of *Purpureocillium lilacinum* at 10 g along with FYM at 500 g or Neem cake at 250 g/plant effectively suppressed the lesion nematode population by 60 and 63%, respectively (Poornima *et al.*, 2004).

ii) Antagonistic bacteria: Reduction in lesion nematode population in soil (67.12 %) and roots (62.00%) and increasing pseudo stem height and girth, number of leaves, root length and weight was obtained by soil application of *Pseudomonas fluorescens* at 20 g/plant (2.5 x 10^8 cfu/g).

Soil application of biofertilizer, *Azospirillum* at 20 g/plant had significant effect in enhancing the plant growth and reduced the population of *P. coffeae* in soil to an extent of 45.5 % over control (Shanthi *et al.*, 2004).

iii) Arbuscular mycorrhizal fungi: The study conducted to determine whether the AMF *Glomus intraradices* was able to incite systemic resistance in banana plants towards *Pratylenchus coffeae* using a split-root compartmental set-up revealed that the AMF reduced nematode population by more than 50%, even when the AMF and the plant parasitic nematode were spatially separated. The results obtained demonstrate that AMF have the ability to induce systemic resistance against lesion nematodes in a root system.

3.4.8.3. Host resistance: Banana cvs./hybrids viz., Kunnan, Pey Kunnan, Then Kunnan, Nattu Poovan, Pidi Monthan, Chirapunji, Singhlal, Sakkarachayam, Malai Kala, Manik Champa, Kartobium Tham,

Vennettukunnan and Hybrid-74 (Matti x Pisang Lilin), H-21, H-55, H-59, H-65, H-84, H-109, H-110 were found tolerant/resistant to *P. coffeae* (Vadivelu *et al.*, 1987; Sundararaju *et al.*, 2004).

3.5. SPIRAL NEMATODE, *HELICOTYLENCHUS MULTICINCTUS*

H. multicinctus is an important parasite of banana throughout the banana growing areas of the world.

3.5.1. Distribution

Helicotylenchus multicinctus, especially important as a pest of banana, is widely distributed in almost all tropical and warm temperate banana growing countries including India. In tropical soils, this nematode often occurs along with *R. similis* but in cooler dry soils *H. multicinctus* alone is prevalent.

3.5.2. Crop losses

This nematode causes serious decline of bananas in the Jordan valley, Israel and was responsible for about 20% loss in yield. It is responsible for 33.83 per cent loss in yield, 55.88 per cent loss in number of fruits per bunch and delayed fruiting by 134 days (Vadivelu *et al.*, 1987).

3.5.3. Symptoms

The above-ground symptoms are not specific and resemble damage caused by other nematode pests of banana. The spiral nematode incites discrete relatively shallow, necrotic lesions on banana roots. Symptoms of damage inflicted by *H. multicinctus* include discoloration of the root epidermis where small reddish lesions can be observed in the superficial cortical region also in rhizome tissue of infected plants along with reduction in the number of lateral roots (Jones and Milne,1982). Under severe infestations, the smaller lesions enlarge and coalesce leading to extensive necrosis in the outer cortical region of root and even root dieback sets in (Quénéhervé and Cadet, 1985). Toppling of plants can be observed due to poor anchorage under severe infestation (Gowen and Queneherve, 1990). No distinct biotypes or races have been reported in *H. multicinctus* (Gowen and Quénéherv, 2005).

In Israel, *H. multicinctus* causes extensive root necrosis, dieback and dysfunction leading eventually to debility of the entire plant.

3.5.4. Host range

This nematode is known to infect banana, cocoa, sweet potato, citrus, sugarcane, cassava, coffee, maize, mango, oil palm, rice, rubber, tea, yam, avocado, Bermuda grass, and grapevine.

H. multicinctus hosts range from most edible banana and plantain to various alterative host plants, such as pigweed (*Amaranthus* spp.), purslane (*Portulaca oleracea*) and ornamentals (Gowen and Quénéherv, 2005).

3.5.5. Life cycle

Groups of 8-26 eggs were observed in discolored cortical tissues. Eggs hatch in 48-51 hr. in tap water at 30 °C and the first molt is believed to take place outside the egg shortly after hatching. The second stage larvae are recognized by their ventral digitate tail process. In the early third stage larvae and at the third molt, the female gonad primordium is 2 and 6 celled, respectively. During the fourth molt, the male and the female gonads complete their development and the vulva and vagina are seen in female which perhaps complete entire life cycle within the roots.

3.5.6. Spread and survival

The nematodes easily get introduced into virgin land with banana soil and suckers (rhizomes), usually brought from old infested plantations. The nematodes survived 4 months without host plants.

3.5.7. Management

3.5.7.1. Physical methods: Removal of black necrotic tissue from the banana suckers and dipping them in hot water maintained at 53-55°C for 20 min. is recommended to eliminate nematodes.

4.7.2. Cultural methods

i) Crop rotation: Rice or green gram after banana suppressed the population of *H. multicinctus.*

ii) Intercropping: Intercropping of banana with sun hemp was found effective and reduced *H. multicinctus* population by 38.4%, followed by marigold and cowpea which recorded a 29.0 and 22.3% reduction, respectively.

iii) Organic amendments: Application of Distillery sludge at 2.5 kg + Vermicompost 1 kg + Neem cake at 1 kg + Poultry manure at 2.5 kg/plant at 3, 5, and 7 months after planting significantly reduced *H. multicinctus* population. It was on par with Distillery sludge at 2.5 kg + Neem cake at 1 kg treatment.

3.5.7.3. Chemical methods: In recent years, emphasis has been shifted to use of granular nematicides (Oxamyl/Aldicarb/Carbofuran/Fensulfothion/Ethoprophos/Fenamiphos) for the management of the spiral nematode.

3.5.7.4. Biological methods: Reduction in spiral nematode population in soil (58.87 %) and roots (56.54%) and increasing the plant growth parameters (pseudo stem height and girth, number of leaves, root length and weight) was obtained by soil application of *Pseudomonas fluorescens* at 20 g/plant (2.5 x 10^8 cfu/g).

Soil application of biofertilizer, *Azospirillum* at 20 g/plant had significant effect in enhancing the plant growth and reduced the population of *H. multicinctus* in soil to an extent of 48.9% over control (Shanthi *et al.*, 2004).

3.5.7.5. Integrated methods: Jonathan and Cannayane (2002) reported significant reduction in population of *H. multicinctus* and increase the plant growth parameters (pseudo stem height and girth, number of leaves, root length and weight) was observed in banana plants treated with *P. fluorescens* at 2g/ *Glomus mosseae* at 25g along with Press mud at 3 kg/plant.

3.6. ROOT-KNOT NEMATODES, *MELOIDOGYNE* SPP.

Root-knot nematodes are present in all banana-growing regions and can be especially damaging to the cultivars of the Cavendish subgroup. The root-knot nematodes (*M. incognita, M. javanica, M. arenaria* and *M. hapla*) are found to be commonly associated with banana.

3.6.1. Crop losses

The average yield loss due to this nematode is estimated to be 27.3% (Anon, 2009). Jonathan and Rajendran (2000) reported 30.90% loss in banana fruit yield due to *M. incognita. M. incognita* is responsible for annual yield loss of 15% in banana amounting to Rs. 9710.46 million in India (Walia and Chakraborty, 2018).

3.6.2. Symptoms

The symptoms on aerial plant parts caused by *Meloidogyne spp. include* reduction in size and chlorosis of leaves, stunting of plants, yield reduction, and delay and unevenness in fruit ripening. Leaf symptoms include a progressive dieback of older leaves from the leaf tip. Symptoms exhibited by severely infected banana plants include stunting of plants, thinning of pseudo stems, and discolored greenish-yellow bands on the leaf blades.

The most characteristic symptoms of root-knot nematode include galling on both primary and secondary roots. In severe infestation, the root galls are enlarged and distorted. Galls vary in size, reaching up to 15 mm in diameter. In bananas, *M. incognita* infection typically results in the formation of small galls on the roots or fine dilations in thicker fleshy roots. Rhizome upper cell layers are colonized by different species of root-knot nematodes in a single gall. The nematode cause reddish-brown discoloration in the vascular system. When the infected root is longitudinally split open, females in different stages of development can be observed along with black haloes (female chambers). The following are the other symptoms:

- There is not normal growth, the leaves become hard and brittle
- The fields look in patches because of abnormal growth due to distribution of nematodes.
- The yield is reduced finally.
- Heavily infested plants have a much reduced root system with large elongated galls in root-knot infested plants.
- Small galls along feeder roots and large galls on the main roots, are found in root knot infested plants.
- Rabie (1991) reported that the combination of root-knot nematodes and other stress factors, such as drought and cold temperatures, are responsible in inflicting symptoms of 'False Panama Disease,' which is similar to 'Panama Disease'.
- In the former case transverse sections of rhizomes of infected plants showed reddish brown to brownish purple discolored vascular tissue while in later case, reddish brown ring like symptoms were observed in cross section of pseudo stem
- Leaf symptoms include progressive dying back of older leaves, starting at the tips. Galls occur on the primary and secondary roots, whilst distortion of roots and sometimes bifurcation occurs after heavy nematode infections.

3.6.3. Biology and life cycle

The nematode is soil-borne and sedentary endoparasite. After penetration, the J2 larvae feed endoparasitically on conducting vessels and develop to adult females which are spherical in shape and lays egg masses in compound galls. The males are vermiform in shape which leaves roots to enter the soil. During summer, the nematode completes its life cycle in about 25 to 30 days, while it takes longer during winter months.

3.6.4. Survival and spread

Root-knot nematodes survive on dicotyledonous plants, which are usually present in most soils in which bananas are growing. Survival and spread also occurs with the planting material on infected roots and corms. The nematodes spread through infected planting material, irrigation water, and agricultural implements.

3.6.5. Management

3.6.5.1. Regulatory methods: Measures to prevent introduction and spread include the use of certified planting material which is an important management strategy.

3.6.5.2. Physical methods: The suckers can be pared to a depth of 1 cm and treated in hot water at 55°C for 20 min (Inomoto and Monteiro, 1989).

3.6.5.3. Cultural methods: To effectively control root-knot nematodes in bananas, special attention should be paid to weeding during fallowing or crop selection for rotation (Willers *et al.*, 2001).

Application of Neem cake at 250 g/plant twice (once at planting and second after four months) helps in suppressing the nematode population.

Intercropping with Marigold at early growth stages in basins and incorporated into the soil around the plants was found effective.

Growing antagonistic plants namely *Tagetes erecta, T. patula* and *Crotalaria juncea* as intercrops in banana field drastically reduces the root-knot nematode population.

Bello *et al.* (2004) obtained promising levels of control of *M. incognita* and *M. javanica* in banana plantations in the Canary Islands with biofumigation using urban waste and urea, or manure and banana residues.

3.6.5.4. Chemical methods: Application of Carbofuran 3G at 20 g/plant or Rugby at 5 g/plant or Caldan at 5 g/plant applied twice once at planting and second application after three months of planting is known to reduce the nematode population and increase the yield significantly.

Application of either Carbofuran or Cartap or Carbosulfan @ 0.3 g a.i. /m^2 proved beneficial (Anon, 2012).

3.6.5.5. Biological methods: The promising biological control agents such as *Purpureocillium lilacinum, Trichoderma viride, Pochonia chlamydosporia*, endomycorrhiza (*Glomus fasciculatum*) and bacterium, *Pasteuria penetrans* are recommended to control root-knot nematodes on banana.

The biocontrol agent, *Pochonia chlamydosporia* parasitized the eggs of root-knot nematode and prevent the hatching significantly.

Pastueria penetrans @ 20 g/m^2 ($5x10^7$ spores/g root powder) mixed with 200 ml sand and evenly distributed per plant at planting and 45 days after planting effectively reduced *M. incognita* population in banana besides increasing the yield. Karuna *et al.* (2004) reported that the plants treated with *P. penetrans* (3.5 kg/plant) gave higher yield when compared to control plants (2.1 kg/plant).

Prior colonization of banana suckers with *Glomus mosseae* was found to reduce the root and soil population of *M. incognita.*

3.6.5.6. Host resistance: In Tamil Nadu, banana cvs. /hybrids such as Chakia, CO 1, Pisang Lilin, H 74, H 94 and H 110 were found resistant to root-knot nematodes (Jonathan, 1994; Sundararaju *et al.*, 2004).

3.6.5.7. Integrated methods: Infected corms can be disinfected of root-knot nematodes by paring the planting material followed by immersing it in hot water at 50-55° C for 30 minutes.

Dipping of pared suckers in mud slurry mixed with Carbofuran (3G) at 20 g/L of slurry followed by drying suckers in shade before planting was found effective.

Paring and pralinage of suckers with Farm yard manure/Vermicompost enriched with *Trichoderma harzianum* + *Purpureocillium lilacinum* at 2 kg/plant at planting and three times afterwards at 3 months' interval proved beneficial.

In heavily nematode infested area, paring banana suckers and dipping in hot water @ 55 °C for 20 minutes and two split applications of Neem cake at 1 kg/plant (at planting and three months later)/Carbofuran 0.5 g a.i. /ha was found effective for the control of root-knot nematode in banana.

Significant reduction in the root-knot nematode population by 76% as compared to control was obtained by integration of soil application of 2 kg FYM enriched with *Pochonia chlamydosporia* (with 1 x 10^6 spores/g) per plant and *Pseudomonas fluorescens* (with 1 x 10^9 spores/g) and at the time of planting and at an interval of 4 months.

Jonathan and Rajendran (2000) reported that application of *P. lilacinum* multiplied on Neem cake at 15-20 g/plant significantly reduced root gall index, egg masses, eggs, females and soil population of *M. incognita* in banana.

Application of *T. harzianum* along with FYM has reduced the root-knot nematode population in soil (from 175/100 ml soil in control plot to 42/100 ml soil in treated plot) and roots (from 17/g root in control plot to 4/g root treated plot). Further, application of the bioagent also reduced *Fusarium* wilt incidence and increased the yield by 4.53 t/ha over control.

3.6.6. Success story in Odisha

Two technologies developed under AICRP on Nematodes for the management of root-knot nematode (*M. incognita*) i.e., (i) application of *Pseudomonas fluorescens*, a biocontrol agent @ 20 g/ m^2 at the time of planting, and (ii) paring of suckers followed by hot water treatment (55° C for 20 minutes) along with application of Carbofuran @ 0.5 g a.i. /plant and Neem cake @ 1 kg/plant were demonstrated, besides the untreated check (farmers' practice) in Shri. Basudeb Behera farmer's field located in Mangalpur village of Puri district. The application of *P. fluorescens* and combined treatment of paring, hot water treatment, Carbofuran, and Neem cake registered average bunch weight of 24.2 kg and 25.8 kg with 17.5% and 25.0% increase in yield in comparison to untreated plants (20.5 kg), respectively. But the maximum benefit cost ratio (11.3) was noticed with application of *P. fluorescens* with a net profit of Rs. 52,000/- per hectare. The farmers were well convinced with the benefit of this sustainable, ecofriendly and low cost technology of bioagent application for nematode management.

3.7. ROOT-KNOT AND LESION NEMATODES AND PANAMA WILT DISEASE COMPLEX

3.7.1. Interaction

The root-knot (*Meloidogyne incognita*) and lesion (*Pratylenchus coffeae*) nematodes along with Fusarium wilt (*Fusarium oxysporum* f. sp. *cubense*) are known to interact and cause disease complex in banana. The plants receiving inoculum of either *M. incognita* or *P. coffeae* or both inoculated before inoculation of *F. oxysporum* f. sp. *cubense* resulted in the early onset and increased severity of wilt disease complex and maximum reduction in plant growth parameters (Table 3.5) (Sundararaju and Thangavelu, 2009).

Table 3.5. Effect of *M. incognita, P. coffeae* and *F. oxysporum* f. sp. *cubense* on plant growth and disease complex in banana

Treatment	Plant height (cm)	Root lesion index	Root-knot index (1-5 scale)	Wilt index (1-5 scale)
P. coffeae (*Pc*)	110	4	1	1

M. incognita (Mi)	115	1	5	1
F. oxysporum f. sp. *cubense* (*Foc*)	118	1	1	4
Pc + Foc	105	4	1	3
Mi + Foc	110	1	4	5
Pc + Mi + Foc	90	5	3	5
Control	125	1	1	1

3.7.2. Management

3.7.2.1. Cultural methods: Application of compost and neem cake was most effective in reducing the disease intensity by *Meloidogyne* sp. and *F. oxysporum* f. sp. *cubense* on banana (Roy *et al.*, 1999).

3.7.2.2. Integrated methods: Dipping of banana suckers in 0.2% bavistin solution for 45 min (Roy *et al.*, 1998a); flooding and application of inorganic fertilizers at N-225, P-75, K-60 kg/ha (Roy *et al.*, 1998a); and soil application of bavistin at 0.2% (Roy *et al.*, 1998b) were effective in increasing plant growth, reducing nematode population, and in increasing fruit yield (Table 3.6).

Table 3.6. Management of *M. incognita, F. oxysporum* **f. sp.** *cubense disease complex*

Treatment	Dose	Results	Reference
Bavistin	Dipping of suckers for 45 min	Increased plant growth, reduced nematode population, increased fruit yield	Roy *et al.*, 1998
Flooding, appln. of inorganic fertilizers	N-225, P-75, K-60 kg/ha	Increased plant growth, reduced nematode population, increased fruit yield	Roy *et al.*, 1998
Bavistin	Soil appln. of 0.2% 5, 7 & 9 months	Highest fruit yield of 17.4 kg/plant	Roy *et al.*, 1998

3.8. THE CYST NEMATODE, *HETERODERA ORYZICOLA*

The cyst nematode is an important nematode found on banana cv. Nendran in Kerala.

3.8.1. Occurrence and distribution

The infestation of a cyst nematode, *Heterodera oryzicola* on banana variety Nendran was observed for the first time in India at Banana Research Station, Kannara, Kerala, during July 1983 (Charles and Venkitesan, 1984). These nematodes feed on fibrous roots and cause necrosis. In a subsequent survey, this nematode was recorded in several locations in Trichur, Ernakulam, Quilon and Calicut districts of Kerala especially on banana cultivated in paddy land. *H. oryzicola* also infects paddy and *Cynodon dactylon*. This creates practical difficulties in adoption of crop rotation with banana in rice fields. The infested plants are progressively debilitated leading to substantial yield losses.

H. oryzicola was also reported on banana cultivars Saldathi and Njalippnan from Goa (Koshy *et al.*, 1987).

3.8.2. Economic importance and losses

H. oryzicola has been reported to cause considerable damage to banana in certain pockets of Kerala. An initial inoculum level of 100 and 1000 viable cysts per plant could reduce bunch weight by 20.5 to 56.6 per cent. The nematode infection was observed to affect the quality of fruits by increasing acidity and reducing total sugars (Charles, 1989).

H. oryzicola attacks the minute tertiary feeder roots and not observed to attack the fleshy roots and rhizome portions. The affected feeder roots turn black in color and detach easily (Charles, 1989).

3.8.3. Symptoms

The cyst nematode attacks thin lateral roots and the root weight is considerably reduced (62-87%). The damage caused to the root tissue may suppress the flow of food materials to the various parts of the plant resulting in general decline, stunting and premature defoliation.

3.8.4. Life cycle

Life cycle of the cyst nematode on banana cv. Nendran is completed in 23 days from egg to egg stage. The second stage juveniles took to develop into adult males and white females with egg deposition within 12 and 17 days, respectively (Charles, 1989).

3.8.5. Histopathology

Histopathological studies revealed that the nematode attacked xylem vessels of the root with the head attached to the stele. The cortical cells and epidermis were ruptured at the infection site. Black discoloration was seen on the nearby cortical cells without any cellular change (Charles, 1989).

3.8.6. Ecology, population fluctuation and soil types

Coastal sandy and sandy loam soils significantly favored nematode activities affecting plant growth and nematode multiplication. A temperature of 25 °C, flooded condition of the soil with a pH of 6.7 influenced higher percentage of larval emergence by 50th day (Charles, 1989).

3.8.7. Management

i) Physical methods: The nematode can be controlled by treatment of suckers (by hot water or chemicals) before planting.

ii) Cultural methods: Application of distillery sludge at 2.5 kg + vermicompost 1 kg + neem cake at 1 kg + poultry manure at 2.5 kg/plant at 3, 5, and 7 months after planting significantly reduced *H. oryzicola* population. It was on par with distillery sludge at 2.5 kg + neem cake at 1 kg treatment (Sundararaju *et al.*, 2002).

3.9. RENIFORM NEMATODE, *ROTYLENCHULUS RENIFORMIS*

The reniform nematode probably occurs in all tropical and subtropical areas of the world feeding as sedentary semi-endoparasite on the secondary or tertiary roots of banana

3.9.1. Economic importance

In field plots where *R. reniformis* was the predominant parasitic nematode, yield reduction of 19.3 tons per ha was observed.

3.9.2. Symptoms

Necrotic lesions are produced in the areas where females are attached to the roots. The posterior part of the female body and egg mass are protruded outside the host. High population of the nematodes causes severe necrosis and destruction of feeder roots.

3.9.3. Biology

Only the immature female of this nematode is parasitic. Eggs are laid in a mass covered by gelatinous substance. An egg mass may contain 45 to 66 eggs. After hatching, juveniles (J2s) migrate in the soil and undergo three molts in quick succession without feeding. After the last molt, the female infects roots by penetrating the tissue with its head and leaving rest of the body outside the host tissue. At the same time, the body enlarges and takes a kidney shape. Nurse cells are formed as a prerequisite of feeding. The egg to egg cycle is about 24-29 days at 28-31 °C.

3.9.4. Management

Management tactics include cultural, physical, chemical, biological methods, host resistance and integrated management.

3.9.4.1. Physical methods: Physical methods include removal of infested tissues by paring and disinfecting them in hot water at 55 °C for 20 minutes before planting. In Kerala, the practice of 'sun drying' the rhizome before planting has been found effective in control of nematodes.

3.9.4.2. Cultural methods: Cultural methods include summer plowing, organic manuring, crop rotation and destruction of infected crop residues. Crop rotation with Pangola grass gives effective control of the reniform nematode on bananas.

3.9.4.3. Chemical methods

- » Carbofuran and Phorate at 2 g a.i./plant at planting is the most effective and popular granular nematicides used against banana nematodes. Ethoprop at 2-3g a.i. /plant or carbofuran at 1.2 g a.i. / plant reduce the nematodes population with high benefit: cost ratio.
- » Application of 4 g Carbofuran, Fensulfothion or Oxamyl per plant, applied to the soil at planting followed by repeated treatment four months later decreased nematode population and increased the yield.

3.9.4.4. Biological methods: The promising biological control agents include the fungus, *Purpureocillium lilacinum*, AMF, and bacterium, *Pasteuria penetrans*.

3.9.4.5. Host resistance: A few germplasm like Kadali, Kunnan, Pisang Seribu and Ayiranka Poovan are known to support low nematode populations.

3.9.4.6. Integrated methods: Integrations of paring of suckers and hot water dipping at 55 °C for 20 min., soil application of neem cake at 400 g per plant and carbofuran at 20 g per plant before planting significantly reduce the nematode population and improve the plant growth.

The best way of management of nematode population is obtained by paring (removal of infected part of the suckers) and pralinage (coating of suckers with some nematicide) with carbofuran or fensulphothion at 42 and 17 g per sucker, respectively before planting.

3.10. FUTURE LINES OF INVESTIGATIONS

- » Intensive survey for *Radopholus similis* and *Heterodera oryzicola* in major banana growing regions.
- » Crop loss assessment due to major nematode pests.

- Introduction of internal plant quarantine regulation for the burrowing nematode and disinfection of banana suckers at the sites of indigenous germplasm collections.
- Production of certified banana suckers free from nematodes.
- Research on biological control, crop rotation, cultural practices and sanitation methods.
- Studies on the use of mycorrhizae and other antagonistic fungi/bacteria.
- Studies on different races of *R. similis* in India.
- Role of nematodes in disease complex with fungi and bacteria.
- Developing resistant cultivars.
- Developing integrated management technologies using tolerant/resistant cultivars, parasitic fungi and bacteria, minimal quantity of nematicides, organic amendments and arbuscular mycorrhizal fungi should be the priority areas of research.

3.11. CONCLUSION

Plant-parasitic nematodes cause severe damage to the banana crop and reduce yields considerably. By using the nematode-free tissue cultured planting material, nematode pests seldom become problematic during the first years after establishment of commercial plantations. Besides, farmers should send soil and root samples annually for nematode analysis to monitor plant-parasitic nematode populations and follow management recommendations. This way, the effect of nematode pests in commercial banana production is effectively restricted. However, there will always be a need to investigate alternative strategies to keep nematodes under control. The need for nematode management strategies in the smallholding farming sector is significant and requires more attention.

Chapter - 4

Citrus, *Citrus* spp.

4.1. INTRODUCTION

4.1.1. Origin

Citrus originated in South East Asia. North East region of India is the home of some citrus species e.g. mandarins.

4.1.2. Area and production

Citrus fruits (*Citrus* spp.) rank third in area and production after mango and banana with an estimated production of 14.850 million tons from an area of 1.106 million hectares (2021-22). There are many commercially grown *Citrus* spp. such as mandarins (*Citrus reticulata*), sweet orange (*C. sinensis*), acid lime (*C. aurantifolia*), lemon (*C. limon*), grape fruit (*C. paradisi*), pumello (*C. grandis*), *etc*. Although one or the other species of the citrus fruit is grown in almost all the states of India, but the major citrus growing states are Andhra Pradesh, Maharashtra, Karnataka, Punjab, Gujarat, Madhya Pradesh, Orissa and Manipur. Citrus is a highly sensitive crop and is attacked by number of insect pests, diseases and nematodes. These play major role towards the low productivity of citrus in India with national average of 13.426 tons per hectare, as compared to 25-30 tons per hectare in other citrus growing countries. The major citrus producing countries in the world are presented in Table 4.1.

Table 4.1. Major Citrus fruits (Total) producing countries in the world (FAO, 2016)

Country	Area (million ha)	Production (million tons)	Productivity (tons/ha)
China	2.63	38.39	14.62
Brazil	0.76	19.59	25.78
India	0.94	12.04	12.75
USA	0.30	7.51	25.11
Mexico	0.56	8.11	14.37
Spain	0.36	7.02	19.31
Egypt	0.20	4.81	24.48

Nigeria	0.84	4.06	4.85
Turkey	0.14	4.29	31.65
Argentina	0.15	3.28	22.53
Others	2.47	37.31	15.08
World	9.45	146.43	15.49

The state-wise area, production and productivity of mandarin, sweet orange, lime/lemon fruits in India are presented in Tables 4.2 and 4.3.

Table 4.2. State-wise area, production and productivity of mandarin and sweet orange, (2017-18) (Area - '000 ha, Production - '000 tons, Productivity - tons/ha)

States	Mandarin			Sweet orange		
	Area	**Production**	**Productivity**	**Area**	**Production**	**Productivity**
Andhra Pradesh	-	-	-	82.89	2003.11	24.17
Arunachal Pradesh	32.73	69.74	2.13	-	-	-
Assam	14.95	203.72	13.62	0.14	1.98	14.12
Bihar	-	-	-	0.40	4.40	11.00
Chhattisgarh	-	-	-	0.47	2.37	5.09
Himachal Pradesh	8.71	14.10	1.62	1.81	2.77	1.53
Jammu & Kashmir	3.16	5.48	1.73	3.16	5.72	1.81
Karnataka	3.65	79.07	21.64	1.52	25.77	16.95
Kerala	0.20	0.91	4.51	-	-	-
Madhya Pradesh	12.11	2103.64	17.37	6.43	111.72	17.39
Maharashtra	107.32	797.93	7.43	55.18	684.80	12.41
Manipur	4.46	39.89	8.94	-	-	-
Meghalaya	9.28	45.24	4.87	-	-	-
Mizoram	16.37	44.02	2.69	1.59	4.94	3.11
Nagaland	6.52	47.33	7.26	0.22	1.65	7.55
Punjab	51.65	1208.42	23.40	3.00	25.09	8.35
Rajasthan	23.19	317.68	13.70	0.20	2.69	13.45
Sikkim	13.08	18.99	1.45	-	-	-
Tamil Nadu	1.75	33.72	19.31	1.67	17.53	10.49
Telangana	0.37	5.70	15.42	24.76	368.60	14.89
Tripura	5.62	25.33	4.51	1.12	2.59	2.32

West Bengal	4.14	40.18	9.71	-	-	-
Others	0.03	0.11	3.30	0.08	0.11	1.43
TOTAL	428.31	5104.21	11.91	34.62	3265.83	17.69

Table 4.3. State-wise area, production and productivity of lime/lemon fruits (2017-18)

States	Area ('000 ha)	Production ('000 tons)	Productivity (tons/ha)
Andhra Pradesh	34.88	562.01	16.11
Arunachal Pradesh	0.16	0.11	0.66
Assam	13.30	112.40	8.45
Bihar	18.29	113.19	6.19
Chhattisgarh	13.31	108.30	8.14
Gujarat	46.28	605.62	13.09
Himachal Pradesh	11.55	6.72	0.58
Jammu & Kashmir	4.97	12.74	2.56
Jharkhand	7.16	54.60	7.63
Karnataka	13.10	306.21	23.37
Kerala	0.50	0.46	0.93
Madhya Pradesh	20.29	306.73	15.12
Maharashtra	27.27	250.62	9.19
Manipur	6.24	66.95	10.73
Meghalaya	1.19	4.94	4.15
Mizoram	8.10	25.90	3.20
Nagaland	0.96	7.51	7.84
Odisha	26.95	259.83	9.64
Punjab	0.99	7.62	7.74
Rajasthan	2.69	14.47	5.38
Tamil Nadu	9.65	118.46	12.37
Telangana	13.43	178.25	13.27
Tripura	4.77	24.21	5.07
Others	0.17	0.64	3.80
TOTAL	186.20	3148.47	11.00

4.1.3. Important rootstock species

4.1.3.1. Rough lemon (Jatti khatti), ***Citrus jambhiri*****:.** It is the most widely used rootstock in North India. It needs less water, suits arid conditions. Tree is medium in size, with sharpened strong spines. Leaves are light green in color and of medium in size. The leaf tips are notched; however, small somewhat colored. Fruit round in shape with irregular peel. On ripening, peel turns light yellow. Seeds bold and many up to 32 per fruit, highly polyembryonic may have 4-6 embryos.

4.1.3.2. Kharna Khatta (Khatta), *Citrus karna*: Tree is of medium size spreading than jatti khatti. Leaves are bigger than jatti khatti but are green in color. Petioles are winged. Flowers colored. Fruit medium-sized become orange colored at ripening. Fruits have well developed papilla. Peel thick with rough surface. Seeds 30-34 rarely polyembryonic. There may be 2 or 3 embryos in few seeds.

4.1.3.3. Cleopatra mandarin, *Citrus reshni*: Attractive tree with symmetrical growth. Thorns are absent. Leaves small but dark green in color. Flowers are smaller than jatti khatti and kharna khatta. Fruits small, orange, loose skinned. Average number of seed per fruit 14. Seeds are highly polyembryonic. Very good rootstock for commercial citrus cultivars particularly Kinnow and sweet oranges.

4.1.3.4. Rangpur lime, *Citrus limonia*: Trees on Rangpur lime are very vigorous and adapt well to a great variety of soils, even the weakest sandy ones, always with supplemental fertilization. It is heavy and early bearing, and drought resistance. Resistant to Tristeza but susceptible do Blight and Sudden Death. Fruit are large but internal quality is only average. Sugar production per area is excellent. Rangpur lime induces early maturity to the fruit, when prices are higher.

4.1.3.5. Trifoliate orange, *Poncirus trifoliata*: Dwarfing, resistant to phytophthora and nematodes, suitable to shallow soil and high rainfall areas. It is generally a good rootstock for oranges, grapefruit, most mandarins and lemons but Eureka lemon could be an exception. Cropping efficiency is good. Trees propagated on *Citrus trifoliata* are suitable for planting at high density. Fruit quality is excellent, with high total soluble solids and acid content and smooth thin peel. Fruit size is medium and maturity mid to late. Juice content is high.

4.1.3.6. Swingle citrumelo (*Poncirus trifoliata* x *Citrus paradisi*): The hybrid is tolerant to tristeza, *Phytophthora parasitica*, nematodes and is moderately tolerant to salts. High sugar levels in the fruit, excellent flavor and high sugar levels per box of fruit. Tree growth is less vigorous than that of Rangpur lime. However, due to its disease resistance and longevity, trees on swingle become very large. In spite of that, trees on Swingle are usually planted in higher densities, yielding high crops per unit area. It has shown satisfactory promise for sweet oranges and grapefruit. Valencia oranges on this stock are more cold hardy than on Rough lemon.

4.1.4. Cultivars

4.1.4.1. Mandarins

i) Kinnow (King x Willow Leaf): This is the most important cultivar of mandarin like fruit. It is a cross between *Citrus deliciosa* x *Citrus nobilis*. The plants were planted first At Fruit Research Station, Abohar, Punjab. It performed well and now, it has become the number one commercial cultivar of North India. Tree is medium in growth with glossy and shining leaves. Leaves are lanceolate in shape.

Tree is precocious in bearing and comes to bearing as early as in the third-year of age. Fruits are of medium in size and globose in shape. Rind is normally tight can become loose if kept up to February end on the tree. The color of rind becomes golden yellow. Segments separate with little difficulty. Very sweet and juicy TSS of juice varies from 12-15 percent with 2 to 1 percent acidity. Seeds per fruit varies from 10-25. Sometimes fruits may become seedless due to seasonal change. It ripens in January. However, its fruits are used for juice purpose along with Musambi from November onward.

ii) Local Sangtra: The trees are erect growing with sparse foliage. Fruits are smaller than Kinnow, but have short neck. The rind is loose and easy to peel, yellow in color. The segments get separated easily. Seeds 4-5 sometimes seedless. The fruits ripen in December. T.S.S. varies from 8-10 percent with 2-3 percent acidity.

iii) Khasi mandarin: It is commercially grown mandarin cultivar of Assam area and North Eastern states. There it is known as Sikkim or Kamala. Some plants in citrus orchards are mixed planted in the North also. Tree growth is just like local Sangtra. Fruits globose to ovate. Rind bright yellow to orange. The rind is smooth. The fruits may have short necks, very juicy. Juice is orange colored. Seeds varies from 7-25.

iv) Coorg mandarin: Most important cultivar of South India. It is grown on large scale in Kodagu area. Trees are vigorous and upright with dense foliage. Fruits medium with rind developing a bright orange color at maturity. The fruit shape varies from oblate to globose. Rind thin easily peeled. Very juicy with many seeds ranging from 10-30.

v) Nagpur mandarin: It is very popular mandarin cultivar of India and famous in the world. It is also known as Ponkan. The trees are vigorous, with compact foliage, leaves are lanceolate and usually erect. Fruits medium in size sub-globose. Rind loose develops cadmium color at maturity. Fruits have small neck. Very juicy. Seeds vary from 5-7. Seedless fruits are also there TSS varies from 10-12 percent with 1-2% acidity.

4.1.4.2. Sweet oranges

i) Musambi: Tree is of medium size with sparse foliage. Fruits medium in size and sub-globose in shape. The rind shows longitudinal furrows and a circular ring at the apex. The colour of rind at ripening turns light yellow. The juice is acid-less with 10 percent TSS. Fruits ripen in November, seeds varies from 20-25 per fruit. The plants raised on Pectinifera roots perform better than on jatti-khatti.

ii) Pineapple: Tree is of medium size with spreading habit. Fruits medium-sized, round in shape. Rind color yellow to orange, rind glossy and shining. Fruits are very juicy. TSS: Acid ratio well blended. Seeds varies from 10-20. Fruits ripen in December.

iii) Jaffa: Tree is similar to pineapple. Fruits are bigger in size than pineapple, round in shape. Skin yellowish orange. TSS: Acid well blended. Number of seeds per fruits varies from 9-12. Fruits ripen in December. Plants raised on Cleopatra rootstock perform better than jatti-khatti.

iv) Blood Red: Trees are vigorous than jaffa. Fruits are of medium-sized and round in shape. The skin develops deep orange color at ripening in December. Flesh red when fully ripened TSS 10-13 percent, acidity 1-2 percent. The seed number per fruit ranges between 9-12. Juice reddish in color. Plants budded on jatti-khatti show stem pitting and crease formation. Therefore, cleopatra rootstock is being used to check decline in Blood Red orange.

v) Valencia Late: Tree is vigorous. Heavy bearer, fruits are of medium-size, slightly oval in shape. Skin turns golden yellow at ripening. Juice abundant with good blend of TSS and acidity. It is late ripening cultivar. The ripening time in North India is February. The seeds vary from 2-7 per fruit.

4.1.4.3. Grape fruits

i) Red Blush: The tree is dwarf in nature. The fruits are of medium sized oblate in shape. Smooth rind, glossy and deep yellow with crimson patches, which become visible at ripening. Juice vesicles have deep crimson blush. Juice is less acidic with good TSS. Seeds usually get aborted. There can be 0-8 seeds per fruit. It ripens by the end of November.

ii) Marsh Seedless: Trees are spreading and vigorous than Red blush. Fruits are medium to large in size, oblate-roundish in shape, skin light yellow and smooth. Juice has medium TSS and acidity. Mostly seedless due to seed abortion. It ripens in December.

iii) Duncan: Tree is vigorous and spreading. It bears big- sized fruits of oblate shape. Skin pale light yellow or creamy. Juice acidic and sweet with well-marked bitterness. It contains very high number of seeds, may be up to 50. It ripens in November.

iv) Foster: Fruits are medium to large in size with oblate shape, skin pale yellow. Flesh pink and juice has well blended acidity. Taste is bitter. Seeds 40-50 per fruit. It ripens in November.

4.1.4.4. Lemons

i) PAU Baramasi: Tree is vigorous with dense foliage. The shoots drooping usually touch the ground. Fruits are lemon yellow, round in shape with rounded apex and tapering base. Skin is thin, fruit juicy and mostly seedless sometime containing few seeds. Bears throughout the year.

ii) Fureka: Tree is semi-vigorous. Fruit oblong in shape with nippled apex. Skin-lemon yellow and sooth. Juice abundant and strongly acidic, with excellent flavor. Seeds rarely present. Fruits ripen in August.

iii) Punjab Galgal: Tree is vigorous, foliage light green in color. Fruit is medium is size and oval in shape. Peel yellow smooth and glossy. Juice is very acidic (5-6 percent acidity), with 5-8 seeds per fruit. It ripens in November-December.

4.1.4.5. Lime

i) Kagzi lime: Tree is spreading and spring with dense small leaves and twigs usually attacked by citrus canker. Fruits small round and thin skinned. Juice vesicles green in color, juice is very acidic (7 percent acidity).

4.1.4.6. Sweet lime (Mitha)

i) Local: Tree is spreading with light green foliage. Stout thorns present on twigs. Fruit medium globose to ellipsoid. Skin smooth with distinct aroma. Juice abundant, non-acidic and insipid. Seed 5-6 per fruit. Fruits ripens in September.

4.1.5. Exports

In 2020, India exported worth US$ 68.4 million in Citrus, making it the 25th largest exporter of Citrus in the world. At the same year, Citrus was the 424th most exported product in India. The main destination of Citrus exports from India are: Bangladesh (US$ 50.2M), Nepal (US$ 7.95M), United Arab Emirates (US$ 6.99M), Qatar (US$ 567k), and Oman (US$ 535k).

The fastest growing export markets for Citrus of India between 2019 and 2020 were Bangladesh (US $ 33.6M), United Arab Emirates (US $ 2.05M), and Oman (US $ 234k).

4.1.6. Nematode problems

Nematodes are one of the important limiting factors in citrus production throughout India. Citrus crop being perennial in nature, harbor and encourage the buildup of nematode population throughout the year. The first record of an association between a nematode and citrus appears to be that of Thirumala Rao (1956), who found *Meloidogyne* sp. parasitizing citrus roots in Andhra Pradesh. But it was not until the discovery in 1961 of *Tylenchulus semipenetrans*, on the roots of citrus trees in Uttar Pradesh (Siddiqi, 1961) that a nematode was found to cause a disease condition of citrus, later called "slow decline".

Of the several plant parasitic nematodes reported in association with citrus roots from India, proof of pathogenicity to citrus is available for only eight species viz., *Tylenchulus semipenetrans* (Siddiqi, 1961), *Pratylenchus coffeae* (Siddiqi, 1964), *Hoplolaimus indicus* (Gupta and Atwal, 1972), *Meloidogyne africana* (Chitwood and Toung, 1960), *M. indica* (Whitehead, 1968), *M. javanica* (Mani, 1986), *Radopholus citrophilus* (Suit and DuCharme, 1953) and *Hemicycliophora arenaria* (Van Gundy, 1957). Except for the citrus nematode (*T. semipenetrans*), other nematode species that are pathogenic on citrus have limited geographic distribution.

4.2. THE CITRUS NEMATODE, *TYLENCHULUS SEMIPENETRANS*

Siddiqi (1961) reported the citrus nematode for the first time from India in a paper presented before the Indian Science Congress and observed that about 80 per cent of the citrus trees at Aligarh, Uttar Pradesh, were infested with this nematode. The nematode causes "slow decline" and is considered to be one of the factors responsible for die-back of citrus trees in India (Chona *et al.*, 1965).

4.2.1. Occurrence and distribution

The citrus nematode is known to be widely distributed in major citrus growing regions of the country. It has been reported from Uttar Pradesh (Siddiqi, 1961); Delhi, Punjab, Rajasthan, Maharashtra, West Bengal, Assam (Chona *et al.*, 1965); Himachal Pradesh (Mukhopadhyaya, 1970a), Sikkim (Edward and Rai, 1970); Orissa (Khuntia and Das, 1969); Haryana (Mukhopadhyaya and Suryanarayana, 1969); Bihar (Prasad and Chawla, 1965); Karnataka (Swamy *et al.*, 1973); Kerala (Nair, 1965); Tamil Nadu (Muthukrishnan and Sivakumar, 1973) and Andhra Pradesh (Krishnamurthyrao and Thammiraju, 1975).

4.2.2. Economic importance and losses

Yield increases of 40 to 200 per cent have been obtained following the control of the nematode over a range of growing conditions (Swarup and Seshadri, 1974). *T. semipenetrans* was responsible for 69%, 29% and 19% loss in fruit yield of sweet orange (Baghel and Bhatti), lemon (Mukhopadhyaya and Suryanarayana, 1969) and sweet lime (Mukhopadhyaya and Dalal, 1971), respectively.

The citrus nematode is responsible for annual yield loss of 27% in citrus amounting to Rs. 9828.22 million in India (Walia and Chakraborty, 2018).

4.2.3. Symptoms

'Slow decline' of citrus is a diseased condition of trees with symptoms similar to those caused by drought and malnutrition. Affected trees exhibit reduced vigor, chlorosis and falling of leaves, twig dieback and consequently, reduced fruit production (Prasad and Chawla, 1965). This decline of the tree is gradual and persists until the crop is so small that tree maintenance may become uneconomical. The decline symptoms are often more pronounced in the uppermost portion of the trees. In Israel, the decline symptoms are associated with a critical nematode population of approximately 40,000 larvae and males per 10 grams of roots, above which the tree performance is markedly lowered.

Soil particles usually cling tightly, even after washing to the gelatinous egg masses which cover the protruding part of the female nematode body. In heavily infested roots, the cortex separates readily from the vascular stele.

As the nematode feeds on roots and reproduces, a large proportion of the feeder roots of citrus trees, particularly in the upper soil layers, is inactivated or destroyed, the uptake of water and minerals from the soil is reduced and the symptoms appear on the above ground tree parts. Heavily infested roots are darker in color with branch rootlets shortened, swollen and irregular in appearance than in normal ones (Chona *et al.*, 1965).

4.2.4. Host range

Mainly host range is confined to Rutaceae family. A total of 75 Rutaccous species mainly citrus and citrus hybrids are the hosts of this nematode. Other non Rutaceous hosts reported from India are grapes and loquats (Parvatha Reddy and Singh, 1978a), *Clerodendron inerme* (Nand *et al.*, 1994), *Jasminum sambac* (Bajaj *et al.*, 1988), *Garcina mangostana* (Chawla *et al.*, 1980) and loquats (Chona *et al.*, 1965).

4.2.5. Life cycle

Van Gundy (1958) studied the life cycle in detail. Eggs hatched in 12 to 14 days at about 24 °C and sex differentiation was possible at the second stage. The second stage male larvae developed to maturity within 7 days without feeding. The second stage female larvae required 14 days to locate the root, feed on the epidermal cells and molt. The life cycle from egg to egg required 6 to 8 weeks.

In experiments conducted in Israel, the minimum time required at 24 °C for completion of the life cycle was 14 weeks on *Poncirus trifoliata*, 10 weeks on *Ruta bracteosa* and 7 weeks on *Citrus aurantium* and *C. limettoides* (Cohn, 1965).

4.2.6. Histopathology

Histopathological changes in the root tissue from the time of inoculation with second stage larvae to the establishment of feeding zone are described in citrus rootstocks viz., Karna Khatta (*Citrus karna*), Jamberi (*C. jambhiri*), Hill lemon (*C. pseudolimon*), Rangpur lime (*C. limonia*), Sweet lime (*C. limettioides*) (Misra and Edward, 1977) and Gajanimma (*C. pennivesiculata*) (Edward *et al.*, 1965).

4.2.7. Races/Biotypes

Based on different reports and studies on the infectivity of citrus nematode on different host's world over, four races of *T. semipenetrans* have been proposed by Inserra *et al.* (1980). 1) 'Poncirus Biotype' which reproduces on *Citrus* spp., *P. trifoliata* and their hybrids, grapes, but not on olive; 2) 'Citrus Biotype' which reproduces poorly on *P. trifoliata*, but infects *Citrus* spp., Carrizo and Troyer Citrange, Olive and Persimmon; 3) 'Mediterranean Biotype' which is close to citrus biotype, but does not reproduce on olive; 4) 'Grass Biotype' reproduces only on grass, *Andropogon rhizomatus*. Indian population of the nematode is considered related to Mediterranean biotype (Table 4.4).

Table 4.4. Biotypes of *Tylenchulus semipenetrans*

Host differential	Biotypes		
	Poncirus	Citrus	Mediterranean
Citrus sp.	+	+	+
Poncirus trifoliate	+	-	-
Grape	+	+	+
Olive	-	+	-

Persimmon	?	+	+
Distribution	California Japan Israel	California Italy India?	Mediterranean region South Africa India?

4.2.8. Economic threshold level

Citrus is a perennial crop for which to find out economic threshold level is slightly difficult. This also varies from field to field depending upon the agro practices followed. In India, economic threshold level has not been estimated under field conditions. However, it has been estimated in California that soil stages (juveniles/100 g soil) below 800 represents a non-damaging population level. The orchards with more than 1600 juveniles/100 g soil may respond economically to nematicide treatments and at level more than 3600 during peak population growth period, treatment may improve yield substantially. The females per gram root are also used to define the damage level, with counts of <300, >700 and >1400 representing low, moderate and high ranges, respectively. In Florida, it was estimated that yield was not measurably reduced if populations were below 2000 juveniles/100 cc soil during peak period of the soil population development (Duncan and Cohn, 1986).

4.2.9. Survival and spread

T. semipenetrans is extremely sensitive to lack of moisture and high temperature. Most of the juveniles are killed if infected roots and soil are exposed to sun during summer. Under field conditions, the nematode survives mainly on the left over roots. In the absence of host, at 3 bars moisture stress, 9-27% of *T. semipenetrans* juveniles could survive after 450 days (Gaur and Sehgal, 1988); while at 30 bars of moisture stress at 40 °C, only 5% of the populations could survive after 7 days (Sehgal *et al.*, 1990).

The dissemination of this nematode to newer areas is mainly through infected nursery plants. The irrigation water, particularly, flood irrigation is the prime source for its spread from plant to plant within the orchard. This can also spread through contaminated implements.

4.2.10. Ecology and population fluctuation

The active population development period may vary from place to place depending upon agro climatic conditions. In Haryana, two peaks in October and April have been recorded on *C. sinensis* (Baghel and Bhatti, 1982). In Delhi, maximum populations have been recorded on *C. aurantifolia* in August (Prasad and Chawla, 1965). While contrary to these reports, one peak of nematode population has been recorded in Maharashtra in January-February on *C. reticulata* grafted on *C. jambhiri* (Singh, 1997).

The distribution of *T. semipenetrans* in the rhizosphere of citrus trees is related with the active root zone. The maximum nematode populations are observed at a distance of 90 cm from tree trunk and up to 30 cm depth (Chhabra, 1978; Baghel and Bhatti, 1982), though, *T. semipenetrans* have been observed up to the depth of 180 cm (Chawla and Sharma, 1984). Sharma and Sharma (1977) reported maximum concentration of population at 30 cm distance from tree trunk in top 30 cm soil irrespective of plant age.

The population density is also related with the decline condition of the trees. Usually the population is more on the trees at initial stage of decline as compared to the population at advance stage of tree decline. This is mainly due to reduction in root mass in the advance stage of decline (Chona *et al.*, 1965; Bindra *et al.*, 1967; Mani, 1994). The nutritional level of plant also influences the nematode population. The low levels of N and P result in higher population while higher dose of

K favors low nematode population (Mangat and Sharma, 1981).

4.2.11. Interaction with other pathogens

The role of root-rot fungi in the disease syndrome caused by *T. semipenetrans* is significant. Misra and Edward (1977) reported severe root damage to citrus when the nematode was accompanied by secondary invaders like *Fusarium oxysporum* and *Macrophomina phaseoli*. The nematode helps weakly pathogenic fungus like *Fusarium* to cause secondary infection of citrus roots (Chhabra, 1972).

In the presence of *T. semipenetrans*, significantly more citrus plants become infected by *Fusarium oxysporum* and *F. solani* and the combined activity of the nematode and *F. solani* is thought to be a factor in the debilitation of citrus. *Fusarium* spp. seems to cause more damage in combination with the citrus nematode in cool, wet soils and with *Phytophthora* spp. in warm, wet soils.

4.2.12. Management

4.2.12.1. Physical methods: The nematode can be controlled by dipping the bare roots of citrus seedlings in hot water at 45 °C for 25 min or 46.7 °C for 10 min (Bindra *et al.*, 1967). This treatment leaves no adverse effect on seedlings. During summer, exposing the soil to hot sun can control the nematodes substantially (Sehgal *et al.*, 1990).

4.2.12.2. Cultural methods: Populations of the citrus nematode in soil and roots can be reduced by removing the old feeder roots before the growth flush followed by application of FYM.

The citrus nematode is easily transported to new orchards through infested seedlings and budded plants which are usually carried 'balled' with infested soil. Therefore, nurseries should never be established on or near old citrus orchards and nursery soil should be sterilized before planting. Care should also be taken not to spread infection through tools, machinery and irrigation water used in infested groves.

Commencing from March, the bimonthly irrigation of citrus orchards with dung extract (prepared by mixing 2/3 cow dung + 1/3 water and kept for a period of two weeks with daily shaking; diluted with water in a ratio of 1: 10 before application) improved tree growth and markedly reduced the nematode population.

4.2.12.3. Intercropping: Intercropping of acid lime with marigold and mustard markedly reduced the rate of multiplication of *T. semipenetrans* to 10.03 and 8.59, respectively (Mani, 1988b). Intercropping of onion, garlic or marigold in the citrus orchards not only reduces the nematode population but also provides additional income.

4.2.12.4. Botanicals: Singh and Sitaramaiah (1973) reported that castor cake was highly effective in reducing 50% of the citrus nematode population. Mobin and Khan (1969) have also observed that neem, mahua, groundnut and mustard oil cakes were effective in reducing the nematode population in citrus rhizosphere. Application of neem cake at 20 kg per tree basin at an interval of 4 months can effectively reduce the nematode population of *T. semipenetrans and Meloidogyne javanica* (Mani and Murty, 1986).

Irrigation of citrus orchards with sewage water reduces the citrus nematode population to a very low level.

Castor (*Ricinus communis*), Aak (*Calotropis procera*) and Datura (*Datura strumonium*) chopped leaves applied at 20 and 40 g/kg soil in 9 sq. m. area around tree trunks can result in reduction of nematode population and increase in yield (Baghel, 1995).

4.2.12.5. Chemical control

i) Nursery bed treatment: Soil application of aldicarb or carbofuran at 4 kg a.i. /ha was effective for the control of *T. semipenetrans* (Parvatha Reddy, 1988).

ii) Seedling bare root-dip treatment: The bare root-dip treatment of acid lime seedlings with monocrotophos or chlorpyriphos at 1000 ppm for 45 min. reduced the multiplication of *T. semipenetrans* by 90% (Mani, 1990a).

iii) Main field treatment: Non-fumigant compounds like carbofuran, ethoprophos, phorate, fensulfothion, dimethoate *etc.*, have been tested against citrus nematode under field conditions. Dichlofenthion at 45 l/ha reduced the nematode population by 80% (Mukhopadhyaya, 1970b). Mukhopadhyaya and Dalal (1971) achieved 97.6% control by ethoprophos when applied at 4 kg a.i. /ha and 39.9% increase in yield of sweet lime during second year. Chhabra *et al.* (1977) found fensulfothion very effective at 3 kg a.i./ha in controlling the nematodes on pumello and increased the yield by 68-76%.

Significant reduction in the citrus nematode population (76 to 84 per cent) in sweet orange and increase in yield (61 to 244 per cent) was observed with soil application of aldicarb and carbofuran both at 6 kg a.i. per ha (Baghel and Bhatti, 1983b). The effectiveness of aldicarb and carbofuran was also demonstrated for the control of *T. semipenetrans* infecting acid lime (Parvatha Reddy, 1986) (Table 4.5).

Table 4.5. Effect of different nematicides on *Tylenchulus semipenetrans* infecting different *Citrus* spp.

Citrus spp.	Nematicide & Dosage	Type of application	Effectiveness
Acid lime	Carbofuran - 4 kg a.i. /ha	Soil application	Reduced nema popn. by 65-75%
Grape fruit	Carbofuran, Fensulfothion, Phorate - 5 & 10 kg a.i. /ha	Soil application	Reduced nema popn. by 75-90% at lower conc. & by 90% at higher conc.
Naval orange	Carbofuran, Ethoprophos, Fenamiphos-22 kg a.i./ha	Soil application	Reduced larval popn. in soil, increased fruit yield by 108-137%
Orange	Carbofuran, Carbosulfan – 1.1 kg a.i. /ha;	Through drip irrigation	Reduced female nemas on roots and increased fruit yield
	Fenamiphos – 10 ml/^{m}2	With irrigation water	Reduced larval popn., increased uptake of K, Ca and Cu, gave extra yield of 6.1-7.7 tons
Sweet lime	Ethoprophos - 40 kg a.i. /ha	Soil application	Reduced nema popn. by 98%, checked popn. build-up for 16 months, increased fruit yield at 12 kg/tree
	Carbofuran - 4 kg a.i./ha	Soil application	Reduced nema popn. in soil
	Carbofuran - 6 kg a.i./ha	Soil application	Reduced nema popn. by 76%, increased fruit yield by 152% with cost: benefit ratio of 1:8
	Phorate – 6 kg a.i./ha	Soil application	Reduced nema popn. by 66%
	Oxamyl – 4 μg/ml	Foliar spray	Reduced nema popn. in pots

4.2.12.6. Biological methods

i) Antagonistic bacteria: Application of *Pseudomonas fluorescens* at 20 g/tree three times in a year at 15 cm depth and 50 cm away from tree trunk helps in reducing the nematode population.

Pasteuria penetrans can be readily multiplied on the citrus nematode. The parasite seems to offer potential for the biocontrol of *T. semipenetrans* (Mani, 1988a).

In citrus, a monthly rate of 1.1 kg a.i. per ha of avermectins (*Streptomyces avermitilis*) for 7 months gave maximum increase in yield and reduction of *T. semipenetrans* population (Garabedian and Van Gundy, 1983).

ii) Antagonistic fungi: Several nematophagous fungi such as *Arthrobotrys cladodes, Dactylaria* sp., *Monacrosporium gephyrophagum, Laginidium* sp. *etc.*, can trap the *T. semipenetrans* and could be exploited for nematode control (Gowda *et al.*, 1982). Endoparasitic fungus, *Purpureocillium lilacinum* is a potential biocontrol agent for controlling citrus nematode (Mani *et al.*, 1989).

Since the parasitic fungus, *P. lilacinum* has many attributes of a successful biological control agent, experiments were conducted in nematode infested citrus grove to determine its efficacy in controlling *T. semipenetrans.* It was observed that fruit diameter of the orange cv. 'Valencia' was significantly affected by application of 3 nematicides and *P. lilacinum.* The fruit diameter from fungus treated tree was greater than that from trees treated with temik, vydate and mocap.

Similarly, the number of nematodes in roots and soil around citrus trees inoculated with *P. lilacinum* was significantly lower than that in the non-treated or the nematicide treated trees (Jatala, 1986). Application of *P. lilacinum* in the form of spore suspension checked the multiplication of *T. semipenetrans* by 71.7 to 97.8% (Mani and Murthy, 1989).

iii) Arbuscular mycorrhizal fungi: Arbuscular mycorrhizal fungus, *Glomus mosseae* when added to the soil with *T. semipenetrans* in *Citrus jambhiri,* limited the nematode development and partially neutralized the adverse effect of the nematode (Baghel *et al.*, 1990).

4.2.12.7. Host resistance: Citrus rootstocks which are resistant or highly tolerant to the citrus nematode would offer practical and effective means of control. Trifoliate orange (*Poncirus trifoliata*), Troyer Citrange and Swingle citrumello and its hybrids (Citrumello) were highly resistant to the citrus nematode (Parvatha Reddy and Singh, 1978; Chhabra and Bindra, 1974; Parvatha Reddy and Agarwal, 1987; Mani and Reddy, 1986). The citrus rootstock hybrids evolved at the Indian Institute of Horticultural Research, Bangalore, by crossing Rangpur lime (*Citrus limonia*) with *P. trifoliata* viz., CRH-3, CRH-5 and CRH-41 were highly resistant to the citrus nematode (Parvatha Reddy *et al.*, 1987).

Inter-generic hybrids were evolved, involving the commercial rootstocks Rough lemon, Rangpur lime, Cleopatra mandarin and Orlando tangelo, as seed parents and *Poncirus trifoliata* (the citrus nematode resistant compatible donor) as pollen parent to overcome the citrus nematode problem. Twenty-five promising hybrids and the 5 parents were screened against *T. semipenetrans* by artificial inoculation under glasshouse conditions. Nine hybrids: CRH-57, 47, 10, 35, 17, 5, 41, 3 and 12 were resistant while 2 hybrids, CRH-8 and 24 were susceptible (Prasad *et al.*,1998, 1999). The remaining 14 hybrids were moderately resistant. Among the parents, *Poncirus trifoliata* was resistant while all the others were highly susceptible.

Resistance, based on the number of larvae and adult females on the feeder roots, was evaluated in 57 selections belonging to 4 genera (*Citrus, Fortunella, Poncirus* and *Severinia*) and 24 species/hybrids. All *Severinia disticha* and *P. trifoliata* strains and hybrids except 2 were rated resistant. Pummelo (*C.*

maxima) x *P. trifoliata* and Smooth Flat Seville (*C. aurantium*) x Swingle citrumelo (*C. paradisi* x *P. trifoliate*) exhibited a moderately resistant reaction. It is suggested that Swingle citrumello and *P. trifoliata* strains should be used as citrus rootstock parents. Rangpur lime *(C. limonia)*, grapefruit *(C. paradisi)* and *C. pectinofera* were moderately resistant. Of 13 Rangpur lime (*Citrus limonia*) and rough lemon (*C. jambhiri*) strains tested, none were found to be resistant to *T. semipenetrans*. Rough lemon (Poona), Rough lemon (Madhya Pradesh), Rough lemon 14, Rough lemon 8779 and Rough lemon 8782 were moderately resistant, while all the others were susceptible.

4.2.12.8. Integrated methods

i) Using bioagents and botanicals: Management of the citrus nematode based on the application of neem and castor cakes both at 10 kg per plant along with a nematophagous fungus, *Purpureocillium lilacinum* at 250 g (grown on paddy seeds) per plant 3 times in a year at 15 cm depth and 50 cm away from the trunk was found to be extremely effective in reducing the citrus nematode population with a consequent increase in the growth of acid lime trees. The above treatment also gave highest parasitization of egg masses and eggs of *T. semipenetrans* and increased spore density of *P. lilacinum* in soil (Parvatha Reddy *et al.*, 1991a, 1993).

Trichoderma harzianum in combination with neem oil cake was effective in increasing the growth of acid lime trees and reducing the citrus nematode population both in soil and roots. The parasitization of citrus nematode females with *T. harzianum* increased in the presence of oil cakes (Parvatha Reddy *et al.*, 1996b).

Incorporation *Verticillium lecanii* with neem cake facilitated the effective management of *T. semipenetrans* on acid lime (Parvatha Reddy *et al.*, 1996a).

Integration of neem cake with *Pseudomonas fluorescens* gave maximum reduction in citrus nematode population both in soil and roots and increased plant growth of acid lime (Parvatha Reddy *et al.*, 2000).

ii) Using arbuscular mycorrhizal fungi and botanicals: Inoculation of endomycorrhiza, *Glomus fasciculatum* in the soils amended with neem cake was found effective for the management of *T. semipenetrans* on acid lime. This strategy can help in combating the menace of citrus nematode at nursery stage and also provide highly mycorrhizal seedlings of acid lime for transplanting in the main field for the management of *T. semipenetrans* under field conditions (Parvatha Reddy *et al.*, 1995).

iii) Using two bioagents: Application of bacterial bioagent, *Pasteuria penetrans* (at 2 x 10^9 spores/plant) and fungal bioagent, *P. lilacinum* (at 50 g/plant with 4 x 10^7 spores/g) was effective in reducing the *T. semipenetrans* population and in increasing the parasitization of larvae by *P. penetrans* and eggs by *P. lilacinum* (Parvatha Reddy and Nagesh, 2000).

iv) Using chemicals and botanicals: Neem cake at 1 kg/plant combined with carbofuran at 2 kg/ha reduced 47.1% nematode population and increased the yield by 41.1% (Baghel, 1995).

v) Using bioagents and chemicals: Integration of *P. lilacinum* (at 4 g/plant) with carbofuran (at 30 mg a.i./plant) was found effective in increasing the plant growth parameters of acid lime and in reducing the citrus nematode population both in soil and roots (Parvatha Reddy *et al.*, 1996c).

vi) Biofumigation and solarization: The effect of tissue mulching and irrigation on survival of the citrus nematode, *T. semipenetrans* was measured. Oriental mustard (*Brassica juncea*), yellow mustard (*Sinapis alba)*, faba bean (*Vicia faba*) and a mixture of cereals (rye and triticale) were chopped with a mower at late flowering and sprinkler-irrigated with 30 mm of water within 0.5 hour. Any organic material reduced citrus nematode survival. Citrus nematode suppression was 92% greater after oriental and yellow mustard (except yellow mustard/ plastic) than after cereal or legume, indicating mustard allelochemicals possibly suppressing nematodes.

Biofumigation with yellow and oriental mustard cover crops reduced citrus nematode and *Phytophtora* root rot by 90%. The beneficial fungi such as *Trichoderma* have been found at much higher levels in the mustard treated soils. This does not occur when the soil is fumigated with chemicals. Mustard cover crops are now common place in the Salinas Valley, California.

4.3. THE LESION NEMATODE, *PRATYLENCHUS COFFEAE*

Pratylenchus coffeae is a highly pathogenic nematode reported from citrus in India, Japan and U.S.A. and is causal organism for 'Root-rot of Citrus'. In Florida (USA), it is known to cause 'Citrus slump disease'.

4.3.1. Distribution

P. coffeae is a pest of citrus in India, Japan and the U.S.A. In India, it has been reported from Uttar Pradesh, Punjab and South Western India on citrus (Sethi and Swarup, 1971; D'Souza *et al.*, 1970).

4.3.2. Economic importance and losses

P. coffeae appears to be the most pathogenic species and is responsible for the "Citrus slump disease" in Florida. The nematode root rot of citrus in Uttar Pradesh is principally caused by *P. coffeae* (Siddiqi, 1964). This nematode constitutes the dominant factor in the nematode complex of citrus in Aligarh. O'Bannon and Tomerlin (1973) reported 49 to 80 per cent growth reduction in young trees up to 4 years depending upon different rootstocks.

4.3.3. Symptoms

P. coffeae infected citrus plants exhibit poor growth, general dieback and produce undersized fruits. It is a migratory root endoparasite, the adults and juveniles of which invade the root tip and destroy the growing meristem. As a result, a lateral root usually initiates near the destroyed root tip. The adults and juveniles occur in the cortical tissue of feeder roots where they form small black lesions at their feeding site which expand to girdle the rootlets.

Non-specific above ground symptoms appear in form of stunted plants, general dieback and undersized fruits. Brown black lesions girdle the feeder roots. The infected seedlings show marked reduction in root and shoot weight as compared to healthy ones.

4.3.4. Host range

Citrus limon, C. sinensis, C. reticulata and banana have been found to be hosts of *P. coffeae* in Uttar Pradesh (Siddiqi, 1964). The nematode is transferable from banana to citrus plants and from lemon to orange plants and *vice versa*.

4.3.5. Life cycle

P. coffeae is a migratory endoparasite of the root cortex where it feeds and multiplies. The favored place of entrance is slightly back of the elongating zone in the piliferous region. A colony or 'nest' is established showing brownish lesions on the root. When decay sets in due to secondary invasion by bacteria and fungi, the nematodes quit and migrate through the soil to attack fresh roots. The first molt takes place within the egg and three molts occur outside. *P. coffeae* requires about 45 to 48 days to complete the entire life cycle. Of this time, 15 to 17 days are necessary for eggs to hatch, 15 to 16 days are spent as larvae, and it takes 15 days from maturity to egg production.

4.3.6. Histopathology

Adults and larvae of *P. coffeae* are found in the cortical tissues of the feeder roots where a tiny, brownish-black lesion is formed which gradually expand to girdle the rootlets. When the nematode invaded a root-tip, the meristem often was destroyed and lateral root initiation usually occurred near the destroyed root tip. Inoculation experiments have shown that the young plants of *C. limon* and *C. sinensis* show poor growth when infected with *P. coffeae* (Siddiqi, 1964). A marked reduction in the top as well as root weight of the infected seedlings was noticed. *C. limon* seedlings inoculated with 20 *P. coffeae* had peak populations at 36 weeks with 10,000 nematodes per g root extracted; shoot growth was reduced by 22 per cent. *P. coffeae* extensively invades the cortex of feeder roots, causing cavities and cell necrosis.

4.3.7. Survival and spread

P. coffeae survived best in excised infected roots of *C. jambhiri* stored in soils with initial moisture content near field capacity and at temperature ranging from 10-32 °C and were infective after 4 months. The nematodes did not survive extended storage at temperatures over 38 °C. The optimum temperature for reproduction was 29.5 °C.

4.3.8. Management

4.3.8.1. Cultural methods: Sanitation is important to reduce damage by *P. coffeae*. Citrus nurseries in Florida, South Africa, Brazil and elsewhere have either mandatory or voluntary programs to certify that young trees are free of *P. coffeae*.

4.3.8.2. Chemical methods: Infestation in citrus seedlings of *P. coffeae* was eliminated by treating with fensulfothion and fenamiphos at 4.4 kg a.i. per ha and the stem diameter of the treated seedlings increased by 44 and 46 per cent, respectively after 9 months. The population of *P. coffeae* can be effectively reduced by applying carbofuran at 4 kg a.i. per ha (Baghel and Bhatti, 1983a).

4.3.8.3. Host resistance: Out of 125 selections of citrus, most of the selections were considered highly susceptible supporting population more than 1000 per gram root. Only four selections of *Microcitrus australis* x *M. australasica* hybrid and Rubiboux 70-A5, Trifoliate orange (*Poncirus trifoliata*) had nematode populations of less than 100/g root. They were considered to show some resistance to *P. coffeae* because of low numbers.

4.4. ROOT-KNOT NEMATODES, *MELOIDOGYNE JAVANICA, M. INDICA*

In India, Thirumala Rao (1956) reported that *Meloidogyne* sp. caused considerable damage to citrus in Andhra Pradesh when a susceptible crop like tobacco or okra is grown as an intercrop. This is the first report of nematode damage to citrus in India.

Chitwood and Toung (1960) observed the root-knot nematode resembling *M. africana* infecting *C. sinensis* from Delhi and proved its pathogenicity to citrus. Whitehead (1968) reported *M. indica* on citrus from India.

Mani (1986) reported that *M. javanica* caused severe crop loss of acid lime in Andhra Pradesh and it was pathogenic to acid lime and sweet orange. It is confined to only coastal Andhra Pradesh.

Kagzi lime orchards in North Gujarat, particularly in Banaskantha district were observed to be heavily infected by the root-knot nematode *M. indica*. The citrus root-knot nematode, *M. indica* infection was observed in Banaskantha, Mahesana and Anand districts of Gujarat.

4.4.1. Distribution

The root-knot nematode on citrus has been reported from Andhra Pradesh, Gujarat and Delhi.

4.2. Symptoms

The above-ground symptoms of the citrus root-knot nematode include severe stunting of trees, yellowing of leaves, poor vigor, unthrifty tree appearance, and twig blight leading to citrus decline. The plants were showing yellowing of foliage in initial stage and were completely drying at later stage. On uprooting such infected plants, numerous small to very big sized galls/knots were noticed on the roots, resulting in very poorly developed root system (Patel *et al.*, 1999). They fail to flower and produce fruits even after several years of planting. The roots have conspicuous galls on pioneer and fibrous roots. Big cavities were seen where galls were present earlier during advanced stage. On the surface of roots, thin films can be observed which contain nematode egg masses.

4.4.3. Host range

M. javanica has been reported on *C. aurantifolia* and *C. sinensis* (Mani, 1986), whereas *M. indica* infects *C. aurantium, C. sinensis* (Whitehead, 1968) and *C. aurantifolia* (Parvatha Reddy *et al.*, 1981).

4.4.4. Management

4.4.4.1. Physical methods: Soil solarization is one of the effective ways of suppressing nematode populations and can be employed mostly during hot weather days. The nursery area should be plowed well and leveled after breaking the clods. The prepared field may be covered with polythene sheet of 100 gauge thickness for about 7-10 days before sowing.

4.4.4.2. Cultural methods: At least two to three deep plowings should be given during bright sunny days at an interval of 10-15 days.

It is essential to select root-knot nematode-free planting material to prevent the introduction of nematode inoculum into orchards. It is preferable to avoid growing vegetables as intercrops or as rotational crops in citrus orchards and nurseries. Cereals and millets can be grown in nursery area according to grower's choice.

i) Intercropping: Growing sun hemp in tree basin reduced the population of root-knot nematodes. *Crotalaria* roots exert a toxic effect on *Meloidogyne* population. *Meloidogyne* larvae freely entered roots of *Crotalaria* but failed to survive. The possibility of growing *Crotalaria* in tree basin for the control of root-knot nematodes infecting citrus needs further investigation.

ii) Crop rotation: Once the life of the citrus orchard comes to an end, it is highly preferable to grow cereals and millets for at least 2-3 seasons before initiating new orchard in the same field. This will help in reducing the populations of *Meloidogyne* species and *T. semipenetrans* which are common in citrus orchards.

iii) Botanicals: Field experiments carried out in lime orchards indicated that application of neem cake at 20 kg per tree basin reduced the population of *M. javanica* by 25 per cent three months after application. Repeated applications once in four months significantly reduced the root-knot nematode population. FYM or compost at 20-25 t/ha or oil cakes, particularly neem or karanj cakes at 2-3 t/ha may be incorporated in soil before sowing or transplanting.

4.4.4.3. Chemical methods: Carbofuran at 4 kg a.i. per ha reduced the larval population of *M.*

javanica on acid lime in soil by 62.8 per cent. Chemical should be applied in two split doses at 4 – 6 months interval at the time of formation of new roots to protect them against nematode infection.

4.4.4.4. Biological methods: Use organic amendments enriched with bioagents (*Pupureocillium lilacinum*) for raising seedlings and in standing crop in the field proved beneficial.

The root-knot nematode in sweet orange can be managed by soil application of 100 kg FYM enriched with *Pseudomonas fluorescens* (2 x 10^6 cfu /g) @ 20 kg/ha.

4.4.4.5. Host resistance: It was observed that *C. reticulata* was not infected by *M. javanica.* Sour orange, Lime, Rough lemon, Grape fruit, Troyer citrange, Cleopatra mandarin and Trifoliate orange were highly resistant to *M. incognita* (Race 1 and 3) as well as *M. javanica.*

The response of *Citrus* spp. and related rootstocks to a population of *Meloidogyne javanica* was evaluated by some researchers in a screen-house experiment. Palestine and Rangpur lime, Rough lemon, Sour orange, Sexton and Thentriton tangelos and Volkamer lemon were not infected by *M. javanica*. Galls and tip swellings were observed on the roots of *Poncirus trifoliata* and Troyer citrange. There was no evidence of nematode development. Symptoms induced by the nematode were stelar division, syncytia formation in the vascular tissues and necrotic cells.

4.5. THE BURROWING NEMATODE, *RADOPHOLUS CITROPHILUS*

Perhaps the most spectacular disease with which the burrowing nematode, *R. citrophilus* has been associated is that of 'spreading decline' of citrus in Florida, USA. Suit and DuCharme (1953) have demonstrated that this nematode causes spreading decline in citrus trees in Florida, USA. This nematode is of great economic importance to citrus in Florida and causes reduction in top and root growth. Although attempts have been made to eliminate this pest, several thousand hectares currently under cultivation still remain infested. Spreading decline has been detected only in those citrus groves located on the sandy ridge zone of central Florida. It is responsible for 40-80% loss in fruit yield.

R. citri was described in 1996 (Machon and Bridge, 1996) from roots of citrus seedlings in Indonesia. In pathogenicity experiments, populations of only 1000 nematodes per plant caused severe root degradation and significantly reduced plant growth.

4.5.1. Symptoms

The burrowing nematode infected citrus field show patchy appearance. Trees with heavily infected roots wilt more readily than healthy trees during periods of water stress. The trees appear under-nourished and branches show twig die-back. The leaves are small and sparse. The size of the fruit is small and overall yields are less. Diseased trees are non-thrifty, lack vigor and produce fewer fruits. Affected trees have abundant dead twigs and branches.

R. citrophilus is a migratory endoparasitic nematode which parasitizes the tips of new feeder roots soon after they are formed, resulting in stubby, somewhat swollen root tips. At the nematode infection sites, brown to black lesions are formed and sometimes cankers develop. Extensive feeder root damage takes place 50 cm below ground level in affected trees with only 20 to 30% feeder roots at depths of 25 to 75 cm. Feeder roots are virtually absent at depths below 75 cm in infested orchards.

4.5.2. Life cycle

The burrowing nematode is a migratory endoparasitic nematode. Juveniles and adults penetrate roots in the region of elongation. The nematodes feed on and kill cells forming cavities and tunnels in the cortex and stele of roots. Eggs are laid randomly within the roots. Once inside, the nematodes may spend their entire lives within the root. The life cycle takes about 18-20 days at 24-26 °C. The nematode may survive in soil without feeding for about 6 months.

4.5.3. Histopathology

Females and larvae enter growing feeder roots near their tips in the region of cell elongation and root hair production. The nematode upon penetration, feeds on the cortical parenchyma cells and gradually burrows towards the stele, creating tunnels and cavities in the cortex and stele. Large number of nematodes accumulate in the phloem-cambium ring region. This part of the root is often completely destroyed leaving a cavity filled with nematodes. Cell reactions involve hypertrophy and when the nematode penetrates pericycle, hyperplasia and tumor formation occur. Wound gum accumulates next to the stele in the invaded area, imparting tan to amber color in the older portions of the lesions (DuCharme, 1959).

4.5.4. Biotypes

The nematode appears to have at least 2 biotypes: Biotype 1, which reproduces poorly on Milam lemon roots, and Biotype 2, which reproduces well and significantly damages Milam Lemon and other resistant rootstocks.

4.5.5. Survival and spread

Large root fragments that remain buried in soil after tree removal may help support populations during fallow. The nematode is spread in contaminated rootstock, machinery, subsoil water and it migrates rapidly along developing root systems. The infested area gradually expands in all directions regardless of rows or direction of cultivation. The measurable spread (about 15 meters per year) is characteristic of the disease. The nematode spread is aided by the flow of water in soil.

4.5.6. Interaction with root rot fungi

Spreading decline in citrus tree is due to the combined activity of the burrowing nematode, *R. citrophilus* and other root rot organisms that are always found in the parasitized roots. Although it is recognized that the nematode is the primary cause of disease, pathogenic fungal species such as *Fusarium, Sclerotium* and *Thielaviopsis* are commonly found in lesions soon after nematode invasion. Subsequently, bacteria and cellulose destroying fungi such as *Pencillium* and *Aspergillus* invade the root, causing additional destruction. As the nature of the root tissue changes, the nematodes leave their habitat and migrate to new healthy roots (DuCharme, 1968).

4.5.7. Management

4.5.7.1. Regulatory methods: Since the burrowing nematode that attacks citrus is presently confined to Florida, all other citrus growing areas have stringent quarantines and special regulations to prevent the introduction of *R. citrophilus* via affected planting material.

Sanitation is one of the most important methods to manage the burrowing nematode. The major management tactics consist of 'nematode-free nursery stock' certification program. Citrus nursery certification program in Florida, USA, results in added crop value of more than US$ 17 million per year by slowing the spread of *R. citrophilus.*

4.5.7.2. Cultural methods: Barriers on a small scale are being used to confine *R. citrophilus* within limited areas of infestation (Poucher *et al.*, 1967). These man-made barriers surrounding the infested area consist of a strip of land 3 to 8 meter wide, free of citrus trees, and treated with a nematicide at 6 months intervals to keep the soil in the barrier zone free of roots as well as burrowing nematodes. The surface of the soil is also treated with herbicides to keep the barrier free of all plant growth. A biological barrier composed of a band of resistant citrus trees or a non-host of *R. citrophilus* has been suggested to block the migration of nematodes from infested areas (DuCharme and Suit, 1956; Ford, 1967).

Attempts are being made to eradicate *R. citrophilus* from infested groves of Lakeland fine silt in Florida by the "pull and treat" method (Poucher *et al.*, 1967). This procedure involves destruction of all trees in an area overlapping the infested site by at least 16 meters in all directions and removal of as many roots as possible from the soil. The cleaned soil is treated with a heavy dose of a nematicide, and the area is then kept free of all vegetation for 2 years. After the period of fallow, the site is replanted preferably with trees grafted on a tolerant rootstock.

4.5.7.3. Chemical methods: Various chemicals tried for the management of *R. citrophilus* on citrus as bare root dip treatment is presented in Table 4.6.

Table 4.6. Effect of various chemicals as bare root dip for the management of *R. citrophilus* on citrus

Chemical	Dosage	Duration & type of treatment	Effectiveness
Fensulfothion	1000 ppm	30-60 min. dip	Eliminated infection
Fenamiphos	250-1000 ppm	30-60 min. dip	Eliminated infection
Ethoprophos	1000 ppm	30-60 min. dip	Eliminated infection
Thionazin	1000 ppm	30-60 min. dip	Eliminated infection

Various chemicals tried for the management of *R. citrophilus* on citrus under field conditions is presented in Table 4.7.

Table 4.7. Effect of various chemicals for the control of *R. citrophilus* on citrus under field conditions

Chemical	Dosage	Duration & type of treatment	Effectiveness
Fensulfothion	100-400 ppm	Sub-surface soil drench	Eradicated infection in Rough lemon seedlings
Oxamyl	11.4 kg a.i./ha	As foliar spray 6 times a year at 6 weeks interval	Reduced nematode population
Fenamiphos	11.2 kg a.i./ha	Split application	Reduced nematode population
Ethoprophos	100-400 ppm	Sub-surface soil drench	Eradicated infection in Rough lemon seedlings
	7-25 kg/ha	3 field applications	Increased twig growth
Thionazin	15 kg/ha in 400 liters of water	3 soil drenches around trees + post treat. drench	Increased feeder roots
	100-400 ppm	Sub-surface soil drench	Eradicated infection in Rough lemon seedlings

4.5.7.4. Biological methods: Application of AMF, *Glomus etunicatum* in nursery soil as well as on transplantation increased the plant vigor and reduced the damage caused by *R. citrophilus* on citrus (O'Bannon and Nemec, 1979).

4.5.7.5. Host resistance: Citrus rootstocks such as Milam, Ridge Pineapple, Carrizo Citrange and Algerian Navel were reported to be resistant to the burrowing nematode (Ford and Feder, 1964).

4.6. THE LANCE NEMATODE, *HOPLOLAIMUS INDICUS*

Khan *et al.* (1964) reported *H. indicus* from the soil around the roots of *C. sinensis, C. aurantifolia* and *C. reticulata* from India. This nematode is only known from India where it is widely distributed. It causes damage to citrus in Punjab.

4.6.1. Occurrence and distribution

H. indicus has been reported on citrus from Punjab (Gupta and Gupta, 1966), Delhi, Rajasthan (Sharma *et al.*, 1969) and Uttar Pradesh (Rashid *et al.*, 1973).

4.6.2. Symptoms

H. indicus at 1,000 numbers per 500 g of soil caused some growth reduction of citrus, but a marked reduction occurred at or above an initial population level of 2,000 (1 male; 4 females) (Gupta and Atwal, 1972). None of rough lemon, karna khatta, sweet orange, mandarin, lemon or trifoliate orange was completely resistant to nematode attack at population levels above 4,000 per 500 g of the soil. Lemon proved somewhat more tolerant than other species. The greatest nematode populations were found in roots and soil of trees 10 to 20 years old and in an intermediate stage of decline; smallest populations were associated with trees one to five years old at initial stages of decline and 10 to 20 years old in the final stages of decline.

4.6.3. Host range

H. indicus is reported on sugarcane, banana, pea, guava, tomato, rice brinjal, finger millet, amaranthus, black gram, chillies, cowpea, cucumber, groundnut, okra, maize, mustard, Para grass, cauliflower, harayali grass, mango, peach, wheat, bajra, sorghum and cabbage besides citrus. In citrus, it has been reported on *C. jambhiri, C. karna, C. reticulata, C. sinensis, C. limon* and *Poncirus trifoliata* (Khan *et al.*, 1964).

4.6.4. Life cycle

The first molt occurs within the egg and the development outside the egg consists of 3 larval stages and the adult, with the usual 3 molts. Sex differentiation is indicated early in the second molt by the presence, in the female only, of 4 specialized ventral chord nuclei coinciding the position with that of the vagina. The optimum conditions for maximum population growth are a temperature of 30 °C, a soil pH of 7.0, sandy loam soil (10-20% clay) and 16% moisture content.

4.6.5. Management

Intercropping with non-host/resistant crops reduces nematode population. Chickpea, guar, and tobacco are immune; while beet root, mustard, canola, spring rape, watermelon are resistant to *H. indicus* (Gupta and Atwal, 1971).

Leaving fields fallow for 3 months decreases nematode population.

4.7. THE SHEATH NEMATODE, *HEMICYCLIOPHORA ARENARIA*

H. arenaria was reported on *Citrus jambhiri* from sandy desert region of California, USA (Van Gundy, 1957). *H. arenaria* is considered important on *Citrus* species only.

4.7.1. Economic importance

Growth reduction of citrus ranged from 12% at 25 °C to 37% at 30 °C.

4.7.2. Symptoms

H. arenaria feeds ectoparasitically near root tips of citrus. Root elongation ceases soon after juveniles insert their stylets into cells near root cap. Galls are formed in the meristem region or the zone of elongation by the hyperplasia of pericycle leading to enlarged cortex, some cells being multinucleate. Such galls are always at root tips, and thus can be distinguished from those caused by root-knot nematodes. Lateral root initials are stimulated near root tip, and if feeding occurs in the zone of elongation, abnormally long root hair often develops.

4.7.3. Life cycle

H. arenaria reproduces by parthenogenesis. At 28-30 °C, female laid on an average 6.2 eggs. A gelatinous substance on eggs causes them to adhere to soil particles. The time from egg to egg was 15-18 days at 30 °C (Van Gundy, 1959). The nematode multiplied best in soil containing 90% sand, 5% silt and 5% clay. Adult females can survive in water without feeding for 4-5 weeks.

4.7.4. Histopathology

Histochemical studies by Mc Elroy and Van Gundy (1968) revealed an accumulation of protein in the food cells during the first 6 hours and enlargement of nuclei in the adjacent cells. Afterwards, the food cells which are hypertrophied with distorted nuclei, are ingested. As feeding continues, the contents of neighboring cells are removed and cell walls are thickened. Eventually the food cells are crushed and pushed to the root surface by the new meristem formed by the pericycle, which formed the new food cells.

4.7.5. Management

4.7.5.1. Physical methods: Dipping of bare rooted nursery plants in hot water maintained at 46 °C for 10 minutes eliminated nematode infestation (Van Gundy and Mc Elroy, 1969).

4.7.5.2. Host resistance: Sweet and sour orange, trifoliate orange and citrange are non-hosts to the sheath nematode.

4.8. FUTURE LINES OF INVESTIGATION

» The population threshold level of *T. semipenetrans* on different varieties under different soil types need to be worked out for effective management.

» *T. semipenetrans* is economically important and widely distributed throughout India, but how much economic losses are being caused by this nematode in India need to be worked out.

» The dissemination of nematodes to newer areas is mainly through nursery plants. There is need to develop methodologies for obtaining nematode free seedlings through nematode management at nursery level.

» With the emphasis on drip irrigation system to save the water in citrus plantations, there is need to work out etiology of nematodes under changed irrigation system and also work out the efficient use of nematicides through low volume irrigation system for controlling nematodes.

» The biological control of citrus nematode has not been explored well. Intensive research is needed on this aspect.

- Except citrus nematode, no other nematodes on citrus have been given due attention in India. Systematic survey of parasitic nematodes associated with citrus plants, their densities and pathogenic potential need to be worked out.
- Very often other soil microorganisms are involved with infection of the nematode causing more damage to the plants. *Fusarium* sp., a weak pathogen is reported to cause more damage in presence of citrus nematode, but there is also need to study the associative effects of citrus nematode with *Phytophthora* sp. which is the cause of very severe problem of root rot in citrus.
- There is a need to integrate different methods of nematode control and find out their workability under field conditions.

4.9. CONCLUSION

The most economically important PPN species in India include the citrus *Tylenchulus semipenetrans* due to its wide distribution in most citrus groves and population densities above the economic threshold levels and the root-knot nematode, *Meloidogyne indica* which is a serious problem in Gujarat and Andhra Pradesh. The holistic comprehension of the nematode biology and ecology as well as their reflections on the economic importance of these nematodes should be harnessed for better pest control.

Chapter - 5

Pineapple, *Ananas comosus*

5.1. INTRODUCTION

5.1.1. Origin

Pineapple is said to be originated between Paraguay and southern Brazil. According to MS-Bertone it is considered that Parana-Paraguay river is the origin place of pineapple.

5.1.2. Area and production

The total area under pineapple cultivation in world is 10.98 lakh hectares and total production of 274.03 lakh metric tons (FAO, 2017).

Countries which commercially produce pineapple are Costa Rica, Philippines, Brazil, Thailand, India, Indonesia, Nigeria, China, Colombia, Mexico *etc.* (Table 5.1).

Table 5.1. Major pineapple producing countries in the world (2016)

Country	Area (million ha)	Production (million tons)	Productivity (tons/ha)
Coasta Rica	0.04	2.93	68.15
Brazil	0.07	2.69	39.22
Philippines	0.07	2.61	40.05
China	0.08	2.08	27.48
India	0.11	1.96	17.85
Thailand	24.29	1.81	24.29
Nigeria	0.20	1.59	8.12
Indonesia	0.01	1.40	132.15

Mexico	0.02	0.88	44.94
Columbia	0.02	0.76	41.11
Others	0.34	7.63	22.52
WORLD	1.05	25.81	24.65

The total area under pineapple cultivation in India is 1.1 lakh hectare (FAO, 2017) and production of 18.61 lakh metric tons (FAO, 2017). According to Food and Agriculture Organization (FAO, 2017), Indian is the fifth largest producer of pineapple in world (about 7% of total world production).

The major pineapple producing states of India are Assam, West Bengal, Manipur, Mizoram, Tripura, Nagaland, Meghalaya, Bihar, Tamil Nadu, Karnataka, Kerala and Andhra Pradesh (Table 5.2).

Table 5.2. State-wise area, production, and productivity of pineapple in India (2017-18)

State	Area ('000 ha)	Production ('000 tons)	Productivity (tons/ha)
Andhra Pradesh	4.52	71.33	15.78
Arunachal Pradesh	3.09	23.11	7.49
Assam	16.30	296.52	18.19
Bihar	4.26	115.13	27.03
Karnataka	2.62	163.73	62.42
Kerala	8.22	69.72	8.49
Maharashtra	0.35	11.26	31.88
Manipur	14.16	134.11	9.47
Meghalaya	12.37	144.73	11.70
Mizoram	5.03	27.64	5.50
Nagaland	9.53	132.83	13.93
Odisha	0.97	11.53	11.89
Sikkim	0.15	0.47	3.21
Tamil Nadu	0.83	25.74	31.12
Telangana	0.00	0.05	50.00
Tripura	8.73	127.00	14.55
West Bengal	11.41	345.15	30.26
Others	0.43	5.71	13.26
TOTAL	102.96	1705.76	16.57

5.1.3. Varieties

Kew and Queen are the most suitable varieties for cultivation. The important features of the varieties include:

5.1.3.1. Kew: It is a late maturing variety grown particularly for its canning quality. The plants are vigorous and leaves are long with straight margins. Leaves having small spines at the tip and also at the base near its attachment to the stem; where they are irregularly arranged. Fruit are big sized and weight ranged from 1.5 to 2.5 kg and is oblong in shape, slightly tapering towards the crown. Eyes are broad and shallow, making fruits more suitable for canning. The fruit is light yellow, almost fiberless and very juicy with 0.6-1.2% acid, and 12-16^0 brix TSS.

5.1.3.2. Queen: The plants are characterized by dwarf, compact habit of growth. Foliage is bluish green. The leaves are short, stiff, and very spiny along the margins and thickly covered with a whitish bloom on both surface. Fruit weight ranges from 0.9-1.3 kg. Eyes are small, prominent, deep set. When fully mature, the fruit is golden-yellow and internal flesh is deep golden yellow less juicy than Kew, crisp textured with pleasant aroma and flavor. The TSS content varies from 15 to 17^0 brix and acidity between 0.6-0.8 percent. This variety is suitable for table purpose.

5.1.4. High density planting

The yield of 70-105 tons/ha. may be obtained under high density planting, the increase in yield/ unit area being 45-85 tons/ha. Less weed infestation, protection of fruits from sun burn, increased production of propagules (suckers and slips)/unit area and non-lodging of plants are added advantages of high density planting.

5.1.5. Exports

India exports pineapple worth Rs. 14 crore during the year 2022 to three major countries: Dubai, Qatar, and Bangladesh.

Fresh pineapple export shipments from India stood at 19.6K. India exports most of its fresh pineapple to Qatar, Maldives and United Arab Emirates.

The top 3 exporters of fresh pineapple are Costa Rica with 156,456 shipments followed by Philippines with 27,140 and Guatemala at the 3rd spot with 23,114 shipments.

Export of pineapple has significantly increased in last few years; however, the total quantity exported is negligible. Increase in exports has taken place from 4,194.99 tons in 2007-08 to 3030.66 tons in 2011-12 (Table 5.3).

Table 5.3. Export of pineapple from India in the last five years

Years	Quantity (tons)	Value (in Rs. Lakh)
2007-08	4,194.99	339.76
2008-09	3893.44	367.81
2009-10	2462.33	402.70
2010-11	1705.99	390.28
2011-12	3030.66	603.08

Source: APEDA Database, 2011-12

5.1.6. Nematode problems

`The most common nematodes found to be detrimental to crop production are *Meloidogyne* spp., *Rotylenchulus reniformis*, and the lesion nematode, *Pratylenchus brachyurus*.

5.2. ROOT-KNOT NEMATODES, *MELOIDOGYNE INCOGNITA, M. JAVANICA*

5.2.1. Crop losses

Losses for the first crop were greater than 40% due to the attack of root-knot nematodes.

5.2.2. Symptoms

Root-knot nematodes cause stunting, yellowing and dieback of plants. The root growth is greatly slowed down. Reduction in pineapple plant weight, leaf weight, and fruit weight by 47.5% was observed due to root-knot nematode infection. Root-knot nematodes produce distinct terminal galls on the roots, stopping further root development. The root infested simultaneously by several larvae stops the growth and may produce many secondary roots.

5.2.3. Life cycle

Second stage juveniles penetrate roots in the meristematic region of the root tip and become sedentary after 2-3 days. Vermiform adult males and saccate sedentary females were subsequently developed. Reproduction is by mitotic parthenogenesis and female nematodes produce eggs contained in a gelatinous matrix.

Moyle and Botella (2014) found that *M. javanica* infection life cycle in pineapple takes 10 weeks with the knots appearing on the roots at approximately 4-5 weeks and the production and release of egg outside the roots surface occur at 10 weeks after inoculation.

5.2.4. Survival and spread

Eggs of *M. javanica* contained in galled tissue survived for 20 days at relative humidity of 90%, while the free-living forms survived for 20-24 weeks in desiccated soil without a host. *M. javanica* can survive, although at low levels, as long as 2 years in fallow field soil. The nematode spread over long distances in soil adhering to workers feet, implements and equipment that is moved from field to field. Infected planting stock also serves as a means of nematode spread.

5.2.5. Ecology

Two seasonal *M. incognita* population peaks were observed first in April-May and second during November. Soil temperature range of 21.2 °C-27.5 °C and soil moisture range 14.0-18.6% were found to favor the population development.

5.2.6. Management

5.2.6.1. Cultural methods: Crop rotation is important to avoid build-up of root knot and other nematodes that contribute to large crop losses. Some crops usually included in rotation with pineapples are groundnut and rice.

Yield increases have been observed when planting was done in a field, which was previously had pigeon pea, owing to decrease in infestation by *M. incognita*. The reduction in population of *M. incognita*, *Criconemoides* spp., *Helicotylenchus* spp. and *R. reniformis* and increase in fruit yield in the planted crop (23.5 tons/acre as against 15.1 tons/acre in control) and first ratoon crop (18.4 tons/acre as against 9.3 tons/acre in control) was observed when pineapple was rotated with Pangola grass (Ayala *et al.*, 1967).

The addition of chopped pineapple material to soil at 125-375 tons/ha significantly reduced galling caused by root-knot nematode.

To prepare the land used for pineapple production, green manure plants like cowpea, can be grown and incorporated into the soil prior to planting pineapples. To be effective, green manure crops known to reduce or eliminate root-knot nematode infestation should be planted between pineapple crops.

At 3.00 tons ha^{-1} of poultry manure application, no root-gall damage occurred in any of the pineapple plants. Highest fresh fruit weight obtained on crown propagated plants occurred at 3.00 tons ha^{-1} of poultry manure application (Agu, 2008).

Maintenance of weed-free and host-free fallow period of at least for 6 months was found effective for significant decline in nematode population.

5.2.6.2. Chemical methods: Dipping crowns in solutions of Oxamyl at 2,400-4,800 ppm, or Fenamiphos at 300 ppm provides a good protection against *M. incognita.*

5.2.6.3. Biological methods: The significant suppressive effects of isolates *Trichoderma asperellum* (M2RT4) (ICIPE) (soil application of 120 ml aqueous solution containing 1.0 $\times 10^8$ spores/ml), *Trichoderma* isolate MK4 (soil application of 120 ml aqueous solution containing 1.0 $\times 10^8$ spores/ml) and the *Purpureocillium lilacinum* isolates (KLF2 and MR2) (KBL) (soil application of 120 ml aqueous solution containing 3.0 $\times 10^6$ spores/ml) against *M. javanica* on pineapple. *Trichoderma* spp. colonized the roots of pineapple endophytically (Kiriga, 2013).

Application of *Trichoderma harzianum/viride* and *Pseudomonas fluorescens* for treatment of seed/seedling/planting materials in the nurseries and field proved beneficial.

Addition of *Dactylella ellipsospora* reduced plant injury caused by the root-knot nematode (Linford and Yap, 1939).

5.2.6.4. Host resistance: *Ananas ananasoides* and three of its hybrids are highly resistant to attacks by *M. incognita.*

5.3. RENIFORM NEMATODE, *ROTYLENCHULUS RENIFORMIS*

5.3.1. Crop losses

The reniform nematode can reduce the pineapple plant populations by more than 83% in the fields. Complete failure of ratoon crop could take place due to severe nematode infestation along with moisture stress.

When reniform nematodes are present, however, marketable yields can be reduced up to 38 percent in the planted crop harvest and 60 percent in the first ratoon (Wang and Sipes, 2009).

In Brazil, some reports have demonstrated that the reniform nematode can reduce yield by 60% in the planted crop and around 40% in the second harvest.

5.3.2. Symptoms

Leaves of infected plants are less erect than those of healthy plants, are reddish in color and show poor growth. The reniform nematode reduces the number of lateral and fine feeder roots. Plant collapse and death can result due to severe nematode infestations. Reniform nematode infection inhibits secondary root formation and root systems are poorly developed.

The reniform nematode could be detected by small masses of soil clinging to the gelatinous matrix of females embedded in roots.

5.3.3. Life cycle

Egg hatch is stimulated by root exudates of certain host plants, and second stage juveniles leave the egg and move into the soil. Once in the soil, they undergo 3 molts without feeding, yielding adult males and pre-adult females. Females enter the root system, initiate a feeding site, become sedentary and develop into swollen, mature egg-producing females. Males do not feed.

5.3.4. Survival and spread

The reniform nematode survives extended periods (6 months to 2 years) without a host. The nematode survives fallow periods in the egg stage or as anhydrobiotic juvenile stages, depending on soil moisture.

5.3.5. Ecology

The optimum temperature for development is 25-29.5 °C, and reproduction is limited by temperatures above 36 °C (Rebois, 1973; Heald and Inserra, 1988). Soil temperatures in pineapple growing regions are extremely favorable for the development of the reniform nematode. The optimal pH for reproduction of the reniform nematode in Hawaiian soils is approximately 4.8-5.2 (Rohrbach and Apt, 1986).

5.3.6. Management

5.3.6.1. Cultural methods: Sun hemp is a poor host of reniform and root-knot nematodes, and when incorporated into the soil produces monocrotaline, a compound that inhibits nematode movement and delays nematode development. Thus sun hemp appears promising for managing nematodes in pineapple production systems (Wang and Sipes, 2009). The reniform nematode can penetrate the roots, but their maturation is drastically delayed (Wang *et al.*, 2001). Thus, sun hemp "traps" reniform nematodes and prevents them from entering the state of anhydrobiosis.

Rotation of pineapple with French marigold (*Tagetes patula*), Rhodes grass (*Chloris guyana*), sun hemp (*Crotalaria juncea*), *Desmodium unicatum*, sugarcane or Pangola grass was effective in controlling *R. reniformis.* Sugarcane is generally considered a non-host for *R. reniformis*, so planting pineapple into sugarcane soils may decrease problems with the nematode, provided that weed hosts are not present.

In the greenhouse, cover crops such as sun hemp (*Crotalaria juncea*), rapeseed (*Brassica napus*), and African marigold (*Tagetes erecta*) were incorporated in soil naturally infested with *R. reniformis*. All cover crops suppressed *R. reniformis* numbers when incorporated 3 months after planting. *Crotalaria juncea* treatment increased bacterivorous nematodes and nematode-trapping fungal population densities more than the other treatments. *Crotalaria juncea* could have suppressed *R. reniformis* because it is a poor host and because it enhances nematode-trapping fungi when incorporated into soil (Wang *et al.*, 2002).

Planting of pineapple after Pangola grass reduced the population of *R. reniformis* and increased the fruit yield in the planted crop (23.5 tons/acre as against 15.1 tons/acre in control) and first ratoon crop (18.4 tons/acre as against 9.3 tons/acre in control) (Ayala *et al.*, 1967).

Reduction in the population of *R. reniformis* can be obtained by irrigated fallow, since the nematode prefers low moisture regimes in pineapple crop.

5.3.6.2. Chemical methods: Application of Carbofuran at 2.25 kg a.i. per ha, Oxamyl at 1 kg a.i. per ha and Fenamiphos at 4.5 kg a.i. per ha were effective in controlling *R. reniformis* and in increasing fruit yields. Foliar application of Fenamiphos at 600-2400 ppm gave effective control of reniform nematode. Foliar application of Oxamyl may be as effective as Fenamiphos if applied at twice the rate.

5.3.6.3. Host resistance: Pineapple varieties/species such as Venezolana, *Ananas ananasoides* and several of its hybrids are highly resistant to *R. reniformis* and do not allow the initial infestation to be maintained.

5.3.6.4. Integrated methods: Integrating sun hemp with soil solarization is promising for suppressing reniform nematodes and weeds which is more compatible with this shorter production cycle in pineapple (Wang and Sipes, 2009). This practice can readily be adopted by smaller-scale growers.

5.4. THE LESION NEMATODE, *PRATYLENCHUS BRACHYURUS*

The root-lesion nematode, *Pratylenchus brachyurus*, is prevalent, and of economic importance, throughout the equatorial tropics, such as the Ivory Coast, Uganda, Hluhluwe in Northern Natal (South Africa) and Brazil (Guérout, 1975; Bafokuzara, 1982; Keetch, 1982).

5.4.1. Symptoms

Plants parasitized by *P. brachyurus* had chlorotic leaves. The root lesion nematode invades the outer root tissues, causing black areas (lesions) of dead or injured plant cells on the root surface. These lesions can completely encircle the root. Fresh root weight was reduced in plants parasitized by *P. brachyurus*. It causes delay in leaf emergence, a reduction in fruit weight, and decreased ratoon growth. The lesion nematode damages the cells of the epidermis and the central cylinder of pineapple plants (Costa and Matos, 2000; Ritzinger and Costa, 2004; Sipes *et al.*, 2005), which explains the necrosis found in the roots and the reduced weight of the root system. Plants with a reduced root system have a lower capacity to absorb water and nutrients and have a compromised photosynthetic capacity. Consequently, the plants may become chlorotic, have reduced development and lower fruit weight, and a delay in fruit maturity (Ferreira *et al.*, 2014). *P. brachyurus* suppressed root and leaf fresh weight and decreased the width of leaves and fruit, and reduced the fruit weight.

5.4.2. Biology and life cycle

P. brachyurus is a migratory endoparasite. Males are rare, and reproduction is by mitotic parthenogenesis (Roman and Triantaphyllou, 1969). The life cycle may be completed within the roots. Thus, large populations can develop quickly and cause the rapid destruction of the cortical parenchyma (Guérout, 1975).

5.4.3. Survival and dissemination

Under laboratory conditions, populations of *P. brachyurus* from the Ivory Coast survive from 20 to 22 months in fallow soil (Feldmesser in Wallace, 1963), as long as viable root fragments are present in the soil (Guérout, 1975). If root fragments are absent from the soil, survival without a host is limited to approximately 7 months. After 35 days at 44°C only 25-50% of an original South African population survived (Keetch, 1977).

In the Ivory Coast, *P. brachyurus* is sometimes disseminated when infected suckers are used as seed. Generally, the suckers used as seed are uninfested.

5.4.4. Ecology

The optimum temperature for *P. brachyurus* development is 25-30 °C (Olowe and Corbett, 1976). Approximately 20 mm of rainfall per 10 days is required in the Ivory Coast to maintain high root populations of *P. brachyurus*. Population densities of *P. brachyurus* in roots are higher in acid soils.

5.4.5. Interaction with mealybug wilt

At 8 months after inoculation (MAI), plants affected by mealybug wilt of pineapple (MWP) and parasitized by *P. brachyurus*, were chlorotic with reduced leaf turgor during the hottest hours of the day. Chlorosis increased soon after flowering was induced. The width of the D-leaf was lower in the plants affected by MWP and parasitized by *P. brachyurus*.

At harvest at 19 MAI, in the plants affected with MWP, *P. brachyurus* population density was lower in the soil. Plants affected by MWP and parasitized by *P. brachyurus* had leaves with reduced turgor, that were chlorotic. Necrosis at the tips of the new leaves and death of the older leaves were also seen in this treatment. Plants parasitized by *P. brachyurus* with MWP, had necrotic roots, reduced fresh root and D-leaf weights, and reduced D-leaf width, fruit length and delayed fruit maturity.

At 21 MAI, The decrease in fresh root weight, number of slips per plant and the slips mean weight were reduced in plants affected by MWP and parasitized by *P. brachyurus*.

5.4.6. Management

In the Ivory Coast, six weeks fal1ow can reduce populations of *P. brachyurus* by half (Guérout, 1975). In Nigeria, adding cassava residues or extracts of Neem (*Azadirachta indica*) leaves to soil reduces populations of *P. brachyurus* by 75 and 72%, respectively, (Egunjobi and Larinde, 1975).

Mucuna pruriens established itself as an effective candidate in controlling lesion nematode population build up and improving the yield of pineapple. At harvest, the weed-free Mucuna treatment reduced the populations of *P. brachyurus* by 56% over the bush fallow treatment. The effectiveness of *M. pruriens* was greatest at six months after the application of treatments as weed free-Mucuna treatment reduced the populations of the lesion nematode pest by 341 % over the bush fallow treatment. At harvest, populations of *P. brachyurus* recovered from roots from weed free-Mucuna plots represented significant reduction of 154% over the bush fallow treatment (Osei *et al.*, 2011). The antagonistic activity of *Mucuna* might be due to the production of phytoalexins by the roots (Vargas *et al.*, 1996).

Dipping suckers in 1500 ppm solutions of fenamiphos or carbofuran (1500 ppm) provides control of *P. brachyurus* in the Ivory Coast.

5.5. CONCLUSION

Plant parasitic nematodes identified in pineapple fields were *Meloidogyne* spp., *Rotylenchulus reniformis*, and *Pratylenchus brachyurus*. The most prevalent nematodes were *Meloidogyne* spp. and *R. reniformis* that might have affected the crops manifesting symptoms such as chlorosis, necrosis and stunting. Studies on plant parasitic nematodes in the India are still limited and so does the knowledge of farmers regarding these parasites, thus, further studies with wide sampling sites on the occurrence of plant parasitic nematodes in pineapples is recommended. To further identify the PPN into species level, molecular identification is recommended.

Chapter - 6

Papaya, *Carica papaya*

6.1. INTRODUCTION

6.1.1. Origin

Papaya originated in South Mexico and Costa Rica.

6.1.2. Area, production and yield

Global papaya production amounted to 13.735 million metric tons from an area of 0.452 million ha with an yield of 29.694 tons/ha in 2019.

India is the largest papaya producer in the world with 6.050 million tons production volume per year. Dominican Republic comes second with 1.171 million tons yearly production. India and Dominican Republic produce together more than 50% of World's total. The top 10 papaya producing countries are presented in Table 6.1

Table 6.1. Top 10 papaya producing countries in the world (2019)

Country	Area ('000 ha)	Production ('000 tons)	Yield (tons/ha)
India	149.000	6050	40.604
Dominican Republic	3.861	1171	303.376
Brazil	27.556	1162	42.161
Mexico	18.839	1084	57.594
Indonesia	10.599	987	93.121
Nigeria	101.634	855	8.418
Democratic Republic of the Congo	12.535	212	16.92
Venezuela	10.002	200	19.981
Cuba	6.258	184	29.457

Peru	9.713	1717	17.674
WORLD	452.552	13735	29.694

India is still the world's biggest papaya producer in 2019. Based on FAO Statistical Database in 2019, India's papaya production reached 6.05 million tons. In this country, the center of papaya cultivation is spread over in several states, namely Andhra Pradesh, West Bengal, Karnataka, Orissa, Gujarat, and Maharashtra (Table 6.2).

Table 6.2. State-wise area, production and yield of papaya (2017-18))

State	Area ('000 ha)	Production ('000 tons)	Yield (tons/ha)
Andhra Pradesh	18.01	1687.82	93.72
Arunachal Pradesh	0.23	0.73	3.18
Assam	7.21	147.40	20.44
Bihar	1.90	42.72	22.45
Chhattisgarh	14.40	381.42	26.48
Gujarat	20.31	1256.51	61.86
Himachal Pradesh	0.23	1.21	5.32
Jharkhand	2.66	116.41	43.73
Karnataka	8.75	593.69	67.84
Kerala	16.15	95.03	5.88
Madhya Pradesh	10.55	421.55	39.96
Maharashtra	10.28	408.30	39.71
Meghalaya	0.85	6.82	8.06
Mizoram	1.25	25.00	20.00
Nagaland	1.41	17.18	12.17
Odisha	3.05	70.29	23.05
Rajasthan	0.51	4.37	8.57
Sikkim	0.76	0.57	0.74
Tamil Nadu	1.53	141.58	92.83
Telangana	1.18	79.27	66.92
Tripura	2.61	26.35	10.08
Uttar Pradesh	2.01	96.83	48.08
West Bengal	12.38	365.95	29.56
Others	0.16	1.84	11.31
TOTAL	138.40	5988.83	43.27

Source: Horticulture Statistics Division, Department of Agri. Co-op. & Farmers Welfare.

The major papaya producing districts in different states in India include:

» Gujarat (Kachchh, Tapi, Vadodara, Aravalli, Chhotaudepur *etc.*)

- » Andhra Pradesh (Chittoor, Anantapur, Kadapa, Prakasam, Vizianagaram *etc.*)
- » Karnataka (Chitradurga, Koppal, Ballari, Mandya, Gulbarga *etc.*)
- » Madhya Pradesh (Khargone, Dhar, Sidhi, Barwani, Shivpuri *etc.*)
- » Tamil Nadu (Erode, Dharmapuri, Vellore, Salem, Theni *etc.*)

6.1.3. Major commercial varieties

In India, there are several popular papaya cvs. that are widely cultivated, namely Coorg Honey Dew, Pusa Delicious, Pusa Majesty, Pusa Dwarf, Pusa Giant, Co. 1 to Co. 6, Surya, Pusa Nanha, Mahabindu, Pink Flesh Sweet, Pant 1, Sunrise Solo, Taiwan *etc.*

6.1.4. Advance technology

- » High density planting

6.1.5. Pre-cooling

- » Forced air pre-cooling or hydro-cooling at -1 to 0 ^{0}C and RH – 90-95% are suitable for storing up to 8 months.

6.1.6 Storage conditions

- » Temperature : 10-13 ^{0}C
- » Relative humidity : 85-90%
- » Storage period : 1-3 weeks
- » Freezing point : -2 to -1 ^{0}C

6.1.7. Sensitivity

- » Refrigeration and ethylene

6.1.8. Peak harvesting season

- » Gujarat : January and December
- » Andhra Pradesh : Round the year
- » Karnataka : April to October
- » Madhya Pradesh : January and October to December
- » Tamil Nadu : Round the year

6.1.9. Export from India (2017-18)

- » Quantity: 9.99 Thousand MT
- » Value: 4,150.44 Lakh Rupees

6.1.10. Nematode problems

Root-knot nematodes (*Meloidogyne incognita* and *M. javanica*) and the reniform nematode (*Rotylenchulus reniformis*) have been identified as proven pathogens on papaya.

6.2. ROOT-KNOT NEMATODES, *MELOIDOGYNE INCOGNITA*, *M. JAVANICA*

6.2.1. Crop losses

The root-knot nematode, *Meloidogyne incognita* and the reniform nematode, *Rotylenchulus reniformis* are responsible for 30% annual yield losses in papaya amounting to Rs. 2516.64 million in India (Walia and Chakraborty, 2018).

Ponte (1980) and Taylor *et al.* (1982) reported 10 to 20 per cent reduction in papaya fruit yield due to root-knot nematodes.

Ramakrishnan and Rajendran (1998b) estimated nearly 40% fruit loss in the first crop after planting, with increasing losses likely thereafter due to *M. incognita.*

6.2.2. Distribution

The root-knot nematodes on papaya have been recorded from Delhi, Bihar, Haryana, Karnataka, Madhya Pradesh, Maharashtra, Orissa and Uttar Pradesh.

6.2.3. Symptoms

General symptoms visible in field include poor growth, yellowing of foliage, dropping of leaves, reduction in leaf production, weak vigor and premature dropping of fruits. Papaya orchards infected with *Meloidogyne* spp. show patches of poor growth with many plants missing in the rows.

Roots exhibit typical below-ground symptoms *i.e.* galls of varying sizes on roots. The lateral branching of roots is limited. In heavily infected old roots, adjacent galls join together and form large galls. In mild infestation, root tips become swollen and root growth inhibition is distinctly seen.

6.2.4. Biology and life cycle

The root-knot nematode is soil-borne, sedentary endoparasite. After penetration, the J2 larvae feed endoparasitically on conducting vessels and develop to adult females which are spherical in shape and lays egg masses in compound galls. The males are vermiform in shape which leaves roots to enter the soil. Females lay eggs in masses on root surface on galls, or in advanced stages inside the compound galls. During summer, the nematode completes its life cycle in about 25 to 30 days, while it takes longer during winter months.

6.2.5. Survival and spread

Root-knot nematodes survive as egg masses in infected plant debris and soil or collateral and other hosts like Solonaceous, Malvaceous and Leguminaceous plants which act as sources of inoculums.

The nematodes are disseminated through infected planting material, irrigation water, and agricultural implements.

Loamy light soils are favorable for nematode development and reproduction.

6.2.6. Management

6.2.6.1. Physical methods: Hot water treatment of papaya seedling roots at 50 °C for 10 min proved beneficial. Soil solarization was found effective to raise nematode-free seedlings.

6.2.6.2. Cultural methods: Neem cake at 2.5 t/ha gave reduction in root-knot nematode (*M. incognita*) population in both soil and roots and increased yield by about 200% (10.4 kg/plant compared to 5.5 kg/plant in control) (Nayak, 1986). Application of farm yard manure at 8 kg and Neem cake at 1 kg/

plant was found beneficial.

6.2.6.3. Chemical methods

i) Nursery treatment: Application of carbofuran and fenamiphos each at 2 kg a.i./ha, a day or two before seeding in nursery, is effective in the management of root-knot nematodes (36.2 and 43.4%, respectively) and getting higher healthy transplantable seedlings of papaya (12.9 and 17.3%, respectively). Highest germination (94%) was obtained with carbofuran at 4 kg a.i. per ha. Maximum plant height was recorded in fensulfothion at 2 kg a.i. per ha.

ii) Main field treatment: Carbofuran at 2 kg a.i. per ha was found effective against nematodes and gave highest yield.

Nayak (1986) reported that carbofuran and ethoprophos all at 6 kg a.i. per ha were effective in controlling *M. incognita* on papaya.

6.2.6.4. Biological methods: Nursery bed treatment with *Purpureocillium lilacinum* and *Trichoderma harzianum* each at 10 kg per ha proved beneficial.

Soil application of *Pochonia chlamydosporia* at 5 kg/ha increased plant height (3.68 m as compared to 2.07 m in control) and fruit yield (38.1 kg/plant as compared to 20.5 kg/plant in control) of papaya and reduced nematode population both in soil and roots (Kumar, 2009) (Table 6.3).

Table 6.3. Effect of talc formulation of *Pochonia chlamydosporia* for the management of *Meloidogyne incognita* infecting papaya

Treatment	Plant height (cm)	Nematode population		Yield (kg/plant)
		Soil	Roots	
P. chlamydosporia at 1 kg/ha	22.6	162	12.4	24.7
P. chlamydosporia at 2 kg/ha	23.9	155	11.3	29.6
P. chlamydosporia at 3 kg/ha	22.4	140	9.4	31.8
P. chlamydosporia at 4 kg/ha	26.2	134	8.6	34.4
P. chlamydosporia at 5 kg/ha	36.8	122	7.3	38.1
Control	20.7	305	19.8	20.5
CD (P = 0.05)	4.9	8.1	0.2	2.7

Among several biological control agents (*Pseudomonas putida, P. fluorescens, Bacillus amyloliquefaciens, B. pumilus, B. subtilis, B. megaterium* and *Purpureocillium lilacinum* each @ 50g/m^2) tested against root-knot nematodes in papaya nursey, *P. putida* gave least root-knot index (1.45 as compared to 2.28 in control) followed by *P. lilacinum* (1.57) and *B. amyloliquefaciens* (1.58) (Table 6.4) (Patel and Patel, 2019). Seedlings raised in the beds treated with *P. putida* had maximum fresh seedling weight (1048.33 g) and it was statistically at par with *P. lilacinum* (T7). *B. amyloliquefaciens* (T3) was at third rank.

Significantly higher transplantable seedlings were recorded in treatment of *P. putida* and *B. amyloliquefaciens* (79), followed by *P. lilacinum* (78), *B. subtilis* (77) and *P. fluorescens* (77), statistically all treatments found at par with each other.

Table 6.4. Effect of different bio-agents on plant growth and root galling in papaya

Treatment	Germination count/1.44 m²	Fresh seedling weight (g)	Transplantable seedlings/1.44 m²	RKI (0-5)* (√x+0.5)
T1 (*Pseudomonas putida*)	80	1048.33	79	1.45 (1.59)
T2 (*Pseudomonas fluorescens*)	79	670.67	77	1.94 (3.26)
T3 (*Bacillus amyloliquefaciens*)	80	869.00	79	1.58 (2.00)
T4 (*Bacillus pumilus*)	79	611.00	76	2.05 (3.69)
T5 (*Bacillus subtilis*)	79	673.67	77	1.92 (3.19)
T6 (*Bacillus megaterium*)	78	650.00	75	2.03 (3.62)
T7 (*Purpureocillium lilacinum*)	80	1000.00	78	1.57 (1.96)
T8 (Control)	78	495.33	74	2.28 (4.70)
CD at 5 %	NS	155.97	3.02	0.16

*0 = Free; 5 = Maximum disease intensity. Figures in parentheses are retransformed values of transformation value

6.2.6.5. Host resistance: The papaya cv. Pusa 22-3 was resistant, while Thailand, Waimanalo, Mukund Farm, Pusa 1-15 and Pusa 1-45 were moderately resistant (Parvatha Reddy *et al.*, 1988). Nayak (1986) reported that cvs. CO-2 and CO-3 were found to inhibit the development and reproduction of *M. incognita*.

Khan *et al.* (1992) found that cv. CO-2 was resistant, while Peradeneya, Selection No.7 and Washington were moderately resistant against *M. incognita* Race 1.

Pusa Delicious, Pusa Dwarf and Pusa Giant were reported to be resistant to *M. incognita* (Laqman Khan, 2001).

Papaya cv. Pusa 22-3 to be resistant, and several other cultivars are reported to have varying degrees of resistance to *M. incognita* (Parvatha Reddy *et al.*, 1988; Khan *et al.*, 1995; Ramakrishnan and Rajendran, 1998a, b).

6.2.6.6. Integrated methods: Seed treatment with *P. fluorescens* (10^8 spores/g) combined with soil application of *T. harzianum* (10^6 spores/g) and *P. fluorescens* (10^8 spores/g) at 5 g/kg soil gave significant reduction in *M. incognita* population both in soil and roots, number of eggs per egg mass and hatching of eggs and increased root colonization by both the bioagents in papaya under nursery conditions (Table 6.5) (Rao, 2007).

Table 6.5. Effect of integration of bioagents for the management of *Meloidogyne incognita* infecting papaya

Treatment	No. of nemas/ 10 g roots	No. of J_2/ 100 ml soil	No. of eggs/egg mass	% egg hatch suppression	Root colonization (cfu/g)	
					Th	*Pf*
T1	47	78	389	32	---	8769
T2	42	70	364	45	---	18457
T3	50	65	375	43	22649	---

T4	35	52	358	65	21895	18249
T5	38	56	326	55	---	24531
T6	45	63	347	54	22412	8566
T7	31	45	310	67	20874	24124
T8	68	126	412	---	---	---
CD at 5%	**6.6**	**8.5**	**33.2**	**7.4**	**2365.7**	**2487.2**

* T1 - Seed treatment with *P. fluorescens*, T2 – Nursery soil mixed with *P. fluorescens* (5 g/kg), T3 - Nursery soil mixed with *T. harzianum* (5 g/kg), T4 - Nursery soil mixed with *T. harzianum* (5 g/kg) and *P. fluorescens* (5 g/kg), T5 - Seed treatment with *P. fluorescens* + Nursery soil mixed with *P. fluorescens* (5 g/kg), T6 - Seed treatment with *P. fluorescens* + Nursery soil mixed with *T. harzianum* (5 g/kg), T7 - Seed treatment with *P. fluorescens* + Nursery soil mixed with *T. harzianum* (5 g/kg) and *P. fluorescens* (5 g/kg), T8 – Control.

The root-knot nematode can be managed by soil application of Neem cake at 2.5 tons per ha and Carbofuran at 6 kg a.i./ha. Highest yield of 11 kg per plant was recorded in Carbofuran followed by Neem cake (10.4 kg), whereas control plants yielded 5.5 kg (Nayak, 1986).

Nursery bed treatment with *T. harzianum* and *P. lilacinum* each at 5 or 10 g/kg soil resulted in production of highly vigorous papaya seedlings whose roots were colonized with both the bioagents. There was significant reduction in root galling in the combination treatments (2.9 to 3.2) compared to control (8.9) (Table 6.6) (Rao and Naik, 2003)

Table 6.6. Effect of integration of bioagents for the management of *Meloidogyne incognita* infecting papaya

Treatment*	**Root-knot index**	**Root colonization (cfu/g)**		**Propagule density (cfu/g soil)**		**% egg masses parasitized by *Th* and/or *Pl***
		Th	***Pl***	Th	Pl	
T1	5.6	---	26434	---	22097	59.36
T2	5.2	---	29157	---	23134	57.44
T3	6.4	35765	---	29347	---	58.42
T4	5.2	37464	---	30956	---	62.46
T5	3.0	35124	23301	27437	17104	80.62
T6	2.8	35624	24211	30147	16122	73.84
T7	8.5	---	---	---	---	---
CD at 5%	0.65	2613.46	1905.27	2116.82	1817.43	8.75

* T1 – Nursery soil mixed with *P. lilacinum* (5 g/kg), T2 - Nursery soil mixed with *P. lilacinum* (10 g/kg), T3 - Nursery soil mixed with *T. harzianum* (5 g/kg), T4 - Nursery soil mixed with *T. harzianum* (10 g/kg), T5 – T1 + T3, T6 – T2 + T4, T7 – Control.

Simultaneous application of *P. lilacinum* and *T. harzianum* both at 1 g a.i. per plant in the main field gave maximum increase in plant growth parameters and highest reduction in reproduction factor and root galling in papaya (Khan, 1991).

Integrated management strategy for root-knot nematodes (*M. incognita*) infecting papaya was standardized by nursery bed treatment with Neem based formulation of *P. lilacinum* (1 x 10^6 cfu/g) + *T. harzianum* (1 x 10^9 cfu/g) (each at 5 or 10 g/kg soil) and application of 35 g each of *P. lilacinum* + *T. harzianum*/pit while transplanting. The above treatment increased plant height, plant weight and reduced root galling, number of egg masses/plant and number of eggs/egg mass in nursery (Table 6.7). The integration of both the bioagents also increased root colonization of bioagents, propagule density in soil and parasitization of egg masses under field conditions (Table 6.8).

Table 6.7. Effect of neem based formulations of bioagents on plant growth and management of *Meloidogyne incognita* infecting papaya in nursery.

Treatment/Dose	Plant height (cm)	Plant weight (g)	Root-knot index (1-10 scale)	No. of egg masses/ seedling	No. of eggs/ egg mass
P. lilacinum - 5 g/kg soil	25.35	5.58	5.8	15.45	356
P. lilacinum - 10 g/kg soil	28.87	6.21	5.2	12.86	325
T. harzianum - 5 g/kg soil	23.69	5.76	6.5	17.89	421
T. harzianum - 10 g/kg soil	29.72	6.89	5.4	14.59	398
P. lilacinum - 5 g/kg soil+ *T. harzianum*- 5 g/kg soil	32.89	7.99	3.1	10.33	270
P. lilacinum - 10 g/kg soil + *T. harzianum* - 10 g/kg soil	33.72	8.65	2.8	9.65	262
Control	18.64	4.12	8.7	26.27	412
CD (P = 0.05)	3.67	1.25	0.65	2.38	34.87

Table 6.8. Effect of neem based formulations of bioagents on plant growth and management of *Meloidogyne incognita* infecting papaya in main field.

Treatment/Dose	Root colonization (cfu/g)		Propagule density in soil (cfu/g)		Parasitization of egg masses (%)	
	Th	Pl	Th	Pl	Th	Pl
P. lilacinum - 5 g/kg soil	---	25,876	---	21,765	---	55.89
P. lilacinum - 10 g/kg soil	---	28231	---	23,867	---	61.98
T. harzianum - 5 g/kg soil	34,789	---	28,765	---	56.76	---
T. harzianum - 10 g/kg soil	37,895	---	31,194	---	66.89	---
P. lilacinum - 5 g/kg soil+ *T. harzianum*- 5 g/kg soil	34,853	24,129	26,986	17,459	41.68	35.98
P. lilacinum - 10 g/kg soil + *T. harzianum* - 10 g/kg soil	36,432	23,679	29,457	16,934	40.21	32.49
Control	---	---	---	---	---	---
CD (P = 0.05)	2655	2106	2317	1567	7.89	7.95

Application of 2 kg FYM enriched with *P. fluorescens* (10^9 spores/g) and *P. lilacinum* (1 x 10^6 cfu/g) per plant at the time of planting and at an interval of 6 months significantly reduced root-knot nematodes on roots by 70%. Significant increase in fruit yield (28%) was also observed. Benefit: cost ratio (calculated for marginal cost of biopesticides and returns accrued by application of biopesticides) was 3.2 (Anon, 2012).

The combined inoculation of two AMF species (*Glomus mosseae* or *G. manihotis*) and a *Bacillus* consortium based on three strains of PGPR were able to reduce nematode infection and damage on papaya.

6.3. INTERACTION OF ROOT-KNOT NEMATODE WITH FUSARIUM WILT

6.3.1. Interaction

Increasing inoculum of *M. incognita*, whether present alone or together with *Fusarium solani*, decreased seedling emergence of papaya. The combination of highest inoculum of both pathogens (3000 nematodes and 3g culture of fungus) caused maximum inhibition of seedling emergence and also increased post-emergence damping-off of papaya seedlings (Khan and Hussain, 1990).

Inoculation of nematodes 12 days prior to fungus or simultaneous inoculation of both pathogens resulted in significant reduction in plant height, root length, shoot and root weight of papaya. Root galling was highest in case of nematode alone followed by nematode inoculation 12 days prior to fungus, and simultaneous inoculation of nematode and fungus (Table 6.9) (Kishore *et al.*, 2005).

Table 6.9. Effect of *Meloidogne incognita* and *Fusarium solani* on disease complex in papaya

Treatment/Dose	Plant height (cm)	Shoot weight (g)	Root weight (g)	Root rot (%)	Galls/g roots
M. incognita	98.7	58.2	10.8	12	74
F. solani	106.3	63.3	13.3	48	0
M. incognita + *F. solani* (simultaneously)	65.3	53.6	9.2	64	43
M. incognita (prior) + *F. solani* (later)	73.2	60.3	9.7	94	65
F. solani (prior) + *M. incognita* (later)	83.7	76.6	15.6	52	35
Uninoculated Control	125.7	85.8	70.6	0	0
CD (P = 0.05)	12.66	5.12	2.37	7.15	---

6.3.2. Management

Khan *et al.* (1997) have reported that *P. lilacinum* gave better overall protection of papaya plants against *M. incognita* and *F. solani* disease complex than *T. harzianum*. They reported that application of both the biocontrol agents further limited the damage caused by *M. incognita* and *F. solani* and gave a 35% increase in plant growth compared to individual applications.

6.4. RENIFORM NEMATODE, *ROTYLENCHULUS RENIFORMIS*

The reniform nematode was first reported on papaya from Northern India by Prasad *et al.* (1964).

6.4.1. Distribution

The reniform nematode has been found in the states of Gujarat, Madhya Pradesh, Orissa, Punjab, Tamil Nadu and Uttar Pradesh (Khan and Khan, 1998).

6.4.2. Crop losses

R. reniformis caused 28% loss in fruit yield of papaya (Rajendran and Naganathan, 1981).

6.4.3. Symptoms

Nematode affected plants are stunted in growth; foliage is slightly yellow in color with reduced leaf size. Under severe nematode infection, roots show discoloration with dark brown lesions which are prone to attack by soil pathogens. Egg masses on roots are covered with soil particles.

6.4.4. Biology

The egg masses were collected from *R. reniformis* infested papaya roots and observations were made on embryonic and post-embryonic developmental stages. The embryonic development was completed in 3 to 4 days and the eggs hatched from 3rd day and continued up to 7th day. The second stage juveniles remained in water until they developed into fourth stage juveniles or the pre-adults which were formed after 12 to 14 days. The pre-adults are the infective stages of reniform nematode. The pre-adults were then inoculated in papaya seedlings and developed into matured females with laying eggs on 12th day of inoculation. Hence, the total life cycle was found to be 25 to 26 days from egg to egg in papaya (Rao et al., 1999).

The young females penetrate the cortex of the root and then become sedentary. The portion of the body which remains outside the root enlarges until it resembles a kidney shape. After the female matures, it secretes a gelatinous substance about its body in which about 100 eggs are laid.

6.4.5. Histopathology

Sivakumar and Seshadri (1972) studied histopathological changes in papaya due to the infection of *R. reniformis*. The infected tissue at the site of feeding showed hypertrophy, hyperplasia, thickening of cell walls, granular protoplasm and enlarged nuclei and nucleoli. *R. reniformis* was mainly a phloem feeder.

6.4.6. Ecology

The incidence of *R. reniformis* is higher in wetter soils.

6.4.7. Management

6.4.7.1. Chemical methods: Foliar application of Fenamiphos and Oxamyl on papaya was effective for the management of reniform nematode (Ayala *et al.*, 1971).

Rajendran and Naganathan (1981) reported that Carbofuran at 2 kg a.i. per ha gave significant reduction in soil population of *R. reniformis* and increased papaya fruit yield by 38.4 per cent over control.

6.4.7.2. Host resistance: The inhibition of development and reproduction of *M. incognita* was observed in Papaya cultivars CO-1, CO-2, CO-3, CO-4, CO-5, Thailand and Giant (Nayak, 1986). Patel *et* *al.* (1989) reported that papaya cvs. Solo and Washington were resistant, while Coorg Honey Dew was moderately resistant. Papaya cv. Mammoth was found resistant, while CO-1, Peradeneya, Sel.7 and Washington were moderately resistant (Khan and Khan, 1998).

6.4.7.3. Integrated methods: Application of 2 kg FYM enriched with *P. fluorescens* (10^9 spores/g) and *P. lilacinum* (1 x 10^6 cfu/g) per plant at the time of planting and at an interval of 6 months significantly reduced reniform nematodes on roots by 66%. Significant increase in fruit yield (28%) was also observed. Benefit: cost ratio (calculated for marginal cost of biopesticides and returns accrued by application of biopesticides) was 3.2 (Anon, 2012).

6.5. DAGGER NEMATODE, *XIPHINEMA BASIRI*

6.5.1. Economic importance and losses

The pathogenic capability of *X. basiri* on papaya has been established. Inoculation of 2000 nematodes per 2 kg soil caused significant reduction in plant height (30.4%) and weight (38.6%) on papaya.

6.5.2. Symptoms

The nematode feeds on the roots of papaya and induced stubbiness of roots, thickness at root tips resulting in galling and fish hook formation (Sundarababu and Muthukrishnan, 1990).

6.5.3. Host range

X. basiri is associated with banana, citrus, fig, guava, litchi, mango, mulberry, plum, rose, tomato, okra and castor in India.

6.5.4. Management

Decomposing organic mulch or straw, green manure and trap crops such as *Crotalaria* and *Sesbania* can be used to encourage fungi like *Pochonia* that kill plant parasitic nematodes (Wildmer *et al.*, 2002). Mustard bran and mustard seed meal can be incorporated into soils because these plants contain nematicidal elements (Moseley *et al.*, 2010).

Marigolds (*Tagetes* spp.) can be grown as cover crops and are effective in reducing populations of nematodes. Marigolds produce root exudates (containing alpha-terthienyl) that suppress plant parasitic nematodes, pathogenic fungi, and bacteria (Wang *et al.*, 2007, Krueger *et al.*, 2010).

Chapter - 7

Mulberry, *Morus* spp.

7.1. INTRODUCTION

Mulberry provides delicious and nutritious fruits which are very popular in North and South India. *M. laevigata* varieties have long fruits which are seedless and sweet. Therefore, it has commercial value. *M. alba* produces white fruits, which are very sweet to taste. Fruits are sold from March – May at Mahabaleshwar, Delhi, Dehradun, Rajasthan, Almora and other adjacent areas. Mulberry fruits contain high carbohydrate besides vitamins and minerals and they are used for jam, jelly, juices and other products. Mulberry fruit juice helps to prevent high fever, dyspepsia and melancholia. The mulberry may be exploited as minor fruits.

7.1.1. Origin

White mulberry is native to northern, eastern and central Asia; black mulberry is native to Southwest Asia; while red mulberry is native to Eastern North America.

7.1.2. Area and production

Major mulberry cultivating countries are China and India. Japan, America, Australia, Korea, and Nepal also cultivate mulberry tree.

The mulberry plantation in India was over 242 thousand hectares in financial year 2022.

7.1.3. Varieties

There are of three types:

» White mulberry (*Morus alba*)

» Black mulberry (*M. nigra*)

» Red mulberry (*M. rubra*)

The major mulberry varieties include Knava-2, S-13, S-1, S-34, Victoria-1, China white- (temperate, clonal selection), Goshoerami- (temperate, introduction from Japan), Tr-10, MR- 2, and DD.

7.1.4. Health benefits of mulberries

Mulberry fruits are edible. Like other fruits, mulberries are full of minerals, vitamins, organic compounds and dietary fibers. All of them are beneficial in keeping people healthy, happy and keeping away from illness.

- » Mulberry berries contain dietary fiber which is helpful for the good health of digestive system.
- » The presence of high level of Iron content in the mulberries is beneficial in significantly boosting the red blood production in the body.
- » The Resveratrol present in mulberries is helpful for good circulation of blood. Resveratrol is specific flavonoid that found only in the dark-skinned berries such as grapes.
- » The high content of vitamin A, vitamin C, anthocyanins, and various other phytonutrient and polyphenolic compounds in the berries are beneficial in preventing people from many kinds of cancers.
- » Mulberries are also helpful in improving the eye vision.
- » The vitamin C along with minerals and vitamins present in the berries is helpful in boosting the immunity and keeping the illness away.
- » Mulberries are also an excellent source of vitamin K, calcium, phosphorous, magnesium, and iron which is good for the bone health.
- » Mulberries are also a good anti-aging agent and prevent us from premature aging.
- » The vitamins A and E present in them is helpful in keeping hair healthy and shiny.

7.1.5. Export

Mulberries are seen as superfoods on the European market. Demand for the dried fruits is increasing. Fruits, leaves and their extracts have potential in the market for food supplements, because of their high level of antioxidant activity and chemical composition. The top 10 countries exporting mulberry fruits are presented in Table 7.1.

Table 7.1. Export trends overview of fresh mulberry fruits in top 10 countries (2021)

Country	Export value (Million US$)	Country	Export value (Million US$)
Spain	575.1	Portugal	251.2
Mexico	532.9	Poland	47.9
Morocco	437.6	France	38.3
United States of America	395.9	Germany	35.3
Netherlands	296.9	Belgium	32.0

7.1.6. Nematode problems

Plant parasitic nematodes play an important role in reducing herbage yield and quality of leaves besides the life span of mulberry plants ultimately leading to poor production. Among mulberry nematodes, root-knot nematode, *Meloidogyne incognita* is economically important as it affect the crop quantitatively and qualitatively.

7.2. ROOT-KNOT NEMATODES, *MELOIDOGYNE INCOGNITA*

In India, it has been reported first by Swamy and Govindu (1965) from Mysore and few years later by Mathur *et al.* (1969) from Uttar Pradesh and Haque (1976) from West Bengal. *M. incognita* and *M. javanica* have been reported on mulberry from India (Swamy and Govindu, 1965; Narayanan *et al.,* 1966; Sharma and Sarkar, 1998).

The root-knot nematode, *Meloidogyne incognita* is a major constraint in mulberry cultivation and plays an important role in reducing herbage yield and quality of leaves besides the life span of mulberry plants. Govindaiah *et al.* (1993a) reported that only Race-2 of *M. incognita* is found to infect the mulberry in India.

7.2.1. Economic importance

M. incognita is found to be highly pathogenic to mulberry causing significant reduction in plant growth and leaf yield at 1000 larvae/plant or 150 larvae/250 cc soil under field conditions. The avoidable leaf yield loss is estimated up to 15% (Govindaiah *et al.,* 1991a; Sharma *et al.,* 1998; Sharma, 1999a). Besides reducing the leaf production, it also affects the nutritive value by reducing the protein content.

7.2.2. Symptoms

The severely affected plants showed stunted growth with low water moisture in leaves, and later marginal necrosis and yellowing of leaves. The underground symptoms include the formation of characteristic knots/galls on the root). Galls are spherical and vary in size; young galls are too small and yellowish-white in color, old galls are big and pale brown. The nematode damaged roots do not utilize water and fertilizers as effectively as uninjured roots. Thus, as a result of poor plant growth, about 10 -12 % leaf yield is lost in addition to affecting the leaf quality (Govindaiah *et al.,* 1991).

7.2.3. Biology

During its life cycle, the nematode passes through three stages such as egg, larva and adult. Eggs and larvae are the main sources to spread the disease. The eggs are ellipsoidal with a diameter of 40 - 45 µm and length of 90 - 100 µm while second stage juvenile larvae are vermiform with 15 - 20 µm diameter and 400- 450 µm length. The second stage juvenile larvae invade the roots through root tips by making the hole with the help of their stylets. After entry, they start feeding on the parenchymatous cells. While feeding on the roots of parenchymatous cells, the larvae seem to inject the saliva into roots, which contains the proteolytic enzyme and flows forward into the oesophagus and ejected through the stylet leading to enlargement of the cells. Due to the process of hypertrophy and hyperplasia tumors or cancerous galls are formed on the roots. Usually, four molts occur to complete the life cycle, first molt occurs within the egg, which hatches to form the second stage larva. The larvae inside the roots undergo three molts and develop into mature adults. The male is cylindrical worm-like, transparent, colorless with stripes across the body having length of 1400 - 1450 µm and diameter of 30 - 35 µm. The female is pear shaped, elongated and pyriform with a length of 700- 750 µm and diameter of 400-450 µm (Minamizawa, 1997; Sharma, 1999a). The reproduction takes place mostly by parthenogenesis and very rarely by sexual means, as occurrence of males is very rare. Each female lays 300 - 500 eggs covered with gelatinous substance called as egg mass (Ertian, 2003). The nematode produces six generations within a year and life span is completed within 30 - 40 days. Temperature from 15 - 35°C (optimum being 20 - 30°C), soil humidity of more than 60% and soil pH of 4 - 8 are favorable for the development of the nematodes (Murthy, 1986).

7.2.4. Histopathology

Histopathological observations revealed that due to nematode infection, the vascular tissues and cortex of roots are disorganized hampering the uptake of water and minerals from the soil to the aerial parts of the plant (Shetty and Rudramuniyappa, 1992; Ertian, 2003).

7.2.5. Spread

Electron microscopic study has revealed that two ways of infestation of *M. incognita* occurs *viz.*, primary infestation by second stage juveniles from soil and secondary infestation by juveniles developed within the root tissues (Babu *et al.*, 1996). The disease spreads primarily through contaminated soil, farm implements and run-off irrigation water from infected plot to the other plot. Plantation of infected saplings, cultivation of other susceptible crops along with mulberry, and growth of some susceptible weeds in and around the mulberry gardens acts as the secondary sources of infection (Govindaiah *et al.*, 1989; Teotia *et al.*, 1992; Sharma, 1999a).

7.2.6. Ecology

The root knot disease incidence and intensity were found to be very high in red sandy soils followed by red loamy soils while in black cotton soils, the incidence was found very less. Among the different farming systems, the incidence and intensity found to be very high under irrigated gardens, while under rainfed gardens it was very poor (Govindaiah *et al.*, 1994; Philip et *al.*, 1997; Sharma and Sarkar, 1998).

Temperature between 27-30 °C, soil moisture of less than 40 % and pH of 5 to 7 are favourable for the development of the root-knot nematode. The disease outbreak is throughout the year and more common in sandy soils under irrigated conditions.

There is significant increase in density of nematode population up to the depth of 15 - 30 cm, and gradually decreases up to 75 - 90 cm depth. Further, the distribution of nematode population was very high at rhizosphere region as compared to inter-stump (space between plants) and inter-ridge regions *i.e.*, space between rows (Sharma, 1998). Such information may help to formulate the appropriate strategies for nematode management.

7.2.7. Interaction with other soil-borne pathogens

The root-knot nematode interacts with soil-borne pathogens *viz.*, *Fusarium solani*, *F. oxysporum*, and *Botryodiplodia theobromae* resulting in death of grown up plants.

In India, the interaction between *M. incognita* with *F. solani/Rhizoctonia bataticola* cause disease complex in mulberry and *M. incognita* provides a good site for easy entry of developing hyphae of these pathogens into the roots of mulberry (Bhagyrathy *et al.*, 2000; Sukumar *et al.*, 2000; Nishitha Naik *et al.*, 2003).

7.2.8. Management

7.2.8.1. Physical methods: Hot water treatment of mulberry seedlings is effective at 50-- 53°C for 20 min. before planting to kill the nematodes inside the root (Ertian, 2003). Hot water immersion must be followed up with a cold water treatment.

7.2.8.2. Cultural methods: Intercropping of mulberry with marigold at wider spacing of 30 to 45 cm controlled *M. incognita*, increased leaf moisture and leaf yield of mulberry (Govindaiah, 1990).

Intercropping of plants like marigold (*Tagetes patula*), sesame (*Sesamum indicum*) or sun hemp

(*Crotalaria spectabilis*) at 30 cm distance in between mulberry rows reduces 60 - 65% root galls and egg masses of nematode resulting in marginal increase of 8 - 10% in leaf yield. Mulching of these plants after full growth enriches the soil fertility, in addition to controlling the disease (Govindaiah *et al.*, 1991b, 1997).

The root-knot disease can be easily and effectively controlled by the use of amaranth plants as trap crop for root-knot nematodes intercropped with mulberry plants in the naturally root-knot infested field increasing production. As intercrop, amaranth could be harvested at frequent intervals to keep the nematode population to a minimum level. This way, amaranth could serve as highly effective and beneficial to the farmers in protecting other crops from invading nematode larvae as well as by controlling root-knot nematodes in the naturally infected field, and also through buying and selling of the edible amaranth plant regularly from the intercropped field (Datta, 2005; Datta and Datta, 2007).

Application of neem oil cake (obtained from seeds of *Melia azadirachta* after oil extraction) at the rate of 2 tons/ha/year in four equal split doses around the basin of the plants during intercultural operations/fertilizers application is recommended, which reduces root galls and egg masses of nematode to an extent of 65 - 70 % resulting in 16-18% increase in leaf yield. The neem oil cake should thoroughly be incorporated to the soil by plowing/digging followed by irrigation (Govindaiah *et al.*, 1994, 1997; Sharma *et al.*, 1998).

Other cultural methods that can be used to reduce the root-knot nematode include:

» Deep plowing of infested garden to a depth of 30-40 cm during summer exposes the nematode eggs and larvae to direct sun.
» The higher temperature and lower humidity developed by this practice kill the nematode eggs and larvae present in the soil because they are sensitive to heat (Minamizawa, 1997; Sharma, 1999a).
» Use nematode-free saplings for new plantation.
» Mulching of green leaves of Neem (*Azadirachta indica*) and Pongamia (*Pongamia pinnata*) at 2.5 tons/ha is also effective for the control of root-knot nematode disease (Govindaiah *et al.*, 1989).

7.2.8.3. Chemical methods: Application of Organophosphate and Carbamate nematicides like Carbofuran (Furadan 3G) at the rate of 40 kg/ha/year or Sebufos (Rugby lOG) at the rate of 30 kg/ha/year in four equal split doses at an intervals of 3 months can effectively reduce 70 - 75% root galls and egg masses of the nematode resulting in 12 -16% improvement in leaf yield. The nematicide should be applied during intercultural operations/fertilizers application particularly around the basin of the plants by broadcasting. After broadcasting, it should be well mixed in the soil by plowing/digging followed by irrigation. After 40 - 45 days of nematicide application, the leaves can be used for silkworm rearing (Govindaiah *et al.*, 1993b, 1997).

7.2.8.4. Biological methods: *Purpureocillium lilacinum* was able to infest eggs and egg masses of *M. incognita* race 2. The density of *P. lilacinum* was 680 propagules/g of soil obtained from a mulberry infested plants.

Trichoderma harzianum THN1 treated plants recorded 46 and 52% reduction in root-knot and egg mass indices and a 23% increase in leaf yield which was on par with Carbofuran treatment (Table 7.2) (Sukumar *et al.*, 2004).

Table 7.2. Comparative efficacy of *Trichoderma harzianum* THN1 with other control agents on *M. incognita* under field conditions

Treatment	Root-knot index	Egg mass index	Leaf yield/plant (g)
T. harzianum (Yearly twice)	2.6	2.4	218
P. lilacinum (Yearly twice)	3.0	2.7	224
Carbofuran (Yearly once)	4.6	3.8	182
Carbofuran (Yearly twice)	2.3	2.2	227
Control	4.8	5.0	178
CD at 5%	0.45	0.34	13

7.2.8.5. Host resistance: In Brazil, Campos *et al.* (1974) have screened several mulberry varieties against *Meloidogyne* spp. and variety Calabresa was found to be fairly resistant. In India, 250 mulberry varieties were screened for resistance against root knot (M. *incognita)* and varieties like S-13, S-30, S-1096, RFS-135, V-1 and Assambola were found to be moderately resistant to the disease (Govindaiah *et al,* 1996; Sharma *et al.,* 2001). These results could be utilized in disease resistance breeding programs for evolving nematode-resistant varieties.

Mulberry cvs. Tr10 and V1 Varieties can be selected for the regions with good irrigation/rainfall conditions, while cv. RFS175 can be used for rainfed cultivation (Vijaya Kumari and Sujathamma, 2016).

Among the ten mulberry varieties screened, only one variety viz., RFS 135 showed resistance response against *M. incognita*, while the varieties, C20 and DD showed moderate resistant reaction. A positive correlation was observed between the root-knot nematode resistance and peroxidase enzyme activity in mulberry genotypes (Gnanaprakash *et al.*, 2016).

The mulberry cultivar Yuesang C20 was immune to the disease, varieties Yuesang C8, Yuesang 87 and Yuesang 44 being highly resistant; while Yuesang 85, Yuesang C10, Yuesang C21, Yuesang C19, Yuesang C15, Yuesang C121, Yuesang C43 being moderately resistant (HuDie *et al.*, 2019).

7.2.8.6. Integrated methods: Combined soil application of *P. fluorescens* + *T. viride* each at 10 g/plant as soil application was effective to check the root-knot nematode disease and to improve growth of mulberry with increased leaf yield and reduced nematode population (Table 7.3) (Muthulakshmi *et al.*, 2010a).

Table 7.3. Effect of biocontrol agents on plant growth characters of mulberry in *Meloidogyne incognita* infested field (Pooled data from two experiments) (Muthulakshmi *et al.*, 2010a)

Treatment	Plant height (cm)	Leaf yield (kg/plant)	Gall index	No. of egg masses/g root	No. of eggs/egg mass	Antagonistic population (cfu/g soil)
Pseudomonas fluorescens @10g/ plant	270.45 (+25.42)	1.568	2	13.44 (-48.58)	230.51 (-29.67)	2.2×10^6
Trichoderma viride @ 10g/plant	265.65 (+23.19)	1.546	2	12.93 (-50.53)	256.67 (-21.69)	1.2×10^4
P. fluorescens @ 10g/plant + *T. viride* @ 10g/ plant	290.27 (+34.61)	2.252	2	11.54 (-55.85)	186.98 (-42.95)	3.1×10^{6}* 4.6×10^{4}**

P. fluorescens @ 5g/ plant + *T. viride* @ 5g/plant	275.48 (+27.75)	1.821	2	12.56 (-51.95)	203.94 (-37.78)	2.2x10^{6}* 4.1x10^{4}**
Carbofuran 3G @ 1 kg a.i. /ha	240.58 (+11.57)	0.958	3	14.10 (-46.06)	216.43 (-33.97)	---
Control	215 .63	0.810	5	26.14	327 .76	---
CD (P = 0.05)	0.833	---		0.163	1.532	---

Figures in parentheses are per cent increase (+) or decrease (-) over control

* *Pseudomonas fluroscens* in soil

** *Trichoderma viride* in soil

Integration of moderately resistant variety (V-1/S-13/S-34) and soil application of Bionema (*Pochonia chlamydosporia*) along with neem oil cake in disease prone areas reduces the disease severity to an extent of 95 - 98% with an improvement of 28 - 30% leaf yield (Sharma *et al.*, 2001). In moderately resistant varieties, the penetration of juveniles into the roots is lesser resulting in lesser number of root-knots, egg masses and nematode population (Govindaiah *et al.*, 1996).

When the infection is very severe in mulberry gardens, Bionema along with neem oil cake treatment should be given for 2 - 3 consecutive years. Application of Bionema/ neem oil cake in combination with intercropping of marigold has to be taken in gardens with moderate infection.

Soil solarization during summer season followed by application of Bionema along with neem oil cake and plantation of moderately resistant variety (V-1/S-13/S-34) are more appropriate in disease prone areas.

Talc-based bionematicide (Bionema) produced by a biocontrol agent, *Pochonia chlamydosporia*, is used along with neem oil cake (Sharma, 1999a). *P. chlamydosporia* parasitizes the eggs of *M. incognita* and stops further hatching (De Leij and Kerry, 1991) whereas the compounds present in the neem oil cake (nimbidin, thionemone and phenolics) kill the nematode larvae present in soil (Alam, 1993).

Bionema and neem oil cake application reduces the disease severity by 85 - 90% and increases the leaf yield up to 23%. *P. chlamydosporia* is compatible with neem oil cake, Furadan 3G, Rugby 1OG and chemical fertilizers, VA-mycorrhiza, bacterial biofertilizer *(Azotobacter)* and vermicompost.

Use of bioagents is most preferred for the management of *M. incognita* since it is not advisable to use chemicals as the leaves are fed to the silkworms. Soil application of *Purpureocillium lilacinum* @ 25 g + 100 g neem cake/plant is recommended for the management of *M. incognita* infesting mulberry.

Combined inoculation of AMF and *Pseudomonas fluorescens* had positive effect on root-knot nematode control on mulberry (Kumutha, 2001).

Soil application of 24 g of neem cake + 200 g FYM + 1 g of bionema + 100 g of AMF mixed in 20 kg of FYM applied in three split doses /year gave maximum reduction of root galling (85.30) (19.9 galls/plant in treated as compared to 135.8 galls/plant in control) and egg mass production (86.45%) (5.00 egg masses/plant in treated as compared to 35.7 egg masses/plant in control) in mulberry cv. Tr 10 (Vijaya Kumari and Sujathamma, 2014) (Table 7.4).

Table 7.4. Effect of Integrated control in the management of root knot nematode *Meloidogyne incognita* in different mulberry varieties

Mulberry Variety	No. of galls / plant			No. of egg masses / g root		
	Control (infested)	Treated	Reduction over control (%)	Control (infested)	Treated	Reduction over control (%)
Tr10	135.80	19.90	83.30	36.70	5.00	86.45
DD	166.36	28.1	83.10	47.66	7.40	84.47
V1	178.30	30.73	82.75	56.60	9.20	83.63
RFS175	185.00	35.5	80.80	78.80	14.50	81.55
K2	195.53	40.33	79.37	86.30	16.80	80.36
MS8	205.90	48.66	76.30	101.66	22.00	78.32

7.3. ROOT-KNOT NEMATODE, *MELOIDOGYNE INCOGNITA* AND ROOT ROT, *MACROPHAMINA PHASEOLINA* DISEASE COMPLEX

7.3.1. Interaction studies

The root-knot nematode, *Meloidogyne incognita* and the root rot fungus, *Macrophomina phaseolina* plays an important role in reducing herbage yield and quality of leaves besides the life span of mulberry plants. Disease complex reduces the nutritive value of mulberry leaves and adversely affect the growth and development of silkworm ultimately leading to poor cocoon production. Besides causing direct damage, nematodes are also involved in disease complex with other microorganisms and provide easier entrancc to wcak pathogenic fungi and break the host plant resistance.

Inoculation of *M. phaseolina* prior to *M. incognita significantly reduced the nematode population.* Inoculation of *M. phaseolina* simultaneously with *M. incognita and 15 days before nematode inoculation caused disease incidence and significant reduction in plant growth compared to the inoculation of fungus or nematode alone and fungus inoculation prior to nematode (Muthulakshmi et al., 2010b).*

7.3.2. Management

Best management practices including broad spectrum bio-control agent (consortium of microbes) which can control both RKN and associated root rot causing fungal pathogens to manage existing mulberry gardens. Two fungal bio-control agents, *Trichoderma harzianum* and *T. viride* were found associated in the rhizosphere of mulberry.

7.3.4. Mulberry Ringspot Virus

An RNA-containing virus with isometric particles 22-25 nm in diameter spreading naturally in Japan. It is readily sap-transmissible, has a moderate host range and is transmitted by the nematode, *Longidorus martini.*

7.3.4.1. Geographical distribution

It was found only in Japan (Tsuchizaki *et al.*, 1971: Tsuchizaki, 1975).

7.3.4.2. Symptoms

Found naturally only in mulberry, in which it usually causes mosaic and ringspot symptoms. Some isolates are associated with leaf enations. Systemically infected leaves develop mosaic, ringspots or enations, show filiform and ring post symptoms. This virus infects quickly covering all the newly emerged leaves of the plant (Ganga and Sulochana, 1997).

7.3.4.3. Host range

Besides mulberry, the virus infects *Vigna sinensis cv.* Black Eye (cowpea), *Glycine max* (soybean), *Pisum sativum* (pea), bean, sesame, *Chenopodium quinoa,* and *Nicotiana clevelandiia.*

7.3.4.4. Transmission

Transmitted by the nematode, *Longidorus martini* and *L. macrsoma* (Yagita and Komuro, 1972). Infested soil containing *L. martini* was still infective after storage, in the absence of plants, for 14-17 months at room temperature or for more than 30 months at 0-9°C.

About 10% of progeny seedlings of *Glycine max* cv. Mikawashima are infected (seed transmission) (Tsuchizaki *et al.*, 1971).

7.3.4.5. Management

» Use of certified *virus*-free propagative stock
» Removal and destruction of infected plants
» Washing of hands or sterilizing the tools after handling infected plants.
» Use of disease-free seeds.
» Clean cultivation

7.5. SUCCESS STORY IN KARNATAKA

Management of root-knot nematode (*Meloidogyne incognita*) in mulberry was demonstrated in Shri. Ravi's farm located in Uppinahalli village, Hosakote taluk, Bangalore North District, using soil application of talc formulation of *Purpureocillium lilacinum* @ 20 kg/ha (10^8 spores/g talc) along with Neem cake @ 1 ton/ha. Use of bioagents is most preferred for the management of *M. incognita* since it is not advisable to use chemicals as the leaves are fed to the silkworms. Individual farmer's gain was 60% more leaf yield which amounts to nearly Rs.1,00,000/ha as compared to the poor crop which fetches Rs. 50,000/- to Rs. 67,500/ha.

Chapter - 8

Guava, *Psidium guajava*

8.1. INTRODUCTION

8.1.1. Origin

Guava is native to Central America and is widely cultivated in India, South Africa, Hawaii, and Mexico.

8.1.3. Area, production and yield

Country-wise area and share in production of guava in 2020 are presented in Table 8.1.

Table 8.1. Global production and share in production of guava (2020)

Rank	Country	Production (million tons)	Share (%)
1	India	24.75	45.22
2	Indonesia	3.62	6.61
3	Mexico	2.37	4.34
4	China	2. 37	4.33
5	Pakistan	2.34	4.28
6	Brazil	2.14	3. 01
7	Malawi	1.94	3.54
8	Thailand	1.66	3.03
9	Bangladesh	1.45	2.65
10	Egypt	1.40	2.55

State-wise area, production and yield of guava in India during 2017-18 are presented in Table 8.2.

Table 8.2. State-wise area, production and yield of guava (2017-18)

State	Area ('000 ha)	Production ('000 tons)	Yield (tons/ha)
Andhra Pradesh	9.53	229.78	24.12
Arunachal Pradesh	0.12	0.22	1.89

Assam	4.43	96.69	21.84
Bihar	27.61	427.61	15.49
Chhattisgarh	21.89	197.18	9.01
Gujarat	12.67	169.57	13.38
Haryana	12.09	137.02	11.33
Himachal Pradesh	2.32	2.61	1.12
Jammu & Kashmir	2.34	7.77	3.33
Jharkhand	8.32	89.31	10.73
Karnataka	7.18	140.23	19.52
Kerala	0.16	1.42	8.77
Madhya Pradesh	35.08	686.70	19.58
Maharashtra	9.07	122.83	13.55
Mizoram	0.42	2.55	6.07
Nagaland	0.58	4.75	8.16
Odisha	14.27	105.04	7.36
Punjab	8.69	195.60	22.50
Rajasthan	4.33	55.13	12.73
Sikkim	1.21	17.60	14.53
Tamil Nadu	9.69	155.06	16.00
Telangana	2.56	38.74	15.14
Tripura	0.70	3.40	4.85
Uttar Pradesh	49.53	928.44	18.75
Uttarakhand	3.62	20.37	5.63
West Bengal	16.25	215.20	13.24
Others	0.20	2.73	13.98
TOTAL	264.85	4053.51	15.30

Source: Horticulture Statistics Division, Department of Agri. Co-op. & Farmers Welfare.

The major guava producing states/districts in India are presented in Table 8.3

Table 8.3. Major guava producing states/districts in India

State	**Districts**
Madhya Pradesh	Indore, Khargone, Vidisha, Katni, Singrauli *etc.*
Uttar Pradesh	Budaun, Kasganj, Aligarh, Farrukhabad, Etah, Allahabad, Lucknow *etc.*
Bihar	Nalanda, Rohtas, Bhojpur, Purbi Champaran, Pashchim Champaran *etc.*
West Bengal	24 Paraganas South, Birbhum, 24 Paraganas North, Medinipur West, Nadia *etc.*
Andhra Pradesh	East Godavari, West Godavari, Guntur, Krishna, Ananthapur, Medak, Ranga Reddy, Mahaboob nagar, Prakasham, Khammam

Maharashtra	Satara, Beed, Pune, Ahmed nagar, Aurangabad, Amravati

8.1.3. Varieties

Major commercial varieties grown in different states are presented in Table 8.4.

Table 8.4. State-wise guava cultivars grown in India

State	Varieties
Madhya Pradesh	L-49, Allahabad Safeda, Gwalior-27, Hafshi, Seedless Chittidar
Uttar Pradesh	L-49, Allahabad Safeda, Lucknow Safeda, Apple Color, Chittidar, Red Fleshed, Allahabad Surkha, Sardar, Mirzapuri Seedless, CISH-G-1, CISH-G-2, CISH-G-3
Bihar	Allahabad Safeda, Apple Color, Chittidar, Hafshi, Harijha, Sardar, Selection-8
West Bengal	L-49, Allahabad Safeda, Dudhe Khaja, Gole Khaja, Kabli, Baruipur, Chittidar, Harijha, Sardar
Andhra Pradesh	Allahabad Safeda, Lucknow 49, Anakapalli, Banarasi, Chittidar, Hafshi, Sardar, Smooth Green, Safed Jam, Arka Mridula
Maharashtra	Nagpur seedless, Dharwar, Dholka, Kothrud, L-24, L-49, Nasik, Sindh

8.1.4. Rootstocks

Propagation of uniform material for planting is very important. For the vegetative propagation of guava, rootstock is raised from the seeds extracted from Lucknow-49 (Sardar guava). The seed is extracted from ripe fruits.

8.1.5. Advance technology

High density planting.

8.1.6. Pre-cooling

Forced air pre-cooling or hydro-cooling at -1 to 0 ^{0}C and RH – 90-95% are suitable for cooling up to 8 months.

8.1.7. Storage conditions

- » Temperature : 7-10 ^{0}C
- » Relative humidity : 85-90%
- » Storage period : 2-3 weeks
- » Freezing point : -1 ^{0}C

8.1.8. Sensitivity

- » Refrigeration and ethylene

8.1.9. Peak harvesting season

- » Madhya Pradesh : October to December
- » Uttar Pradesh : December
- » West Bengal : July to August

» Punjab : January, August and December

8.1.10. Export from India (2017-18)

» Quantity : 1.23 thousand tons

» Value : 553.26 lakh rupees

Major importers include the US, the United Arab Emirates, Saudi Arabia, the Netherlands, Kuwait, and Jordan.

8.1.11. Nematode problems

The root-knot nematodes, *Meloidogyne* spp. are the major problems limiting the production of guava in India.

8.2. ROOT-KNOT NEMATODE, *MELOIDOGYNE ENTEROLOBII*

The worst nematode problem affecting guava is the root-knot nematode, *Meloidogyne enterolobii* (=*M. mayaguensis*) which is recognized as a limiting factor in commercial guava production in Central and South America. Guava production has declined during the past quarter century in Cuba, Puerto Rico, Mexico, Venezuela and Brazil due to increasing pressure of *Meloidogyne* spp. (El Borai and Duncan, 2005). In Brazil, *M. enterolobii* was originally detected in guava orchards in 2001 in Pernambuco and Bahia states (Carneiro *et al.*, 2001). Since then, this nematode has been a matter of grave concern in the country because it has been spreading rapidly with infected seedlings, making the cultivation of guava unviable in the heavily infested areas (Carneiro *et al.*, 2007; Siqueira *et al.*, 2009). Unfortunately, the Ministry of Agriculture (MAPA) has not yet established legislative measures to prevent the spread of this pest in Brazil. *M. enterolobii* was reported from regions in China from guava (Xu *et al.*, 2004). The presence of this nematode has increasingly been detected in different geographical regions from a wide range of hosts, including crops carrying genes for resistance to the main *Meloidogyne* spp. (Blok *et al.*, 2002; Brito *et al.*, 2007; Carneiro *et al.*, 2006a).

In India, public and private nurseries have been responsible for the interstate spread of *M. enterolobii* through planting materials of guava; it has already been intercepted from 11 states (Poornima and Walia, 2021), since the first report from Tamil Nadu in 2016 in India (Poornima *et al.*, 2016). Within the Tamil Nadu state of India, the nematode was recorded from all the nine districts surveyed where guava is cultivated (Ashokkumar and Poornima, 2019). Furthermore, Rashid *et al.*, (2019) reported *M. enterolobii* in newly established as well as established guava orchards and nurseries in Madhya Pradesh, Rajasthan and Uttar Pradesh. Recently, *M. enterolobii* has been recorded from the Luni river basin of Rajasthan (Bhati and Parashar, 2020) and Ratlam district of Madhya Pradesh in India (Singh, 2020).

8.2.1. Distribution

Global distribution of *M. enterolobii* is provided and depicts its prevalence in the tropical climatic conditions across five continents (approximately 52° North to 32° South latitude), i.e. Asia, Africa, Europe, South and Central America besides the Caribbean.

Meloidogyne enterolobii has been reported from several countries in North, Central and South America, Africa and Asia (CABI, 2000). Its present distribution in warmer climates suggests that this species will not survive outside greenhouses in northern countries of Europe, but it might be able to establish itself in the Mediterranean region (Castagnone-Sereno, 2012).

M. enterolobii is now recognized as a widespread and economically significant pathogen for many crops, particularly on guava in the tropical countries (Collett *et al.*, 2021; Singh, 2020). *M. enterolobii* was originally described on the 'pacara earpod tree' in China in 1983 (Yang and Eisenback, 1983). However, since its first interception on guava in Brazil (Carneiro *et al.*, 2001), *M. enterolobii* is now recognized as a major problem on guava not only in Brazil but in many other countries in South America, Asia and sub-Saharan Africa (Collett *et al.*, 2021). Furthermore, the interaction of this nematode with *Fusarium solani* in causing 'guava decline' in Brazil (Gomes *et al.*, 2012b) and India (Ganeshan *et al.*, 2019a), which is a cause of concern for guava growers in the tropical parts of the world. Therefore, considering the economic status of *M. enterolobii* on guava, it is proposed to assign it the epithet 'guava root-knot nematode'.

Brazilian states of RJ, CE, RN, PE and BA are also an important seedling producers, responsible for spreading this nematode to other Brazilian states (Carneiro *et al.*, 2007; Siqueira *et al.*, 2009). However, this species was detected in the state of Rio de Janeiro in native areas of Atlantic forest (Lima *et al.*, 2005), suggesting that the nematode has not been introduced in Brazil from other countries, as suggested when it was reported for the first time in the country (Carneiro *et al.*, 2001). In Paraná state, it was also detected in guava without introduction from other states (Carneiro *et al.*, 2006b).

8.2.2. Crop losses

Root-knot nematodes are responsible for 28 % annual yield loss in guava amounting to Rs. 2350.88 million in India (Walia and Chakraborty, 2018).

In the San Francisco (Pernambuco state of Brazil) region, the area cultivated with guava has decreased from 6,000 to 1,669 ha in seven years (2000-2006), a reduction of more than 70% in guava production (Plantec/Codevasf), a reduction of 10% per year was estimated (Carneiro *et al.*, 2007). Pereira *et al.* (2009) verified direct damages of R$ 1127 in guava production in the Brazilian states of RJ, CE, RN, PE and BA. In addition to these damages, 3,703 jobs were lost due to the decline and death of the orchards in these regions.

8.2.3. Symptoms

Meloidogyne enterolobii affects growth, yield, life span and tolerance to environmental stresses of infected plants. Young infected guava trees exhibit symptoms of yellowing, shedding of leaves, drying of branches, reduced vigor and less productivity. Rapid decline in tree health and death of trees results under heavy nematode infestation. Moderate infestations are associated with general chlorosis, nutrient deficiency symptoms and reduced flowering and fruiting. Roots of infected trees show multiple galls and root necrosis, causing a drastic decrease in feeder roots (Cetintas *et al.*, 2007). The root-knot nematode infects the whole root system, from small rhizoids to more lignified roots. The fruit yield and quality are affected by the nematode.

8.2.4. Host range

Similar to *M. incognita*, *M. enterolobii* also has very broad host range that include economically important crops such as bell pepper, (watermelon), coffee, soybean, sweet potato, tomato, tobacco, bean, guava, aborigine, ornamental plants (e.g. *Ajuga, Brugmansia, Clerodendron, Tibouchina*) and wild plants (e.g. *Bidens pilosa*) (Humphreys *et al.*, 2012). Additional host ranges of *M. enterolobii* include species of *Cactus, Ficus, Syngonium, Rosa* and *Vitis* in the Netherlands.

8.2.5. Biology and life cycle

The root-knot nematode is sedentary endoparasite and soil-borne. After penetration, the J2 larvae feed endoparasitically on conducting vessels and develop to adult females which are spherical in shape and lays egg masses in compound galls. The males are vermiform in shape which leaves roots to enter the soil. Under north Indian conditions, the nematode completes its life cycle in 25-30 days during summer months; while it takes 60-90 days in winter.

8.2.6. Histopathology

The infected guava roots exhibited complete disruption of the epidermis, cortex, endodermis, pericycle, and xylem and phloem cells (Poornima *et al.*, 2019). The root sections were crowded by clusters of adult females (sometimes even up to 60 per gall) of *M. enterolobii*, and nearly one-fourth of the sectioned area was occupied by giant cells. The giant cells surrounding the nematode head were in groups, each with 8–10 cells of varying sizes, with granular cytoplasm and multiple nuclei. The surrounding cortical cells were completely damaged. Mature females of *M. enterolobii* were present in the pericycle region. The cortical cells showed granulations, and the cell organelles were condensed in the centre of the cells. Large cavities were seen in the cortex, and many of the cells were distorted. The females may be few in numbers in simple galls and several in compound galls with long necks with a bulb-like posterior, crowding to reach the pericycle area. This enormous crowding seems to cause choking of the stelar region, ultimately disrupting the vascular bundles and their function.

8.2.7. Interaction with other pathogens

The decline and death of guava trees infected by *M. enterolobii* is a complex disease involving other micro-organisms like *Fusarium solani* (Gomes *et al.*, 2010), *Verticillium dahliae*, *Pythium aphanidermatum*, and *Trichoderma* sp. (Avelar-Mejía *et al.*, 2001).

8.2.8. Spread

The root-knot nematode is disseminated through infected planting material (grafted material, ground layers *etc.*). As is the case for other plant-parasitic nematodes, *M. enterolobii*'s own movement is limited at most to a few tens of centimeters in the soil. The nematode is mainly spread through infested planting stock and soil attached to equipment and machinery and irrigation water (Castagnone-Sereno, 2012).

8.2.9. Ecology

Sandy loam porous soils favor the multiplication of *M. enterolobii* due to increased porosity (Vanitha *et al.*, 2018). Peak soil density attained in January corresponded to high rainfall and temperatures. In root populations exhibited two peaks, in February and in August. There was a negative correlation between nematode populations in soil and roots (Almeida *et al.*, 2010).

8.2.10. Management

8.2.10.1. Physical methods: Deep plowing exposes nematodes to sunlight.

8.2.10.2. Cultural methods

- Use nematode-free planting material.
- Use sterilized potting media (e.g. sterilized by steam or roasting).
- Use clean implements and tools.
- Do not use previously infested fields for planting.
- Use recommended fertilizers to improve crop vigor.

» Use trap crops like marigold as intercrops.
» Green manuring with repellent crop before planting sun hemp and mustard.
» Add burnt paddy husk to planting hole.
» Restrict flow of irrigation water from infected fields to healthy fields.
» Add compost to the soil to reduce the effect of nematodes (after infestation).
» Remove weeds and infested crop residues to protect nearby healthy trees.
» Suppress nematodes by biofumigation with crucifers, intercropping with cabbage and mustard and their soil incorporation.
» Add organic soil amendments like neem cake, green manure, FYM, compost, *etc.* to the planting pit.
» Remove infested plants from the field.
» Use of non-host cover crops like peanut (cvs. IAC Oirã, IAC Poitã, IAC Tatuí and Cavalo Vermelho), rye (cv. IPR 89), green gram, and oat (cv. IAPAR 126).
» Growing non-host crops or bare fallow are the most efficient methods for reducing *M. enterolobii* populations. Fourteen fruit trees are non-hosts to *M. enterolobii* (atemoya, avocado, cashew, citrus rootstocks, coconut, and mango) or poor hosts (assai palm, common mulberry, jaboticaba, grape rootstocks, passion fruit, sapodilla, and sour sop). Crops like cabbage, garlic, grapefruit, maize, peanut, sour orange, and welsh onion are non-hosts.
» Growing marigold around guava plants.

i) Intercropping with non-host cover crops: An effective alternative control of this nematode is the use of non-host cover crops. Among 38 cover crops assessed, 26 were resistant: peanut (*Arachis hypogaea*) 'Cavalo vermelho', white oat (*Avena sativa*) 'IAPAR 126', black oat (*Avena strigosa*) 'IAPAR 61', rapeseed (*Brassica napus*) 'CAN 420', 'CAN 401', foxtail millet (*Setaria italica*), millet (*Eleusine coracana*), rattle pod (*Crotalaria anguroides, C. apiclolice, C. grantiana, C. juncea, C. okraeluka*), cowpea (*Vigna unguiculata*), hyacinth bean (*Dolichos lablab*), castor bean (*Ricinus communnis*) 'IAC 80', grey velvet bean (*Mucuna cinerea*), green velvet bean (*Mucuna* sp.), black mucuna (*Mucuna aterrima*), forage radish (*Raphanus sativus* var. *oleiferus*) 'AL 1006', 'Seletina Nova', 'N4', 'Jesuíta', white tephrosia (*Tephrosia candida*), timbó (*Ateleia glazioveana*), triticale (*Triticum aestivum* × *Secale cereale*). Nine species/genotypes of plants were immune: peanut 'IAC Oirã', 'IAC Poitã', 'IAC Tatuí', Italian ryegrass (*Lollium multiflorum*), rye (*Secale cereale*) 'IPR 89', butterfly pea (*Clitoria ternatea*), mungo bean (*Vigna radiata*), cooper (*Glycine wightii*). Only three species of plants were susceptible: hairy vetch (*Vicia villosa*) 'Ostssat', rice bean (*Vigna unbellata*) and jack bean (*Canavalia ensiformis*) (Table 8.5).

Table 8.5. Summer and winter cover crops immune (RF = 0.00) to *Meloidogyne enterolobii*

Crop	**Cultivars**
Peanut, *Arachis hypogaea*	IAC Poitã
	IAC Tatuí
	IAC Oirã
Butterfly pea, *Clitoria ternatea*	---
Cooper, *Glycine wightii*	---
Italian ryegrass, *Lollium multiflorum**	---
Forage radish, *Raphanus sativus* var. *oleiferus**	'N 4'

Rye, *Secale cereale**	IPR 89
Mung bean, *Vigna radiata*	---

* Winter plants; those not marked are summer plants.

The summer and winter cover crops which were immune to *M. enterolobii* are presented in Table 8.6.

Table. 8.6. Summer and winter cover crops resistant (RF = < 1.00) to *Meloidogyne enterolobii*

Crop	**Cultivars**	**Reproduction factor**
Peanut, *Arachis hypogaea*	Cavalo Vermelho	0.03
Timbó, *Ateleia glazioveana*	---	0.02
White oat, *Avena sativa*	IAPAR126	0.02
Black oat, *Avena strigosa**	IAPAR 61	0.04
Rapeseed, *Brassica napus**	CAN 420	0.36
	CAN 401	0.36
Rattlepod, *Crotalaria anagyroides*	---	0.02
Rattlepod, *Crotalaria apioclice*	---	0.15
Rattlepod, *Crotalaria grantiana*	---	0.03
Sun hemp, *Crotalaria juncea*	---	0.50
Rattlepod, *Crotalaria ochroleuca*	---	0.02
Field bean, *Dolichus lablab*	---	0.03
Finger millet, *Eleusine coracana*	---	0.05
Black velvet bean, *Mucuna aterrima*	---	0.27
Green velvet bean, *Mucuna* sp.	---	0.06
Grey velvet bean, *Mucuna cinerea*	---	0.01
Forage radish, *Raphanus sativus* var. *oleiferus**	AL 1006	0.80
	Seletina Nova	0.04
	Jesuíta	0.01
	IPR 116	0.01
Castor, *Ricinus communis*	IAC 80	0.13
Foxtail millet, *Setaria italica**	---	0.03
White tephrosia, *Tephrosia candida*	---	0.20
Triticale, *Triticum aestivum* × *Secale cereale**	IPR 111	0.07
Cowpea, *Vigna unguiculatta*	Australian	0.04
	---	0.60

* Winter plants; those not marked are summer plants.

Freitas *et al.* (2017) suggested that these non-host fruit species (avocado, cashew nut, citrus, coconut, grape, mango, mulberry, papaya, passion fruit, sapota, and strawberry) can replace eradicated guava trees in fields severely infested by *M. enterolobii* and provide an economical option for growers.

Growing of onion and garlic in guava basins proved to be better 'repellant crops' than marigold and sun hemp against *M. enterolobii* (Vanitha *et al.*, 2018).

Some crop systems alternating non-host plants in summer and winter or only planting in the summer may work to control *M. enterolobii*, taking into account different climatic regions where this nematode causes damage (Table 8.7).

Table 8.7. Reaction of 'Paluma' guava to *Meloidogyne* spp.

Populations	RF[1]	Status[2]
Meloidogyne incognita race 1	0.03	R
Meloidogyne incognita race 2	0.05	R
Meloidogyne arenaria race 2	0.0	I
Meloidogyne javanica	0.0	I
Meloidogyne enterolobii	182.4	HS

[1]RF = Final population/Initial population. [2]Status: I = Immune (RF=0), R = Resistant (RF1.0), HS = Highly susceptible.

Six genotypes of corn ('NB-7361', 'SHS-5080', 'GNX-1020', 'GNX-3010', 'BRS-1031' and 'BM-1115') were highly resistant (FR >1.0) to *M. enterolobii* and can be used in crop rotation systems to reduce or eliminate nematode populations (Dias *et al.*, 2010). Field experiments are necessary to validate these greenhouse results.

Growing non host crops or bare fallow are the most efficient methods for reducing *M. enterolobii* populations. Unfortunately, the list of non-host plants is very small, including cabbage, garlic, grapefruit, maize, peanut, sour orange and welsh onion.

ii) Raising nematode-free planting material: Use soil-less media instead of field soil for hardening of rooted cuttings. The main advantage of using a soil-less medium is the complete alleviation of nematodes and other soil-borne pathogens.

- Alternatively, virgin soil free from vegetation can be used for filling polythene bags. The sources of virgin soil can be sand dunes or soil/silt collected from dried-up water tank areas, river beds and ponds, *etc.*
- The use of solarized soil seems to be the best practical solution. The soil to fill polythene bags should be spread on a cemented floor and covered with 25 μm transparent polythene sheet for about a month during summer. After filling the solarized soil in the polythene bags, the unused bags can be kept covered with polythene sheets for further solarization.
- Using air layers that are hardened in nematode-free soil ensures nematode-free planting material.
- Bio-hardening guava seedlings with effective endophytes that would deter the entry of root-knot nematode and wilt-causing fungus.

8.2.10.3. Chemical methods: Effective management of guava root-knot nematode can be achieved by soil application of Carbofuran at 15 g a.i./ m^2 followed by irrigation.

Since 2018, two new chemical molecules, i.e. Fluensulfone 2GR (granular formulation) [5-chloro-2-{(3,4,4-trifluoro-3-buten-1-yl) sulfonyl} thiazole] (trade name NIMITZ® by Adama) and Fluopyram 400SC (liquid formulation) (Pyridinyl ethyl benzamide) (trade name Velum Prime® by Bayer CropScience) have been registered in India as nematicides. Protocols have been developed, and the efficacy of the new molecules has been demonstrated to the guava growers, particularly in new plantations that are severely damaged by *M. enterolobii* in the states of Rajasthan, Uttar Pradesh, Gujarat and Haryana in India (Khan *et al.*, 2022).

8.2.10.4. Biological methods: Incorporation of egg parasitic fungi, *Pochonia chlamydosporia* and *Purpureocillium lilacinum* mixed in FYM into the pits before planting.

Application of *Purpureocillium lilacinum/Psudomonas fluorescens*) at 20 g/m^2 (about 9 m^2 around the tree trunk) in the rhizosphere region at the time of flowering gave effective control of *M. enterolobii.*

Talc-based formulations (TNAU Coimbatore, India) of *P. lilacinum* and *P. chlamydosporia* are being used for field application after multiplying them in moistened Farm Yard Manure (1:100 ratio) for two to three weeks. Declining guava plants showed revival symptoms after treatment in many instances.

A commercial formulation of *T. harzianum* (higher than 1 x 10^6 spores/ml) effectively controlled root-knot nematodes in guava and stimulate adventitious root growth in plants in Thailand (Jindapunnapat *et al.*, 2013).

In field studies, soil application of 20 g each of *Trichoderma asperellum* + *Bacillus amyloliquefaciens* + *Purpureocillium lilacinum* per plant resulted in a significant reduction of disease incidence (decline) in guava (Ganeshan *et al.*, 2019b).

8.2.10.5. Host resistance: Use resistant varieties and rootstocks such as Horana White, Horana Giant, Pubudu, and Kanthi (all are moderately resistant).

The susceptibility of 46 and 112 wild *P. guajava* to *M. enterolobii* was observed in our study in Brasília and Petrolina, respectively. The same was detected by other authors for *M. enterolobii* (Almeida *et al.*, 2009; Carneiro *et al.*, 2007; Maranhão *et al.*, 2001; Scherer, 2009). In disagreement with this work, Cassasa *et al.* (1997) evaluated 7 accessions of *P. guajava* in relation to *M. enterolobii.* The resistance was observed only in *P. guajava* accession S3.

Milan (2007, 2010) also reported resistance in guava accessions. In the first work, *P. guajava* "B-12" was classified as resistant to *M. enterolobii* (FR=0.88) having the potential to control the nematode, so that it could be used as a rootstock for commercial guava clones. A total of more than 2000 F1 seedlings were produced in the hybridization program and the F1 seedlings are in the nursery stages for further evaluation. The results also showed that *P. longipes* and *P. arayan* showed potential resistant characters to the root-knot nematode, but their potential as rootstocks for guava cultivation is still questionable. In the second work, this author observed that three accessions of *P. guajava* (K-10, A-06, and J-16) were resistant to *M. enterolobii*, with IG more than 2. Unfortunately, the RFs were not measured. These accessions are graft-compatible with local commercial clones and thus could be used as a rootstock. No more information about the reaction of 2000 F1 seedlings to *M. enterolobii* was available.

P. friedrichsthalianium (Costa Rican wild guava) was considered resistant to *M enterolobii* (=*M. mayaguensis*), *M. incognita* race 1, *M. arenaria, M. javanica*) in different countries: Venezuela (Casassa *et al.*, 1997, 1998; Molero *et al.*, 2010), Colombia (Villota and Agudelo, 1997), and Brazil (Carneiro *et al.*, 2007).

The performance of guava grafted onto tolerant *P. guajava* rootstock and onto resistant *P. friedrichsthalianium* was evaluated. In the nursery, 100% successful grafts were obtained with guava and Brazilian wild guava (*P. guineense*) rootstocks, while success with *P. friedrichsthalianium* was 90% ('Arrayán', 'Criollo'). Initial vegetative growth was significantly higher in the guava rootstock treatment. In the field, this treatment also showed more vigorous growth. 'Arrayán' rootstock clearly showed a strong dwarfing effect while the 'Brazilian' and 'Criollo' wild guava rootstocks exhibited an intermediate effect on vegetative vigor. Number and weight of fruits was higher when guava was used as a rootstock and no differences were observed with wild guava rootstocks (Bogantes-Arias and Newcomer, 2010). In order to evaluate the susceptibility of the rootstocks to nematodes of adult plants, an additional control treatment was included in the field, which consisted of the application of nematicides to the guava rootstock treatment. The results indicated that the root-knot nematode was able to colonize the roots of the guava rootstocks with and without nematicides, as well as colonizing the roots of Brazilian wild guava. No nematodes were detected in the roots of the 'Criollo' and 'Arrayán' rootstocks, confirming the resistance of *P. friedrichsthalianium* (Bogantes-Arias and Newcomer, 2010).

Only Costa Rican wild guava *P. friedrichsthalianium, P. cattleyanum* (yellow wild guava), *A. sellowiana* and *P. rufum* showed resistance or immunity to *M. enterolobii. P. acutangulum* and *P. guineense* (Brazilian guava) were susceptible. However, Lee *et al.* (1998) classified the only *P. guineense* accession tested as immune. The same results were observed by Costa *et al.* (2012). Almeida *et al.* (2009) classified *P. friedrichsthalianum* (probably an incorrect identification) as susceptible to this pathogen (FR=13.03). Additionally, these authors verified that three non-identified *Psidium* spp. accessions and were resistant to *M. enterolobii*, with FRs less than 1. *P. friedrichsthalianum* is not originally from Brazil, and this Brazilian rootstock is probably another *Psidium* species. Ten plants of *P. friedrichsthalianum* and nine of *P. cattleyanum* were compatible with 'Paluma' guava when grafted by budding technique in greenhouse conditions. These grafted plants were maintained in field conditions. One year later, five Costa Rican plants were still alive and producing buds, flowers and fruits, while the plants grafted on *P. cattleyaum* presented an incompatibility reaction (enlargement in grafting region); consequently all nine plants died during the first year.

Carneiro *et al.* (2012) reported that only one accession of *P. friedrichstalianium* (Costa Rican wild guava) was resistant and compatible as rootstock with *P. guajava* 'Paluma', in field conditions.

Abdul Karim *et al.* (2002) found that the Red strawberry guava (*P. littorale* var. *longipes*) which showed apparent tolerance.

Cuadra and Quincosa (1982) found that Costa Rican guava (*P. friedrichsthalianum*) was highly resistant to *Meloidogyne* spp. in Cuba.

Milan (2007) found that only *P. guajava* Acc. B-12 has the potential to control the root-knot nematode and graft-compatible with commercial clone of guava because they are from the same species. *P. guajava* Acc. B-12 had the potential to control the nematode and could be used as a rootstock for guava commercial clones.

8.2.10.6. Integrated methods: The guava growers are advised to avoid nematode infected saplings, removal and destruction of nematode affected trees, application of farm yard manure at 100 kg, Neem cake at 250g, and nematode egg parasitic fungus *Purpureocillium lilacinum* at 25g and Carbofuran 3G at 60 g/tree at early stage of nematode infestation to overcome the nematode problem and to enhance fruit yield.

Effective management of guava root-knot nematode can be achieved by soil application of 1 kg Neem cake enriched with bioagents + Carbofuran at 7g a.i. /m^2 just before flowering.

8.3. INTERACTION OF ROOT-KNOT NEMATODE WITH FUSARIUM ROOT-ROT

The root-knot nematode, *M. enterolobii* interacts with the sol-borne fungus, *Fusarium solani* in causing root rot disease complex in guava. The nematode pre-disposes the roots for fungal infection.

In Brazil, *M. enterolobii* has become a menace for guava production. Nationwide, a fourth of the cultivated area is infested, for which the prospect is the decimation of the orchards. Because infected plants have their roots entirely rotten as the disease progresses, a study was set to investigate whether a soil-borne pathogen could be involved. In the field, guava seedlings were inoculated with *M. enterolobii* only or a combination of the nematode with one of 11 *Fusarium* spp. isolates. One hundred-fifty days later, half of the plants inoculated with *M. enterolobii* and *Fusarium* sp. had started to decline. Variables related to the plants´ shoot and root growth were negatively affected by the fungus, especially when root rooting was intense. The six most damaging isolates were identified as *F. solani*. The extent of the root rotting and damage caused by *F. solani* when associated with *M. enterolobii* indicates that a complex disease involving these organisms may be responsible for the widespread decimation of orchards throughout Brazil.

8.3.1. Interaction

M. enterolobii and *F. solani* have a synergistic effect on disease complex development. Whereas *M. enterolobii* alone was able to induce galling and a limited amount of root rot, *F. solani* did not cause root rot when in isolation. When combined, nematode and fungus led to drastic root rot and typical guava decline symptoms, which suggests that guava decline is a disease complex between *M. enterolobii* and *F. solani* (Gomes *et al.*, 2010, 2012).

No root rot and virtually no effect on all variables were observed in the seedlings inoculated with the fungus isolates, with or without physical injury. *M. enterolobii* is also not highly aggressive to guava when inoculated alone (Gomes *et al.*, 2011; Almeida *et al.*, 2011). Major root rot and a negative effect on all variables were observed in the seedlings inoculated with *M. enterolobii* and all four *F. solani* isolates. The penetration and initial development of the fungus were reinforced by the morpho-physiological alterations caused by *M. enterolobii* infection in guava plants, which increased the susceptibility to *F. solani*. In this complex disease, parasitism by *M. enterolobii* predisposes guava plants that are immune to *F. solani* to extensive root deterioration caused by this fungus, which leads to loss in production and to death of plants in a short period of time. This characterizes guava decline as a complex disease caused by the synergistic effect of these organisms, in which parasitism by the nematode predisposes the plants to root decay caused by the fungus.

The effect of *Meloidogyne enterolobii* and *Fusarium solani* was studied in greenhouse on development of the root and aerial part of guava plants with root system divided into two plastic container (I and II), in greenhouse, inoculated with 500 eggs and second-stage juveniles of *Meloidogyne enterolobii* and/or 5 g of wheat grains colonized by *Fusarium solani* isolate UENF/CF 163. The evaluation took place 120 days later, and the results are presented in Table 8.8.

Table 8.8. Effect of *Meloidogyne enterolobii* and *Fusarium solani* on development of the root and aerial part of guava plants with root system divided into two plastic container (I and II), in greenhouse

Treatments applied to the root system		Fresh shoot mass (g)	Root system volume (cm^3)		Fresh mass of root system (g)	
I	II		I	II	I	II
Non-inoculated	Non-inoculated	93.6a	25.8a	24.4a	27.8a	28.0a
Non-inoculated	*M. enterolobii*	98.9a	20.8a	20.4a	23.2a	26.9a
Non-inoculated	*F. solani*	95.5a	23.8a	24.0a	22.8a	22.5a
Non-inoculated	*M. enterolobii* + *F. solani*	74.7b	20.4a	8.3b	21.2a	10.4b
M. enterolobii	F. solani	92.8a	18.6a	23.6a	25.6a	28.4a
CV%	---	23.4	23.8	18.4	21.2	16.2

Values followed by the same letter in the column are not different according to T test at 5%.

8.3.2. Symptoms

Declining trees show chlorosis, scorching of margin, wilting and shedding of leaves (Anon, 2011). Gomes *et al.* (2008) have shown that these symptoms are associated with foliar deficiency of nitrogen, potassium and phosphorus, a near-deficiency of calcium and magnesium, and an imbalanced accumulation of manganese, chlorine and sodium. Declining trees develop progressive rotting of the entire root system. Usually, declining trees do not recover, and death occurs within weeks to a few months.

The onset or worsening of shoot symptoms is often associated with (i) an increase in the number of root galls in the entire root system, (ii) following a high production season or (iii) following drastic pruning, which is used by some growers to synchronize production.

Significant ($P < 0.05$) root rot and damage occurred when *M. enterolobii* and *F. solani* were combined. The synergistic effect of *M. enterolobii* and *F. solani* in causing guava decline is clear, and this combination reflects the complex nature of the disease.

8.3.3. Management

Several methods have been employed for the management of *Meloidogyne-Fusarium* disease complex in guava. However, the search for a nematode-resistant rootstock is most favored because of the perennial nature of the crop and being environmentally benign. Chemical nematicides have been used routinely, but the use of bioagents has also found favor among the other methods.

8.3.3.1. Cultural methods: Soil amendments (cow manure, sugarcane fiber cake, poultry compost) were applied at 20 or 40 kg/tree in the young and mature orchards, respectively. The combined applications were successful to manage nematodes in the moderately infested orchards only, and proved unsuccessful in orchards with heavy infestation of *M. enterolobii.*

8.3.3.2. Chemical methods: Propiconazole (0.2%) was found significantly superior as compared to other chemicals. Trees treated with Propiconazole (0.2%) took 105 days to recover, while trees treated with Difenconazole (0.2%) took 120 days (Sonyal, 2010).

Being the only registered nematicide in India, carbofuran 3 G (Furadan 3 G) has been used by farmers extensively to manage nematodes in different crops, including guava.

8.3.3.3. Biological methods: Singh *et al.* (2003) reported bioagent, *Aspergillus niger* was most effective in controlling the wilt disease followed by *Trichoderma viride*. *Aspergillus niger* was found very fast growing, easy to propagate and most effective in controlling the wilt disease in field. Besides this quality, it is also growth enhancer and the plants treated with *A. niger* developed faster with more height, more thickness and more numbers of leaves (Misra *et al*., 2000).

8.3.3.4. Host resistance: In wilt sick plot that was artificially inoculated repeatedly with *Gliocladium roseum*, *Fusarium solani* and *Fusarium oxysporum*; F1 population of *Psidium molle* x *P. guajava* was free from the above wilt causing pathogens. The rootstock besides being resistant to wilt pathogens, was also found graft compatible with *P. guajava* (Misra *et al*., 2003)

8.3.3.5. Integrated methods: Integrated eco-friendly approach for the control of guava wilt was suggested by Misra *et al*. (2004) and Misra (2005) utilizing bioagent *Aspergillus niger* strain AN17, using resistant rootstock (*P. molle x P. guajava*), intercropping of guava with turmeric or marigold as well as proper cultural practices to minimize losses due to the disease complex.

Integration of *T. harzianum* + *P. fluorescens* each at 1 x 10^6 gave effective management of the disease complex (Sonyal, 2010).

8.4. INTERACTION OF SPIRAL NEMATODE, *HELICOTYLENCHUS DIHYSTERA* WITH WILT, *FUSARIUM OXYSPORUM* F. SP. *PSIDII*

It is a very destructive disease which is widely distributed in almost all guava growing regions of India. It is caused by the fungus, *Fusarium* along with spiral nematode. Guava wilt is prevalent in Uttar Pradesh, Delhi, West Bengal, Bihar, Madhya Pradesh, Andhra Pradesh, Haryana, Punjab and Rajasthan. The disease complex recorded 5 -10% loss annually in 12 districts of Uttar Pradesh which was estimated around Rs. 0.8 to 1.0 million. In West Bengal, guava wilt reduced the yields by 80%, from 11.35 tons/ha in healthy plantations to about 1.82-2.27 tons/ha in affected plantations.

8.4.1. Symptoms

Symptoms appear during rainy season. The leaves turn to brown color, dry and drop. The infected leaves show deficiency of nitrogen and zinc. The infected branches and twigs show typical die-back symptom. As the disease advances, the entire plant dies in three to four weeks. The cross section of root shows typical blackening of xylem vessels.

8.4.2. Biomanagement

8.4.2.1. Antagonistic bacteria: *Bacillus subtilis* was found relatively effective by controlling 50% of the anthracnose disease on fruits at post-harvest stage.

8.4.2.2. Antagonistic fungi: Biological control with *Aspergillus niger* AN 17 (Pusa Mrida), *Pencillium citrinum*, *Trichoderma harzianum*, *T. virens* and intercropping with marigold and turmeric gave good control of wilt. *A. niger* was found most effective which can be mass multiplied on FYM and applied every year as it controls wilt and also provide nutrition to guava plants (Table 8.9).

Table 8.9. Evaluation of bioagents against guava wilt.

Bio-agent	Average wilting (%)	Control
Pencillium citrinum	47.33	52.67
Aspergillus niger AN 17 (Pusa Mrida)	21.66	78.34
Trichoderma harzianum	24.66	75.34
Control	100.00	0.00

Pencillium citrinum, which can be easily multiplied on rotten amla fruits and other cheaper substrates locally available with the orchardists, is a potential biocontrol agent against guava wilt.

Out of 5 AMF inoculants, *Glomus aggregatum* was found to show highest potential in colonizing the roots of guava and improve its performance. It also suppressed the population of *Fusarium oxysporum* f. sp. *psidii* in roots.

8.5. FUTURE CHALLENGES

- » There is an urgent need to strengthen and enforce domestic quarantine mechanism to suspend spread of plant parasitic nematodes with vegetative propagules, especially through seed potatoes and rooted plants – along with soil, from nurseries/ sick plots/ hot-spot areas to disease free niches.
- » Develop nematode screening methods with more *Psidium* spp. with correct rootstock botanical classification.
- » Improve nematode screening methodology, using the Reproduction Factor (RF) as an obligatory parameter.
- » Improving grafting techniques in *Psidium* spp.
- » Develop interspecific hybridization with *Psidium* spp. to obtain nematode-resistant rootstocks.
- » Study *in vitro* production of resistant plants by somatic embryogenesis.
- » Study the mechanisms and molecular characterization of resistance in *Psidium* spp.
- » Develop molecular markers related to resistant genes in *Psidium* spp.

8.6. CONCLUSION

M. enterolobii has emerged from a 'minor' to major nematode problem due to its wide host range and global distribution in tropical and sub-tropical regions with the potential to establish in cooler regions, as is evident from stray reports from Europe. Further, *M. enterolobii* is known to reproduce on crops carrying nematode-resistant genes. *M. mayaguensis* is now considered a junior synonym of *M. enterolobii*; molecular tools are now available for accurate diagnosis of the species.

The nematode has spread to far and wide geographies along with planting materials. This aspect attains significance in developing and under-developing countries where phytosanitary measures are not thoughtfully implemented and trained quarantine staff are hardly available.

Chemical nematicides to manage the disease are not standardized but seem to be the only alternative under high infestation levels. The initial field efficacy trials with environmentally benign new nematicidal chemicals such as fluopyram and fluensulfone *etc.* are very encouraging. Resistant rootstocks like *P. cattleianum* and *P. friedrichsthalianum* seem to be most promising; however, grafting techniques need to be perfected. Bioagents can be effective in mild and low infections and provide the best protection with organic amendments, but severe infestations warrant chemical interventions only. Cultural practices like intercropping with profitable antagonistic crops (such as country onions, garlic, marigold *etc.*) and trap crops (cowpea) may be promoted.

Raising nematode-free planting materials should remain the primary focus. Popularizing the already available technique of 'air layering' in guava will solve the problem of nematode infection; the air layers may be further hardened in autoclaved/sterilized potting media to raise nematode-free planting materials of guava.

Chapter - 9

Grapevine, *Vitis vinifera*

9.1. INTRODUCTION

9.1.1. Origin

Grapes originated in Central Asia (Region between Black sea and Caucasian sea). Armenia near Caspian Sea.

9.1.2. Area and production

Based on data from the Food and Agriculture Organization Corporate Statistical Database, the estimated total world production for grapes in 2020 was 78,034,332 metric tons, up by 1.3% from 77,000,008 tons in 2019. China was the largest producer of grapes, accounting for 18.9% of global production. Italy came second at 10.5%, followed by Spain at 8.7%.

Area, production and productivity of the major grape growing countries during 2016 are presented in Table 9.1

Table. 9.1. Major grape producing countries in the World (2016)

Country	Area (Million ha)	Production (Million tons)	Yield (tons/ha)
China	0.84	14.84	17.60
Italy	0.67	8.20	12.28
USA	0.41	7.10	17.31
France	0.76	6.25	8.25
Spain	0.92	5.93	6.45
Turkey	0.44	4.00	9.19
India	0.122	2.59	21.23
Chile	0.20	2.47	12.18
Iran	0.21	2.45	11.82

Argentina	0.22	1.76	7.85
Others	3.35	21.67	6.47
WORLD	7.1	77.44	10.91

Grape is one of the important fruit covering an area of 155.30 thousand hectares occupying 2.24 % of the total area in 2020-21. Major grape-growing states are Maharashtra, Karnataka, Tamil Nadu, and Mizoram. Maharashtra ranks first in terms of production accounting for more than 71 % of total production and the highest productivity in the country during 2020-21. Karnataka is the second largest producer of Grapes with a share of 24% in 2020-21.

Area, production and productivity of the major grape growing states in India during 2017-18 are presented in Table 9.2

Table. 9.2. State-wise area, production and yield of grapevine in India (2017-18)

Country	Area ('000 ha)	Production ('000 tons)	Yield (tons/ha)
Andhra Pradesh	0.80	15.92	20.00
Haryana	0.04	0.09	2.3
Himachal Pradesh	0.08	0.13	1.59
Jammu & Kashmir	0.33	0.89	2.69
Karnataka	26.61	524.20	19.70
Kerala	0.01	0.18	17.50
Madhya Pradesh	1.28	1.28	15.00
Maharashtra	105.50	2286.44	21,67
Mizoram	2.45	18.00	7.35
Nagaland	0.22	0.54	2.43
Punjab	0.29	8.23	28.67
Rajasthan	0.00	0.00	1.33
Tamil Nadu	2.16	58.93	27.27
Telangana	0.35	5.27	15.27
TOTAL	138.91	2920.09	21.02

Source: Horticulture Statistics Division, Department of Agri. Co-op. & Farmers Welfare.

The major grapes producing districts in different states in India include:

- Maharashtra (Nashik, Sangli, Solapur, Pune, Osmanabad *etc.*)
- Karnataka (Bijapur, Chikballapur, Belgaum, Bagalkot, Bangalore Rural *etc.*)
- Tamil Nadu (Theni, Dindigul, Coimbatore, Tirunelveli, Erode *etc.*)

9.1.3. Major commercial varieties

Different types of grape cultivars grown in India are presented in Table 9.3.

Table 9.3. Different types of grape cultivars grown in India

Different types	Cultivars
Colored seeded	Bangalore Blue, Gulabi (Muscat)
Colored seedless	Beauty Seedless, Sharad Seedless, Flame Seedless, Manjri Naveen
White seeded	Anab-e-Shahi, Dilkhush(clone of Anab-e-Shahi)
White seedless	Perlette, Pusa Seedless, Thompson Seedless and its clones, Tas-a-Ganesh, Sonaka, Kishmish Rozavis White, and Manik Chaman

9.1.4. Advance technology

» Hardwood cutting and bench grafting with machine.

9.1.5. Pre-cooling

» Forced air pre-cooling or hydro-cooling at -1 to 0 ^{0}C and RH of 90-95% are suitable for storing up to 8 months.

9.1.6. Storage conditions

» Temperature : 0-1°C

» Relative humidity : 90-95%

» Storage period : 5-12 weeks

9.1.7. Sensitivity

» Freezing

9.1.8. Peak harvesting season

» Maharashtra : March, April, May

» Karnataka : March, April and November

» Tamil Nadu : May and November

9.1.9. Export from India

The country is also a major exporter of fresh Grapes to the world. The country has exported 263,075.67 MT of Grapes to the world, worth Rs. 2,302.16 crores/ 305.66 USD Millions during the year 2021-22. Major export destinations (2021-22) include The Netherlands, Bangladesh, Russia, U.K., United Arab EMTs and Germany.

9.1.10. Nematode problems

The root-knot (*Meloidogyne incognita*), the reniform (*Rotylenchulus reniformis*), and dagger (*Xiphinema index*) nematodes are the most important grapevine pests. The nematodes of minor importance are the lesion nematode (*Pratylenchus* spp.), the citrus nematode (*Tylenchulus semipenetrans*) and *Paratrichodorus minor*. All these nematode species reduce the effective root system through killing of many feeder roots.

9.2. ROOT-KNOT NEMATODE, *MELOIDOGYNE INCOGNITA*

9.2.1. Distribution

In India *M. incognita* and *M. javanica* are prevalent in grape cultivating regions of Tamil Nadu,

Karnataka, A.P., Haryana, Punjab and Maharashtra.

9.2.2. Crop losses

The root-knot nematode is responsible for 30 per cent annual yield loss in established vineyards of grapevine amounting to Rs. 3940.44 million in India (Walia and Chakraborty, 2018).

The root-knot nematode is responsible for 25 to 50 per cent yield loss in established vineyards of `Muscat Hamburg' in Tamil Nadu.

9.2.3. Symptoms

The affected vines show decline in vigor with inadequate sprouting after pruning, reduction in leaf size, marginal drying of leaves and perceptible reduction in yield. The nematode produces small galls on the roots. Heavy infections may completely destroy the root systems of young plants. The plant roots cannot absorb water and nutrients from soil due to disruption of conducting tissues resulting in death of roots beyond site of nematode infection.

Root-knot nematodes have endoparasitic sedentary mode of parasitism. The freshly hatched second stage juveniles penetrate through young feeder roots, form giant cells around nematode head and feed upon the cell sap thus devitalizing the cells. As a result of nematode feeding, numerous fine rootlets are formed at the infection site, thus leading to a 'hairy root' condition.

9.2.4. Biology and life cycle

Root-knot nematodes are sedentary endoparasites and soil-borne. Second-stage juveniles infect the roots; the nematode develops endoparasitically feeding on the vascular tissues. The adult females are spherical in shape and lays egg masses in compound galls. The males are vermiform in shape which leaves roots to enter the soil. During summer, the nematode completes its life cycle in about 25 to 30 days, while it takes longer during winter months.

9.2.5. Histopathology

Studies conducted to investigate histological changes in root-knot nematode resistant grape (*Vitis rotudifolia*), due to *M. incognita* revealed that juveniles entered the roots but were not able to induce nurse cell formation. The only host plant reaction observed was the dissolution of middle lamellae in some cells surrounding nematode head. Investigations on the impact of *M. incognita* revealed that the physiological efficiency of vine is reduced. Fewer nematodes become established and fewer eggs were produced in roots of nematode resistant vine than in susceptible vine. The production was more on resistant vines as less energy reserve was required against nematode infection, reproduction and to repair the damage caused by nematodes. *M. incognita* did not affect the concentration of reducing sugars at nematode feeding sites on French Colombard (susceptible) and Thompson Seedless (moderately resistant) cultivars; whereas non-reducing sugars increased in susceptible and decreased in moderately resistant cultivar, indicating that there was more translocation of photosynthates to the feeding sites of susceptible cultivar than that of resistant cultivar. This explains why nematode caused more damage to susceptible cultivars and fewer eggs are produced in resistant cultivar.

9.2.6. Ecology

Grapevines are attacked and seriously injured by root-knot nematodes only when the vines are grown in light, sandy or sandy loam soils. Vines in fertile loams and heavier soils are seldom seriously damaged.

9.2.7. Spread

Root-knot nematodes are disseminated through infected transplanting material, irrigation water, and agricultural implements.

9.2.8. Interaction with disease pathogens

A disease complex involving *M. incognita* and *Rhizoctonia solani* was associated with stunting of grapevines in a field nursery. Nematode reproduction was occurring on Colombard (*Vitis vinifera)* and Ramsey (*V. champini*) cultivars of grapevine. Root rotting was higher when grapevines were inoculated with both *M. incognita* and *R. solani* and was highest when nematode inoculation preceded the fungus. Shoot weights were lower when vines were inoculated with the nematode 13 days before the fungus compared with inoculation with both the nematode and the fungus on the same day. It was concluded that both the *M. incognita* population and some *R. solani* isolates were virulent against both Colombard and Ramsey cvs. of grapes (Walker, 1997).

9.2.9. Management

A well planned control program should include the use of nematode-free planting material, nematode resistant rootstocks and the selection of the most effective soil treatment for the root-knot nematode infestation.

9.2.9.1. Physical methods: Hot water treatment for root-knot nematode control is necessary if roots are lightly infested with nematodes but otherwise have made good growth. The roots are submerged in hot water maintained at 51.7°C for 5 minutes and cooled immediately before planting.

9.2.9.2. Cultural methods: Use of nematode-free planting materials and application of Neem cake @ 200 g/vine was found effective.

Tagetes patula used as intercrop, reduced 39-44% nematode population and significantly increased vine yield. Intercropping with marigold, asparagus or sun hemp gave effective control and resulted in increased fruit yield. Under field conditions, 27.0 to 38.7% soil population and 30.1 to 35.7% root population of nematode on grapevine was reduced when intercropped with marigold, asparagus or sun hemp. Marigold was found most effective (Baghel and Gupta, 1986). Cultivation of onion or garlic in the basin area brings additional remuneration besides reducing the nematode population.

9.2.9.3. Chemical methods: Application of Carbofuran 3G at 13 g/m^2 in 9 m^2 basin area of vine after 3 weeks of pruning followed by flood irrigation proved beneficial.

Application of carbofuran, benfurocarb or phorate at 6 kg a.i. /ha, in *M. javanica* infested vineyard gave 63-68% nematode reduction and improved yield (120-133%). Carbofuran at 13 g/m^2 at bud breaking stage is recommended for controlling *M. javanica* (Anon, 1989).

9.2.9.4. Biological methods

i) Antagonistic bacteria: Sundarababu *et al.* (1999) observed that commercial formulation of *Pseudomonas fluorescens* reduced root-knot nematode population effectively and enhanced yield in grapevine. Soil application of talc formulation of *P. fluorescens* containing $15x10^8$ cfu/g at 4 g/vine around root-knot infested grapevine at 15 cm depth in the basin at the time of pruning (during July) significantly reduced root galling due to *M. incognita* (39%), number of egg masses (250%) and increased fruit yield (166%) (Shanthi *et al.*, 1998) (Table 9.4). Natural suppression of *M. incognita* by *Pasteuria penetrans* has been reported on grapevine in Tamil Nadu (Mani, 1996).

Table 9.4. Field evaluation of biocontrol potential of *Pseudomonas fluorescens* against *Meloidogyne incognita* infecting grapevine.

Treatment	No. of galls /5 g roots	No. of egg masses/5 g roots	Root colonization (cfu/g + 10^8)	Yield (tons/ha)
P. fluorescens - 1 g/vine	428c	94c	24	12.07b
P. fluorescens - 2 g/vine	390b	85b	36	15.41c
P. fluorescens - 4 g/vine	326a	72a	58	22.07d
Carbofuran 3G - 60 g/vine	300a	67a	---	31.65e
Control (Untreated)	535d	180d	---	8.33a

Figures with different letters are significantly different from each other at 5% level by analysis of variance test.

ii) Antagonistic fungi: *Purpureocillium lilacinum* when applied at 20 g/m^2/vine reduces *M. incognita* infestation by 33% on grapevine.

iii) Arbuscular mycorrhizal fungi: Endomycorrhizal fungus, *Glomus mosseae* adversely affected *M. javanica* on grapevine cv. Perlette. Mycorrhizal colonization was not affected by nematodes. Growth suppressive effect of nematode was neutralized by the fungus.

9.2.9.5. Host resistance: In soils where the root-knot nematode infestations are heavy, it may be impossible to grow grapevines without the use of nematode resistant rootstocks.

Use of nematode resistant root-stocks has an added advantage over other methods of control. The cost factor is reduced to minimum. In grapevine, Dogridge (*Vitis champini)*, Salt creek (*V. champini*), 1613 (*V. solanis* x Othello), Harmony (1613 x Dogridge), St. George (*V. rupestris*), and A x G1 (Aroman x Ganzin1), are well known nematode-resistant rootstocks. Lake Emerald and Tompa are recently recognized as resistant rootstocks (Mortensten, 1985). Banquabad, Cardinal, Early Muscat, Joazbeli, Loose Perlette and Reisling (all commercial varieties) have been recognized as resistant to *M. javanica* (Anon, 1989). Black Champa, Dogridge, 1613, Salt Creek, Cardinal, Banquabad were recognized as resistant to *M. incognita.* Deshmukh *et al.* (2004) reported that grape varieties Black Champa, Dogridge and Degrasset were resistant, while Pandhri Sahebi and Champanel were moderately resistant to *M. incognita.*

The grapevine rootstocks 1613, Dogridge and Saltcreek are resistant, while 1616 and Talaki 5-A are moderately resistant to the root knot nematode (Parvatha Reddy and Singh, 1984).

9.2.9.6. Integrated methods

i) Two bioagents: The combined application of bioagents *Purpureocillium lilacinum* and *Pasteuria penetrans* was most effective in increasing the fruit yield of grapevine besides checking *M. incognita* with benefit: cost ratio of 3.3 (Sivakumar and Vadivelu, 1999).

ii) Cultural and chemical: Soil application of Carbofuran 3G @ 7 g/m^2 at flowering and growing garlic as an intercrop was found effective.

Pruning (during July) and soil application of 4g talc formulation of *P. fluorescens* (containing 15 x 10^8 cfu/g) around each root-knot infested grapevine at 15 cm depth in the basin significantly reduced root galling due to *M. incognita* (39%), number of egg masses (250%) and increased fruit yield (166%) (Shanthi *et al.*, 1998).

iii) Botanical and chemical: Application of Neem cake (200 g/plant) + Carbofuran (10 g/plant) reduced the nematode populations in the soil and roots. Combination of intercropping garlic and application of Carbofuran at 0.1 g a.i. /m^2 reduced nematode population (44.2-47.6%) and increased yield (88.3-105.6%) over control with benefit: cost ratio of 4-5 (Anon, 1993).

The nematode can be effectively managed by integration of reduced doses of Neem cake and nematicide followed by intercropping with onion and garlic which is profitable.

9.3. ROOT-KNOT, *MELOIDOGYNE INCOGNITA* AND WILT, *FUSARIUM MONILIFORMAE* DISEASE COMPLEX

9.3.1. Interaction

In a study carried out under glasshouse on wilt disease expression in the presence of *Meloidogyne incognita* and *Fusarium moniliforme* on grapevine cv. Muscat Hamburg revealed that growth parameters, viz., shoot length, shoot weight, root length and root weight of vines, were found to be significantly affected by the inoculation of *M. incognita* and *F. moniliforme* alone and in combination. Simultaneous inoculation of nematode and fungus resulted in the greatest reduction of shoot length (40.6%) and shoot weight (63.9%). Wilt disease was observed when both *M. incognita* and *F. moniliforme* were inoculated or when the fungus was inoculated alone. The per cent wilt disease incidence (35%) was higher in vines inoculated with nematodes a week prior to fungal inoculation (Table 9.5) (Senthilkumar and Rajendran, 2003).

Table 9.5 Interaction of *Meloidogyne incognita* and *Fusarium moniliforme* on grapevine cv. Muscat Hamburg

Treatment	Final soil nema popn./250 g	Root gall index*	Wilt disease incidence (%)
F. moniliforme alone	--- (-0.3)[a]	--- (-0.3)[a]	16.0
Inoculation of *F. moniliforme* one week prior to *M. incognita*	2898.8 (3.5)[b]	2.3 (0.4)[b]	25.5
Inoculation of *M. incognita* one week prior to *F. moniliforme*	3353.8 (3.6)[c]	3.8 (0.6)[cd]	35.0
Simultaneous Inoculation of *M. incognita* and *F. moniliforme*	3583.8 (3.6)[d]	3.3 (0.6)[c]	30.0
Uninoculated control	--- (-0.3)[a]	--- (-0.3)[a]	---
CD (P = 0.05)	0.06	0.00	---

Figures in parentheses are per cent decrease over control

*Figures in parentheses are log (x+05) transformed value.

9.3.2. Management

9.3.2.1. Integrated management: Soil application of *Pseudomonas fluorescens* at 100 g and FYM at 20 kg/vine gave effective management of disease complex and improved the plant stand by reducing the final soil nematode population (56.9%), root gall index (1.8) and disease incidence (15.67%). This treatment also increased the number and weight of fruit bunches (17.83 and 155.40%, respectively) and fruit quality (more TSS – 13.53 Brix, TSS: acid ratio – 14.87, lower acidity – 0.91%). The bunch weight of grapevine increased by 155.4% compared to untreated control (Table 9.6) (Senthil Kumar and Rajendran, 2004).

Table 9.6. Efficacy of Pseudomonas fluorescens and *Trichoderma viride* against *Melodogyne incognita* and *Fusarium moniliforme* in grapevine cv. Muscat Hamburg under field conditions.

Treatment	Root gall index	Incidence of wilt (%)	No. of bunches	Bunch wt. (kg)*
FYM (20 kg)	3.2	54.2	49.2	13.5 (10.4)
FYM (20 kg) + *Tv* (100 g/vine)	2.8	32.6	57.4	17.2 (51.0)
FYM (20 kg) + *Pf* (100 g/vine)	1.8	24.8	72.8	29.1 (155.4)
FYM (20 kg) + *Tv* (50 g/vine) *Pf* (50 g/vine)	2.4	35.4	60.2	20.4 (79.5)
Carbofuran 3G (60 g/vine)	1.8	51.2	53.6	15.0 (31.6)
Control	4.6	56.0	45.6	11.4
CD (P = 0.05)	0.72	1.64	1.83	0.52

FYM - Farm Yard Manure: Tv - *Trichoderma viride*; Pf - *Pseudomonas fluorescens*.
*Figures in parentheses are per cent increase over control.

9.4. RENIFORM NEMATODE, *ROTYLENCHULUS RENIFORMIS*

The reniform nematode is a serious problem on Anab-e-Shahi variety of grapevine.

9.4.1. Symptoms

Feed on and damage secondary and feeder roots. Affected roots show brownish discoloration. As a result, nutrient uptake is affected and vine appears sick.

9.4.2. Interaction with crown gall disease, *Agrobacterium tumefaciens*

Lele *et al.* (1978) described an interaction between *R. reniformis* and *A. tumefaciens* in grapevine. Initial damage by the reniform nematode facilitated entry, establishment and disease development.

9.4.3. Management

9.4.3.1. Chemical methods: Soil application of Metham Sodium at 40 liters per ha is effective and increased the bunch weight by 63 per cent.

9.4.3.2. Host resistance: Grapevine cultivars such as Dakshi, Joazbeli and Mukchilani are resistant to the nematode.

9.5. ROOT-LESION NEMATODES, *PRATYLENCHUS* SPP.

Pratylenchus vulnus commonly found on grape is economically most important among lesion nematodes and its occurrence and severity is more in heavier soils.

Simultaneous inoculation of *P. vulnus* and *Xiphinema index* at 1000 and 500 nematodes, respectively per vine (cv. Thompson Seedless) enhanced the stunting of shoot and root as compared to individual inoculation of either nematode alone.

9.5.1. Distribution

The nematode exists in almost every grape producing region. Different species of this nematode prevail in grapes in different parts of the globe. The most prevalent are *P. vulnus, P. scribneri, P. brachyurus* and *P. minyus*. *Pratylenchus coimbatorensis* has been reported from South India.

9.5.2. Symptoms

Plants infected with *P. vulnus* show loss of vigor and reduction in fruit production. Infected young vines remain very weak, often fail to establish root system and eventually die. Below-ground symptoms on roots distinctly shows lesions, which are initially brown and later turn black. In severe infection, black lesions combine and girdle the roots.

All the stages of these migratory endoparasitic forms are infective. The nematodes remain confined to cortical tissue. Affected plants exhibit non-specific symptoms like poor growth and low yields. Young vines if infected, are unable to develop a root system and ultimately die. The roots system of matured vines is badly affected giving rise to a matted, witch's broom effect in some parts of the root system.

9.5.3. Histopathology

Infected plants have generally reduced root system. Nematode penetration has been found restricted to few layers of cortical cells only. Adult female and eggs were not found in meristematic and vascular tissues. Inoculation of 1000 *P. vulnus* per vine has reported to be pathogenic and resulted in suppression of root and shoot growth. *P. vulnus* infection resulted in reduction in potassium and zinc uptake, which was measured in leaf blade and petiole of vines.

9.5.4. Management

- » Post-plant treatments with Nemacur (Fenamiphos) can be effective.
- » Hot water treatments (5 min at 51.7 ^{0}C) eradicates this species from grapevine roots.
- » Some walnut rootstocks are reputed to show tolerance. e.g., hybrids of *Juglans hindsii* and *Juglans regia* - cv Paradox.

9.6. RING NEMATODE, *MESOCRICONEMA XENOPLAX*

9.6.1. Symptoms

In grape, the nematodes feed on young roots, usually near the root tip, often in large numbers. The feeding results in local darkening and destruction of the root tissue, and leads to breakdown of the roots. The root systems of plants attacked by these nematodes are visibly smaller than those of nematode-free plants. Significant reductions in top growth can occur (Klingler, 1975).

The other symptoms include:

- » Reduction of vine growth associated with loss of feeder roots.
- » Responsible for significant root damage and related vigor and yield reduction.
- » Yield reductions are significant when populations approach 6-8 nematodes/g soil.

9.6.2. Life cycle

The female produces 3-5 eggs/day. The first molt occurs inside the egg. Observations of egg maturity in 10-12 days at 20-22 ^{0}C (Seshadri, 1965) and hatch in 15 days at 20 ^{0}C (Thomas, 1959). All stages feed and the life cycle is complete in 24 to 30 days.

9.6.3. Management

Ferris *et al.* (2004) found that spring and fall nematicide applications to orchards on both Lovell and Nemaguard rootstock reduced the nematode degree-days experienced by the trees by about 70%. Spring and summer applications of fenamiphos appeared to be most effective in moderating the rate of increase of *M. xenoplax* dosage. The applications of fenamiphos in the spring and fall are more effective in reducing ring nematode populations than applications in the fall alone (Ritchie, 1989). Other management strategies include:

- » Grow non-host cover crops – Sudan, barley, vetch and alfalfa.
- » Use resistant rootstock GRN-1. Population densities of *M. xenoplax* were lowest in vines on 420A Mgt (≤0.08 g^{-1} soil) and 101-14 Mgt (≤0.24 g^{-1} soil) rootstocks. Rootstock 420A Mgt produced good vine vigor and yield and showed the highest degree of *M. xenoplax* resistance Pinkerton *et al.*, 2005).

9.7. DAGGER NEMATODE, *XIPHINEMA INDEX, X. AMERICANUM*

9.7.1. Distribution

This nematode is a pest of grape vine worldwide. Though number of *Xiphinema* spp. have been found from grape vines, only two viz., *X. index* and *X. americanum* are found to be of pathogenic significance. While *X. index* has been reported from Europe, South Africa, Victoria, USA, North Africa, Iraq and South America; *X. americanum* has been recorded from Indian vineyards.

9.7.2. Symptoms

The dagger nematode, *X. index,* feeds on root tips causing swelling and bending as well as many dead feeder roots. If population levels are high; multiple prolonged attacks can result in darkened, necrotic spots that spread through the root tip. The above-ground symptoms include poor growth, stunting of the plant, yellowing or wilting of the foliage, and reduced root systems (root necrosis, lack of feeder or secondary roots, and occasional tufts of stubby rootlets).

The nematode is highly damaging to grapevine seedlings where it feeds as a migratory ectoparasite. *X. index* causes characteristic 'Curly tip' symptom on grape seedlings. Infested roots are necrotic, lack laterals, have root tip swellings, show cellular hypertrophy and multinucleate cells near nematode feeding site. This results in reduced root system.

9.7.3. Histopathology

Epidermal and outer cortical cells collapse at feeding site and show necrosis. Multinucleate enlarged cells have been found beneath the layer of necrotic cells. *X. index* rarely preferred feeding on root tip.

9.7.4. Management

9.7.4.1. Cultural methods: Certified, nematode-free rootings should be grown for new vineyards or replants. Intercropping of marigold (*Tagetes patula*) reduces nematode population and improves the yields significantly. Old root system should be removed.

i) Biofumigation: Results from vineyard with a high incidence of grapevine-fanleaf virus and its vector nematode species *Xiphinema index* (Bello *et al.*, 2004), suggested that biofumigation may be useful to reduce the fallow period necessary to eliminate vector nematodes in uprooted vineyards.

In the fall, the seed bed is planted with Brassica crops high in glucosinolate. In the spring, Brassicas are green manured 10 to 15 cm deep in the soil. A follow-up irrigation cycle speeds up the Brassica decomposition and aids microbes in the decomposition process. Gases released from the plant material can kill nematodes, weed seeds, and other disease organisms.

Vineyard test results are yielding healthier vines and fewer dagger nematodes. The holistic farming plan is showing good vine growth, low or no presence of dagger nematodes, and noticeably improved soil structure. Biofumigation is elevating soil health while providing the vine with nutrients for improved pest tolerance and resistance.

ii) Crop rotation: In France, to save grapevines from attack by *X. index* and subsequently infection by grapevine fan leaf virus, a 7-year rotation with cereal and alfalfa crop has been recommended (Lamberti, 1981).

9.7.4.2. Integrated methods: The most effective control of both *Xiphinema index* and *Meloidogyne javanica* was obtained with Carbofuran at 0.2 g/L. Integration of Neem seed powder at 4.08 g/pot + Fertinemakil fertilizer at 2.5g/pot and Marigold + Fertinemakil is effective against *X. index* and *M. javanica.*

9.8. INTERACTION OF DAGGER NEMATODE WITH GRAPEVINE FANLEAF VIRUS

Grapevine fanleaf virus degeneration is the oldest known disease of *Vitis vinifera*. Records in Europe date back 200 years and specimens in herbaria displaying symptoms predate the introduction of American rootstocks (Martelli and Sovino, 1988). The observation that *Xiphinema index* transmits grapevine fanleaf virus (GFLV) was the first record of virus transmission by a nematode (Hewitt *et al.*, 1958). Grapevine fanleaf virus (GFLV), which is transmitted by the ectoparasitic nematode *X. index*, is causing $1 billion losses to the grapevine industry in France.

GFLV is naturally transmitted by a single nematode species, *X. index* (Andret- Link *et al.*, 2004). *X. index* transmits the fan leaf yellow mosaic vein banding virus (GVFV-GVYMV) disease of grape. Leftover roots of dead and old plants in vineyards act as source of viral inoculum of dagger nematode year after year. *X. americanum* has been found transmitting tobacco mosaic virus, grape yellow vine virus, peach rosette mosaic virus and tomato ring spot virus.

The virus is lost during the molt of the cuticle between life stages. The virus is not passed through the egg stage. Consequently, the virus is re-acquired by feeding of each vermiform life stage of the nematode. GFLV causes reduced vigor, lack of fruit set, and reduced yield of grapevines. Of great importance in the design of control programs for the *X. index*/GFLV problem is that portions of the grape root system can remain alive and serve as a reservoir for virus and nematode for at least 5 years after vine trunks have been removed (Raski *et al.*, 1965).

The grapevine fanleaf virus is a Nepovirus with isometric particles of 30nm in diameter. It is transmitted by *X. index* and *X. italiae*, with *X. index* the more efficient vector (Argelis, 1987; Martelli and Savino, 1988).

The virus can be acquired in less than 15 minutes of feeding on an infected plant and is retained up to eight months in the absence of feeding (Taylor and Robertson, 1975).

The retention site for GFLV in *X. index* is the cuticular lining of the esophagus, including the area surrounding the odontophore (Taylor and Robertson, 1975). As with other viruses vectored by Longidorid and Trichodorid nematodes, it is selectively and specifically adsorbed at the retention site, indicating a specific association between protein coat of virus and cuticular surface. Dissociation with cuticle probably occurs as glandular secretions pass forward into the plant cell (Taylor and Robertson, 1975).

The virus does not pass through the egg stage. It is not retained through a molt when the esophageal lining and odontostyle are also shed. In *Xiphinema*, although the lining of the esophagus is not shed at a molt, it undergoes structural changes and virus particles may pass into the intestine (Taylor and Robertson, 1975).

The virus does not replicate in the nematode vector.

9.8.1. Detection

From a management standpoint, it is important to determine whether *X. index* is present in a site intended for a vineyard, and whether the nematodes are viruliferous.

Virus particles can be detected in extracts from single nematodes by immunosorbent electron microscopy (Roberts and Brown, 1980). However, as indicated by Esmenjaud *et al.* (1993) the procedure is complex and not readily adapted to routine assays.

Enzyme-linked immunosorbent assay (ELISA) methods are routinely used to determine GVFL in plant tissue (Walker *et al.*, 1994). When applied to nematodes, the techniques appear sufficiently sensitive to detect the virus if more than ten viruliferous individuals are used, but are unreliable for single individuals (Esmenjaud *et al.*, 1993).

9.8.2. Symptoms

Martelli and Savino (1988) describe three distinct syndromes or symptoms of grapevine fanleaf degeneration:

9.8.2.1. Malformations: Leaves are variously distorted, asymmetrical, with wide petiole sinuses and abnormal vein arrangement that results in the appearance of an open fan. Leaf symptoms may also include chlorotic mottling. Foliar symptoms are most evident in the spring and through the vegetative growth period. Shoots show abnormal branching, with double nodes and short internodes, while forming angles at internodes. Bunches are small with poor fruit set and shot berries. They ripen irregularly.

9.8.2.2. Yellow mosaic: Bright yellow discoloration is observed on leaves, shoots, tendrils and inflorescences in the spring. Foliage and clusters are not malformed, but clusters are small.

9.8.2.3. Vein banding: Yellow flecking along main veins and associated interveinal areas of mature leaves appear in mid to late summer. Leaves are not malformed, but fruit set is poor and yield is low.

Several closely related nepoviruses found in grapevines are apparently related to GFLV but are serologically different. They are sap-transmissible, and where vectors have been definitively established they are species of *Xiphinema* and *Longidorus*. They include arabis mosaic virus, raspberry ringspot virus, tomato blackring virus, grapevine chrome mosaic virus, Strawberry latent ringspot virus, artichoke Italian latent virus, and grapevine Bulgarian latent virus (Argelis, 1987; Stellmach and Goheen, 1988).

One or more of ten species of *Xiphinema* are present in all major grape-growing regions of the world. They are *X. algeriense*, *X. americanum*, *X. brevicolle*, *X. diversicaudatum*, *X. index*, *X. italiae*, *X. mediterraneum*, *X. pachtaicum*, *X. turcicum*, and *X. vuittenezi* (Raski, 1988). With *X. index* a notable exception, most species of *Xiphinema* have a wide host range and are adapted to a wide range of soil textures.

X. americanum is actually considered a group of some 39 species, which include *X. brevicolle* and *X. pachtaicum*. Molecular techniques (Vrain, 1993), biological evaluations (Halbrendt and Brown, 1993) and virus transmission studies (Brown *et al.*, 1993) are continuing to underscore the variability and similarities within the *X. americanum* group. These studies are adding to the body of knowledge that will be required for an eventual taxonomic revision. The group also includes *X. californicum*, which is probably associated with grape in California (Lamberti and Ciancio, 1993).

9.8.3. Management

In France, to save grapevines from attack by *X. index* and subsequently infection by grapevine fan leaf virus, a 7-year rotation with cereal and alfalfa crop has been recommended (Lamberti, 1981).

9.8.3.1. Management guidelines for GVFLV - pre-plant

i) Avoidance: The wide host ranges of ectoparasitic nematodes make pest avoidance difficult to achieve. One exception is *X. index* and GFLV. The use of virus-free and nematode-free rootings provide the best method of controlling the complex and certification programs have reduced its dissemination through vegetative propagation (Esmenjaud *et al.*, 1993). The wide host range and current abundance of *X. americanum* in U.S. soils is an indication that quarantine of the nematode is not possible, but the certified virus-free nursery stock program continues to have merit. On-farm nurseries should be maintained nematode-free and located away from old vineyard sites.

ii) Cultural methods: By their nature, ectoparasitic nematodes can survive in soil or persist on weed hosts for long periods of time. Fallowing will only reduce populations, not remove them. The persistence of old grape roots in replant sites is a problem. New grape roots tend to follow the channels in the soil left by old roots. GFLV can be moved from these reservoirs to healthy planting stock through exploratory feeding probes by *X. index.* It may be difficult to develop 'immune' rootstocks on which the nematode cannot reproduce and never even attempts to feed. Replanting of a virus-infected vineyard with a rootstock that is resistant to *X. index* within 6-7 years after removing a virus infected vineyard will not be successful in the long term unless the nematode does not attempt to feed. The site will probably not be safe until 1-2 years after the last old root is dead. Such time intervals may not be economically feasible. Resistance to GFLV will be an important solution to the problem (Esmenjaud, 1986).

Old roots are most easily killed before the old vine is removed by using systemic herbicides. Once the vine is removed, the only reliable method of killing roots in the top 2 m of soil is by soil fumigation. Soil ripping to that depth on 30 or 60 cm centers may increase the speed of root death, but this tactic has not been tested in the field. Heat is a useful root killing agent but we are unable to deliver the required heat using current methods. Soil flooding for months will not kill old grape roots or nematodes. Selected rotation crops will reduce population levels of certain nematode species over time. This will have little value for endoparasites but should be tested with ectoparasites.

Certain *Sorghum bicolor* x *Sorghum sudanense* hybrids grown in summer have negative effects on ectoparasites but the number of years required to eliminate a nematode problem is unknown. Addition of manure, composts and ammonia does not kill old roots, but there are various reports of the benefits of ammonia as a nematode control agent (Mojtahedi and Lownsbery, 1976).

9.8.3.2. Management guidelines for GVFLV - Established vineyards

i) Avoidance: Movement of contaminated grape harvesting equipment and tractors from a site infected with *X. index* should be avoided.

ii) Cultural methods: The diseases may be prevented by planting vineyards with virus disease-free grapevines in clean soil, and by replanting diseased vineyards only after a long fallow period of more than ten years.

By reducing vine stress through more frequent irrigation the damage caused by nematodes can be reduced.

The use of grassy cover crops in vineyards infested with *X. index* should be studied. However, legume cover crops in vineyards should be monitored to assess population levels of *Mesocriconema xenoplax,* which may increase. Populations of *X. americanum* will also increase on most cover crop selections. The difficulty in choice of cover crops is that the host range of most ectoparasites is quite broad.

Nematode species with long bodies tend to be in shallow and non-disturbed sites of a vineyard, so placement of any treatment is important. Tillage and soil disturbance can reduce population levels for short periods of time, but root surface area is also reduced.

Drip-irrigation-applied fertilizers that release ammonia may reduce population levels of ectoparasitic nematodes when applied repeatedly. Fifteen kg/ha of nitrogen in urea salt when re-applied three to five times at 30 to 45 day intervals can reduce population levels of most ectoparasites by half. More field testing of these strategies is necessary. Since grapevines do not have a high nitrogen requirement and some vineyards already have excess nitrogen, this technique should be tested and implemented with caution.

iii) Chemical methods: Organophosphate and carbamate nematicides are lethal to ectoparasitic nematodes when used as single treatments at high rates via drip irrigation. Treatments with currently-available commercial nematicides only reduce populations about 50% for 6 to 8 months after treatment. Multiple treatments with low rates of fenamiphos (1 kg/ha) are ineffective against ectoparasites. When multiple treatments are used against endoparasites for several years, population levels of ectoparasites such as *X. americanum* can be observed to increase above the non-treated population levels.

In the case of ectoparasites, the value of systemic nematicides will be minimal unless the toxicant is available during feeding or leaks out into the rhizosphere (Edwards, 1991).

iv) Biological methods: The addition of biological agents to soil has been inconsistent and usually ineffective. It may be possible, however, with the advent of drip irrigation systems, to apply microbe-produced toxins directly to the soil in irrigation water. Zoosporic fungal parasites have been isolated from *X. rivesi* and *X. americanum.* Such fungi require free water for their movement through soil. Their occurrence suggests research to investigate water management protocols that will enhance fungal parasitism of dagger nematodes.

v) Host resistance: Grapevine rootstocks 039-16 (*V. vinifera* x *V. rotundifolia*) and 043-43 have been patented by the University of California and are available through grape rootstock nurseries for use in GFLV-infested vineyards (Table 9.7). Their long-term use is still under study, but preliminary data suggest that they be recommended for use and distribution as rootstocks providing field resistance to infectious degeneration. Resistance to infectious degeneration is complex and could act at several levels. It could result from the root's ability to prevent nematode feeding, or root cells' ability to limit or stop GFLV spread. An ideal rootstock would be resistant not only to the feeding of dagger nematode but also to GFLV, providing two levels of resistance. Then, even a chance probing and transmission of

GFLV by the nematode would not result in disease because the rootstock would be able to prevent or limit spread of the virus. Researchers at University of California, Davis, are continuing to look for resistance to GFLV with the hope of crossing GFLV-resistant plants with known sources of dagger nematode resistance to produce the next generation of rootstocks to combat infectious degeneration.

Table 9.7. Performance of Cabernet Sauvignon scion on selected rootstocks growing on GFLV-infested site in NAPA Valley

Grapevine rootstocks	Fruit yield/plant (lbs)	Resistance to GFLV
039-16	26.4	Resistant
043-43	24.95	Resistant
AxR#1	16.0	Susceptible
Harmony	9.1	Susceptible
St. George	6.3	Susceptible

vi) Future research: Except for *X. index,* ectoparasitic nematodes have not received the same research attention as endoparasites. Long-term experiments on the damage potential and management of ectoparasitic nematodes in vineyards are needed.

The apparent increase in incidence of *X. index* associated with reduced use of conventional soil fumigants is important and should be monitored. Other problems with ectoparasitic nematodes may similarly emerge. New pre-plant control strategies will be needed in the near future. Resistance and tolerance of rootstocks will continue to be important areas of study.

9.9. TOMATO RINGSPOT VIRUS IN GRAPEVINE

9.9.1. Symptoms

Tomato ringspot virus degeneration differs geographically, with the disease more severe in colder regions. Leaves are mottled, with oak leaf patterns, new growth is weak and buds are susceptible to cold damage. Internodes are shortened, leaf area is reduced, and clusters are small. The grape yellow vein symptom, common in infections with this virus in California, consists of yellow flecking along the veins and associated lamina. There is reduced fruit set and slow decline of the vine (Gonsalves, 1988).

Symptoms are difficult to diagnose early in the season unless vines are severely affected, in which case they have many winter-killed buds and weak, stunted shoot growth. By about 9 weeks after the start of vine growth, shoot and foliage symptoms are conspicuous on one or more shoots. Leaves develop ringspots and mottling, are reduced in size, and rosetted due to the shortening of internodes (Yang *et al.*, 1986). Fruit clusters are reduced in size with many berries aborting. Removal of bark from trunks and stems of diseased vines may reveal thickened, spongy phloem tissue with numerous necrotic pits (Uyemoto, 1975).

9.9.2. Management

The use of resistant cultivars (e.g. for grapes) and the use of healthy planting material can reduce the disease. In addition, it is necessary to achieve good control of weeds.

9.10. PEACH ROSETTE MOSAIC VIRUS IN GRAPEVINES

Peach rosette mosaic virus decline is a disease of *V. labrusca* grapevines that occurs only in Michigan in the U.S. The virus also causes a disease in peach, as indicated by the name (Ramsdell, 1988).

9.10.1. Symptoms

Symptoms of peach rosette mosaic virus appear 3 to 4 years after infection. The plant canopy is umbrella-like with shortened and crooked internodes. Leaves are misshapen with a flattened base. Clusters are poorly formed and may shell berries. Bud burst is sometimes, but not always, delayed. Infected vines exhibit reduced vigor and are predisposed to winter injury and death (Ramsdell, 1988). Vines may die after several years. Over time, a circular pattern of missing or dead vines is commonly observed in mature vineyards. This disease has also been called grapevine degeneration, grapevine decline, berry shelling disease, and delayed budding disease.

9.10.2. Transmission

This nepovirus is a strain of TRSV transmitted by *X. americanum* and *Longidorus diadecturus* with the characteristics of other nepoviruses transmitted by these genera. The virus may be endemic in vineyard weeds, which complicates exclusionary measures.

9.10.3. Management

Plant healthy stock. Use planting material certified free of peach rosette mosaic nepovirus (PRMV). Clones of most rootstocks and cultivars that have been tested and found to be free of all known viruses are available.

There is no way to cure an infected vine. Remove and destroy virus-infected vines. Top-grafting is not advisable, as rootstocks will be infected.

Clean pruning equipment with a 1/10 dilution of household bleach between vines.

9.11. FUTURE THRUSTS

The following priorities for research concerned with grapevine nematology are proposed.

- Consolidation and standardization methods for sampling vineyards, and for extracting and identifying nematodes in the samples. Such methods should then be adopted universally.
- Establishment of better defined relationships between population size and damage potential for various nematodes. This will involve collection of information on the influence of factors such as climate, characteristics of host plants and soil features (including soil microbial status).
- Assessment of nematode distribution, especially in new grape growing regions.
- Collection of information on variation in pathogenicity between and within *Meloidogyne* species, with an emphasis on investigation of the incidence, distribution and likely spread of virulent, or 'resistance-breaking', populations.
- Further assessment of rootstocks presently used in for their resistance or tolerance to the range of nematodes encountered, especially *Meloidogyne* spp. and the more common *Pratylenchus* spp., using local nematode isolates.
- Extending the presently available range of nematode resistant rootstocks by local breeding and by importation, taking into consideration essential rootstock characteristics other than nematode-resistance.
- Exploring and utilizing the potential of genetic engineering for introducing and strengthening resistance in rootstocks or even in *V. vinifera*.

» Enhancing natural suppressiveness of soils to plant parasitic nematodes and develop methods to sustain such suppressiveness.
» Identifying parasites and predators that limit nematode reproduction, and develop commercially viable methods of using promising organisms as inundative biological control agents.
» Expanding knowledge of nematode-suppressive and nematode-promoting plants to establish guidelines for manipulating inter-row ecology.

9.12. CONCLUSION

Much of the information already available relates to root-knot nematodes, perceived as the most significant nematode pest of grapevines. The dagger nematode, *X. index* has also received some attention, less due to its direct pathogenic effect than to its role in transmitting GFLV. Little is known about most of the other nematode species in regard to both their incidence and potential effects on the grapevine. Lesion nematodes (*Pratylenchus* spp.) are perhaps the group that most deserves future attention, as they are common on grapevines and are frequently associated with decline and replant problems.

Chemical control of nematodes is not a sustainable option in any viticultural situation. Even in grapevine nurseries, a move to use of containers for growth of seedlings is preferable to using relatively ineffective nematicides. Utilization of host-plant resistance imparted by rootstocks continues to be the best form of control, although there are also significant gaps in our knowledge of grapevine and nematode genetics, mechanisms of resistance and species differences in the grapevine-nematode interaction. This lack of knowledge may threaten the durability of rootstock resistance.

There is a need for proper nematode identification coupled with analysis of rootstock responses to nematode infection, to facilitate the selection of tolerant or resistant cultivars and help to lessen dependence on nematicides. The performance of rootstocks previously reported to possess a degree of resistance to some species of *Meloidogyne* must be evaluated under the growing conditions prevailing in the region, and in response to locally occurring nematode species or isolates, before they can be considered resistant and are planted extensively.

Chapter - 10

Passion Fruit, *Passiflora edulis*

10.1. INTRODUCTION

10.1.1. Origin

The edible commercial species of passion fruit originated on the edges of South American rain forests in the Amazon region of Brazil and possibly in Paraguay and Northern Argentina.

10.1.2. Health benefits

The juice is delicious and has an excellent flavor. It is known for its blending quality. Fruits contain good proportion of reducing and non-reducing sugars and acids. Fruits are rich in Vitamin A (1300-2500 IU/100 g pulp), Vitamin C (30 –50 mg/100 g pulp) and minerals such as sodium, magnesium, sulfur and chlorides. Passion fruit is used for urinary infections and as a mild diuretic, digestive stimulant, heart tonic and to treat asthma, whooping cough, bronchitis, coughs, gastric cancer *etc.* There is currently a revival of interest in the pharmaceutical industry, especially in Europe, in the use of the glycoside, passiflorine, especially from *P. incarnata*, as a sedative or tranquilizer. Italian chemists have extracted passiflorine from the air-dried leaves of *P. edulis*. The juice of passion fruits is given as a digestive stimulant and treatment for gastric cancer.

10.1.3. Area and production

Passion fruit is grown in most tropical and subtropical parts of the world, and it is particularly important commercially in Australia, Columbia, USA, Indonesia, Australia, Mexico, Vietnam, Ecuador, Philippines, and India. The total global supply of passion fruit is estimated at 13.71 million tons. Brazil is the world's foremost producer of passion-fruit, with about 60% of the production.

In India, though passion fruit was introduced in early part of twentieth century, but its cultivation was limited to few districts of Karnataka, Kerala and Tamil Nadu. Since last one decade, its cultivation started in some parts of northern India, especially North East states. The area under this crop is rapidly increasing in Manipur, Nagaland, Kerala and Mizoram and there is a great potential for its cultivation in these states. The estimated area and production of passion fruit is 19.01 thousand ha and 123.94 thousand tons, respectively. Manipur and Nagaland are the leading states in passion fruit production contributing 90 percent area and production of the country (Table 10.1). The average productivity of passion fruit in India is around 6.519 tons/ha which is much lower than countries like Brazil, Australia, Colombia *etc.* where productivity is 30-35 tons/ha. There is immense potential for passion fruit cultivation in North Eastern states, parts of Karnataka, Kerala and Tamil Nadu. These regions have all sets of climate and sufficient rainfall for growing passion fruit successfully.

Table 10.1. Passion fruit production and share of production in different states in India (2014-15)

Rank	State	Production ('000 tons)	Share (%)
1	Manipur	99.16	76.71
2	Nagaland	21.25	16.44
3	Kerala	6.75	5.22
4	Mizoram	2.11	1.63
TOTAL		129.27	100

Source: Anon (2014-15).

In Nagaland; Kohima, Mokokchung, Wokha, Phek are main passion fruit producing districts (Table 10.2).

Table 10.2. Area, production, and productivity of passion fruit in Nagaland (2013-14)

State	Area (ha)	Production (tons)
Kohima	1400	2551
Mokakchung	1420	2588
Tuensang	525	957
Wokha	1070	1950
Zunheboto	690	1257
Phek	1225	2232
Mon	280	510
Dimapur	30	55
Peren	262	477
Longleng	155	282
Kiphire	483	880
TOTAL	7540	13741

Source: Jt. Director of Horticulture, Nagaland, Kohima

10.1.4. Varieties

There are more than 500 species of *Passiflora* in the family Passifloraceae, but only one *Passiflora edulis*, is known as passion fruit. There are two distinct forms of *Passiflora edulis*, the standard yellow (*Passiflora edulis* f. *flavicarpa*) and the purple (*Passiflora edulis* f. *edulis*), differing in acidity and starch content. The yellow are more acidic and less starchy while the purple are less acidic and more starchy. Both varieties, viz., purple (*P. edulis*) and yellow passion fruit (*P. edulis* var. *flavicarpa*) are of commercial importance. The hybrids of these two have also been developed for cultivation.

'Kaveri' hybrid passion fruit variety is a high yielding hybrid developed from Central Horticultural

Experiment Station, Chethalli, Karnataka (Regional Station of ICAR-Indian Institute of Horticultural Research, Bangalore), in 1986. This is a cross between purple and yellow varieties. Each plant bears 40-60 fruits per annum and produces 200 tons yield/ha over a three-year cropping period. Fruits are ovoid to round and purple dotted. Fruits contain 25-30 per cent juice, 11.5-12.0 percent sugars and 3.0-3.5 mg citric acid/100 ml juice. This is tolerant to Alternaria leaf spot, Fusarium collar rot and root-knot nematodes.

10.1.5. Export from India

Passion Fruit is exported to over 75 countries. In the year 2020-2021 (Apr-Nov), India has exported passion fruit worth of USD 2.3 million.

The top 5 trading partners of India are United Arab Emirates (USD 0.37 million), United Kingdom (USD 0.32 million), Nepal (USD 0.26 million), Thailand (USD 0.24 million), and Viet Nam (USD 0.22 million). The total export value of passion fruit in these countries is USD 1.41 million. These top 5 countries account for over 61.3% of the total passion fruit export from India.

10.1.6. Nematode problems

Although a number of plant parasitic nematodes are reported associated with passion fruit, only root-knot and reniform nematodes are known to cause economic damage. Both nematodes can severely limit fruit production and plant longevity.

10.2. ROOT-KNOT NEMATODES, *MELOIDOGYNE INCOGNITA, M. JAVANICA*

The root-knot nematodes are the major limiting factors in the successful production of passion fruit. In South Africa, *M. javanica* is considered a serious pathogen for both yellow and purple passion fruit, making plant more susceptible to droughts (Villiers and Milne, 1973). This nematode is an emerging problem on passion fruit in Mizoram state of north-eastern region of India.

10.2.1. Economic importance

Root-knot nematode control in South Africa resulted in an increase of up to 42 % in yield of passionfruit (Villiers and Milne, 1973).

10.2.2. Symptoms

Characteristic symptoms of infestation by root-knot nematodes are formation of galls or knots on roots, yellowing of leaves, stunting and eventual wilting of the affected plants. Attacks by *Meloidogyne* spp. are characterized by the formation of root-knots resulting from the toxic action of certain substances produced and injected by the nematode; the root system also becomes deficient, making the absorption of water and nutrients by the plant more difficult. Consequently, plants show lesser growth and foliar yellowing with reduced productivity and longevity. Host differential tests confirmed that the race 2 of *M. incognita* is infecting passion fruit (Khan *et al.*, 2017).

Root-knot nematode (*Meloidogyne incognita*) infected plants displayed stunting, yellowing, root galling and declined symptoms and poor fruit yield on purple passion fruit. The minimum of 10 second stage *M. incognita* juveniles (J2) /200 cc soil can cause reduction of plant growth parameters and highest reproduction (R=87.27) (Khan *et al.*, 2017).

10.2.3. Life cycle

Studies on development of *M. javanica* and *M. incognita* were made by studying their life cycle by cuticle molting. Results showed that the 2nd, 3rd and 4th molts occurred on 22nd, 24th and 26th days, respectively after inoculation. Full-grown females were observed by the 27th day while the eggs of both species were noted on the galls by the 31st day after inoculation (Kanyagia, 1993).

10.2.4. Histopathology

In the histopathological studies, results demonstrated that *M. javanica* and *M. incognita* caused damage to cortical and vascular tissue of passion fruit. The roots were susceptible to both nematode species whose feeding activity induced the formation of multinucleate giant cells by the root around the nematode head (Kanyagia, 1993).

10.2.5. Interaction with other pathogens

The purple passion fruit is more susceptible to the disease complex involving the root-knot nematode and *Fusarium solani* fungus.

M. incognita and *M. javanica* interact synergistically with *F. nirenbergiae* and *Neocosmospora* sp., worsening plant damage (Mangeiro *et al.*, 2022).

10.2.6. Management

The use of healthy seedlings, crop rotation with plants that are not hosts or poor hosts to nematodes, solarization, fallow and nematicides are recommended measures to control nematodes (Fischer *et al.*, 2008).

10.2.6.1. Cultural methods: Crop rotations should be useful for nematode control (Milne, 1982a), since the vine is relatively short lived and seedling establishment is of great importance. Rotation with cassava, cereals, maize, baby corn, sweet corn, sweet potato, onions, cabbage and kale was found effective.

10.2.6.2. Chemical methods: Soil treatment with Furadan (3.2 kg a.i. /ha), Nemacur (3.2 kg a.i./ ha) and Rugby (6.4 kg a.i. /ha) caused significant reductions in nematode population and significant increase in fruit yield (Kanyagia, 1993).

10.2.6.3. Host resistance: Passion fruit cvs. Yellow and Kaveri were found highly tolerant to *M. incognita*.

Use of rootstock such as *Passiflora caerulea*, which is tolerant to root-knot nematodes, has been suggested (Milne, 1982a; Terblanche *et al.*, 1986).

Grafting of purple passion fruit scion on yellow passion fruit rootstock can be used for the management of root-knot nematode.

10.3. INTERACTION OF *MELOIDOGYNE INCOGNITA* WITH *FUSARIUM SOLANI*

10.3.1. Interaction

Meloidogyne incognita is responsible for decreases in vegetative development of plants of the genus *Passiflora* (El-Moor *et al.*, 2009). *Fusarium solani* is considered the main causal agent of collar rot disease in passion fruit plants. According to Ponte *et al.* (1998), this fungus is present in all passion fruit producing states of Brazil, causing death of the plants even before they reach two years of age. Plants infected with both *Meloidogyne incognita* and *Fusarium solani* resulted in early death of plants in passion fruit orchards in northern Minas Gerais, Brazil (Dariva, 2011). Several studies on the interaction of *F. solani* with *M. incognita*, confirmed that the nematode makes the plants more prone to infection by the fungus (Rocha, *et al.*, 2018, Gomes, *et al.*, 2011, Silva and Pereira, 2008).

10.3.2. Management

The management of diseases caused by soil phytopathogens, such as *F. solani* and *M. incognita*, is done mainly by using resistant cultivars. However, the interaction of *F. solani* with *M. incognita* can change the resistance reaction in genotypes of passion fruit plants, making the use of these materials difficult and causing low crop yields. The presence of this nematode can increase damages in species susceptible to fungi and changes the phenotype of resistance in resistant species to a phenotype of susceptibility (Fischer *et al.*, 2010).

Wild *Passiflora* spp. are commonly the present sources of resistance to *Meloidogyne* spp. (Castro *et al.*, 2010) and could be used as rootstocks for commercial varieties, allowing for the control of this pathogen. A higher resistance to the collar rot disease has been observed in *P. alata, P. giberti, P. quadrangularis, P. macrocarpa, P. caerulia*, and *P. nitida* (Fischer *et al.*, 2010).

The genotypes BRS Sol do Cerrado, M-19-UFV, *P. alata*, *P. cincinnata*, and *P. giberti* were found resistant to the nematode, had 100 % plant survival, regardless of the inoculation with *F. solani* or *M. incognita* race 3; while the genotypes *P. ligularis, P. morifolia, P. mucronate, P. sharp*, and *P. setacea* were found susceptible. The genotypes M-19-UFV and *P. cincinnata* have potential for use as rootstocks for passion fruit plants for the control of *M. incognita* race 3 and *F. solani* (Rocha *et al.*, 2021).

Table 10.3. Plant survival of genotypes of passion fruit plants (*Passiflora* spp.) inoculated with *Fusarium solani* (FS) and *Meloidogyne incognita* race 3 (Mi) (Rocha *et al.*, 2021).

Genotypes	Plant survival (%)		
	F. solani	*M. incognita*	*F. solani* + *M. incognita*
BRS Sol do Cerrado	100.00	100.00	100.00
M-19-UFV	100.00	100.00	100.00
P. alata	100.00	100.00	100.00
P. cincinnata	100.00	100.00	100.00
P. giberti	100.00	100.00	100.00
P. ligularis	66.67	0.00	0.00
P. morifolia	100.00	0.00	0.00
P. mucronata	100.00	0.00	0.00
P. sharp	100.00	0.00	0.00
P. setacea	100.00	0.00	0.00

10.4. RENIFORM NEMATODE, *ROTYLENCHULUS RENIFORMIS*

10.4.1. Symptoms

Reniform nematodes cause chlorosis and wilting of leaves, stunted growth, and poor root system, which may present rotted rootlets. Seedlings and mature plants may succumb to heavy parasitism (Silveira and Araújo, 2011). Passionfruit seedlings growing in naturally infested soil were smaller, had chlorotic leaves and darker roots than plants growing in steamed soil in pot studies.

10.4.2. Interaction with other pathogens

R. reniformis interact synergistically with *F. nirenbergiae* s (= *F. oxysporum* f. sp. *passiflorae*) and *Neocosmospora* sp. (= *F. solani* f. sp. *passiflorae*), worsening plant damage (Mangeiro *et al.*, 2022).

In Brunei, nematode has been quoted as an agent predisposing plants to crown rot caused by *Phytophthora* spp. (Perregrine and Yunton, 1980).

Based on field observations in Fiji, Kirby (1978) suggested that *R. reniformis* could be involved in the decline in the productivity of passion fruit orchards, which was initially attributed only to *Phytophthora cinnamomi* and *P. nicotianae* f. sp. *parasitica*. Field observations in Brunei, Brazil, and Venezuela, also led to suggestions that *R. reniformis* increases the severity of *P. nicotianae* f. sp. *parasitica*, *F. nirenbergiae*, and *Fusarium* sp. (Peregrine and Yunton, 1980; Liberato, 2002; Castellano *et al.*, 2012).

10.4.3. Management

10.4.3.1. Chemical methods: The plant life is doubled when infested soil is treated with fenamiphos prior to planting.

Chapter - 11

Litchi, *Litchi chinensis*

11.1. INTRODUCTION

11.1.1. Origin

Litchi appears to be native of southern Province of China and northern Vietnam.

11.1.2. Area and production

The world litchi production is estimated about 3.5 million tons of which 80% (2.8 million tons) of litchi was produced in China in 2018. The other major producers are India (0.67 million tons), Vietnam (0.38 million tons), Madagascar (0.10 million tons), and Thailand (0.048 million tons). Other countries growing litchi include Bangladesh, Australia, South Africa, Reunion, Brazil, and Israel.

India ranks second in the world next to China in litchi production. Litchi is very specific for its climatic requirements and Tripura is the most suitable for its cultivation. So litchi cultivation is coming on in Dehradun of Uttarakhand, North Bihar, Jharkhand, Tripura, West Bengal, Assam, Odisha, and Punjab. Tripura is the second largest producer after Bihar in India.

The area under litchi cultivation is expanding every year from 49,300 hectares in 1991-92 to 93,300 hectares in 2016-17. Total production of litchi also boosted from 243,800 to 568,200 metric tons in the same time period. However, the productivity of litchi is stagnant about 6.1 t ha^{-1}.

State-wise area, production and productivity of litchi in India during 2021-22 (Third advance estimate) is presented in Table 11.1.

Table 11.1. State-wise area, production and productivity of litchi in India (2021-22 Third advance estimate)

State	Area ('000 ha)	Production ('000 tons)	Productivity (tons/ha)
Arunachal Pradesh	0.07	0.09	1.10
Assam	6.13	60.58	9.89
Bihar	36.67	308.06	8.40
Chhattisgarh	8.48	58.91	6.95

Haryana	0.27	3.32	12.47
Himachal Pradesh	6.35	6.18	0.97
Jammu & Kashmir	1.12	2.24	2.00
Jharkhand	4.15	65.37	15.75
Manipur	0.06	0.32	5.75
Mizoram	0.45	1.85	4.11
Nagaland	0.57	3.39	5.91
Odisha	4.37	24.18	5.53
Punjab	3.65	60.39	16.53
Sikkim	0.29	0.97	3.30
Tripura	0.95	3.28	3.45
Uttar Pradesh	4.52	39.37	8.72
Uttarakhand	10.72	24.72	2.31
West Bengal	10.35	74.00	7.15
India	99.17	737.20	7.43

11.1.3. Improved varieties

Shahi, China, and Purbi are ruling varieties for north Bihar and eastern U.P.; Rose Scented for Uttaranchal and adjoining area; Shahi, Ajhauli, China, Swarna Roopa and Purbi for Jharkhand; and Bombaiya, Bedana and Rose Scented in eastern parts of West Bengal are the commercial varieties of importance.

A large number of varieties are grown in different parts of country viz. and they are classified as early, mid and late.

11.1.3.1. Early: Muzaffarpur, Rose scented, Shahi, Swarna Roopa, Early Bedana, Dehra Rose.

11.1.3.2. Mid: China, Purbi, Calcuttia, Bombai, Bedana, Deharadun.

11.1.3.3. Late: Late Bedana, Longia, CHES-2, Gulabi.

The brief description of improved varieties of litchi are as presented in Table 11.2.

Table 11.2. Brief description of improved varieties of litchi India

Varieties	Brief description
Bombai	This is an important cultivar of Tripura having vigorous canopy attaining a height of 7-8 m and spreads 8-9 m. This cv. is early maturing (last week of May to first week of June) and yield about 80-100 kg per tree. Fruits are large in size (3.4 cm long and 3.2 cm diameter) obliquely heart shaped, and weighing 14-21 g. The color of ripe fruit is attractive carmine red with uranium green background.
Shaahi	It is a commercial and early variety, which is going popular day by day. The fruits of Shahi are round deep red color and matured during first week of May. The aromatic pulp is the characteristic feature of the fruit. The maintenance of proper soil moisture in soil regime is very important for fruit development.

China	This is late maturing cultivar and plants are comparatively dwarf. The fruits of China are large, conical and free from cracking problem. It is much popular in litchi growing areas due to deep red color and high ratio of pulp.
Swarna Roopa	It is developed through selection from Horticulture and Agroforestry Research Program at Ranchi. This is the most suitable for tilla land, medium time maturing cultivar and free from fruit cracking. The fruits are attractive and deep pink in color with small size nut.
Calcuttia	Fruits are large, attractive and ready to harvest in third week of June. It is a heavy bearing crop with excellent quality. Fruits are moderately juicy having good flavor.
Dehradun	Early and regular bearing variety. Fruits are ready to harvest in June second week. Fruits are of attractive color but prone to splitting. Fruits are sweet, soft and moderately juicy having good flavor.
Seedless Late	Contain greater portion of flesh. Fruits are deep carmine red, sweet in taste and juicy. Fruits get mature in third week of June.
Rose Scented	Popular mid-season variety. Fruits are of medium to large size. The fruits are juicy and sweet in taste and the pulp is greyish in color.

The improved varieties of litchi grown in different states of India are presented in Table 11.3.

Table 11.3. Varietal distributions of litchi in different states in India

States	Varieties
Bihar	Deshi, Purbi, China, Kasba, Bedana, Early Bedana, Late Bedana, Dehra Rose, Shahi, Manragi, Maclean, Longia, Kaselia and Swarna Rupa
Uttar Pradesh	Early Large Red, Early Bedana, Late Large Red, Rose Scented, Late Bedana, Calcuttia, Extra Early, Gulabi, Pickling, Khatti, Dehra Dun, Piyazi
West Bengal	Bombai, Ellaichi Early, China, Deshi, Purbi and Kasba
Haryana /Punjab	Early Seedless, Late Seedless, Seedless-1, Seedless-2

11.1.4. Exports potential

China's main fresh litchi export markets are Hong Kong, USA and Malaysia. A small amount of litchi (about 150 tons) was exported from India in 2018. South Africa produced about 11,000 tons of litchi in 2017/18 of which 56.4% was mainly exported to Europe.

Currently, Indian litchi is exported mainly to UAE, Nepal, Thailand, France, Saudi Arabia, Canada and several other countries. Litchi in the form of pulp is exported to various countries for further value addition. Even litchi stored in syrup is exported for consumption. The quantity and value for the export of litchi to different countries during 2020-21 are shown in Table 11.4.

Table 11.4. Country-wise quantity and value of the export of litchi products (2020-21)

Country	Quantity (Metric tons)	Value (Rs. '000)
Nepal	88.28	3399
Qatar	0.55	121
United Arab Emirates	0.99	73

Maldives	0.19	7
Total	90.01	3600

11.1.5. Nematode problems

Detailed information on economic nematode damage to litchi is available only from South Africa. Milne (1982) recognized *Xiphinema brevicolle* and *Hemicriconemoides mangiferae* as major nematode pests of litchi, causing a severe tree decline syndrome. *Trichodorus* spp. have also been found associated with nursery seedlings. Typical above-ground symptoms were the presence of many bare twigs and branches, leaf chlorosis, leaf-tip burn, poor flowering and excessive fruit drop, and in some orchards up to 40% of the trees died. Root symptoms were severe stubby root and darkening of the roots, leading eventually to loss of a large proportion of the feeder root mass and consequent interference in the uptake of nutrients and water.

11.2. RING NEMATODE, *HEMICRICONEMOIDES MANGIFERAE*

11.2.1. Distribution

H. mangiferae has been reported from Australia, Brazil, Egypt, Fiji, Ghana, India, Israel, Korea, Mexico, Nigeria, Pakistan, Philippines, South Africa, Sudan, Thailand, United States of America, Vietnam, Venezuela, and the West Indies.

11.2.2. Symptoms

H. mangiferae, which causes extensive destruction of the cortical tissue, is considered as the severe pathogen. Populations as high as 40,000 *H. mangiferae*/cm^3 of soil and roots were recorded.

1t was found that *H. mangiferae* commonly penetrates deep into the root, very often being totally embedded within the cortex. On other occasions, it was found that mass invasion occurred, thereby disrupting the epidermis and causing collapse of the exodermis and part of the cortical parenchyma at the site of infestation. Attack by this nematode was also frequently found to be associated with the development of adventitious roots and this may well explain the "stubby-root" appearance of litchi roots infested with this species.

In South Africa, litchi trees may be killed by *H. mangiferae* and many exhibit bare twigs and branches, leaf chlorosis, leaf-tip burn, poor flowering, and excessive fruit drop. Young replanted trees in decline areas are slow to become established and begin to show decline symptoms after 7 or 8 years. The nematode may cause 'stubby roots' symptoms (Milne *et al.*, 1971). In South Africa, about 40% of litchi trees are known to have been killed by a combination of the nematodes *H. mangiferae* and *X. brevicolle* (Milne *et al.*, 1971).

11.2.3. Life cycle and reproduction

The juvenile is a J1-J4, the J2 is what hatches out of the egg. J4 stage is where sexual differentiation occurs. Adult males do not feed or have a stylet. They are rare and not needed for reproduction. Parthenogenic reproduction may be the rule.

11.2.4. Management

Fruit yields can be maintained if infected trees are well irrigated and fertilized. High organic content in the rhizosphere can be used to have a temporary decline for the nematode population. The nematodes effects can be easily overcome by having an efficient and good irrigation and fertilization program to overcome fruit yield loss.

INM includes application of selective chemical products such as cadusaphos or fenamiphos together with compost, manures and organic material as the latter amendments enhance root growth and soil health.

11.3. ROOT-KNOT NEMATODE, *MELOIDOGYNE JAVANICA*

11.3.1. Symptoms

Below-ground damage of the root system is noticeable in a less dense root system and while healthy roots are mostly whitish, damaged roots have stubby, darkened appearance with limited development of feeder roots. Galls are seldom visible. Nematode damage triggers excessive leaf drop, resulting in bare twigs and branches ending in a tree with a more open canopy where weeds grow underneath the tree canopy, in contrast to healthy trees where the canopy is very dense and no sunlight is able to reach the soil. Furthermore, infected trees experience poor flowering and bear small, less abundant fruits compared to healthy trees. The nematode is, however, not able to kill the entire tree, even over a period of several years, and the symptoms are known as litchi slow decline (Daneel, 2017).

Meloidogyne javanica infects litchi roots in orchards, but galls are generally inconspicuous. Galls produced eight weeks after artificial inoculation of nursery plants, are fairly small and indicate the typical *Meloidogyne* behavior of entering just behind the root tip. In formation of the gall, there is some degree of hyperplasia, however, and the developing female and egg-mass cause distortion of the cortical tissues and epidermis.

11.3.2. Histopathology

Within the vascular bundles, large giant-cells (syncitia) develop as a result of feeding by *Meloidogyne* and these cause disruption and blocking of both the xylem and phloem. The giant cells are typically multinucleate, thick-walled cells. The cause of such a multinucleate condition is at present a controversial matter, and the present study did little to shed light on the problem (Milne *et al.*, 1975). A certain amount of circumstantial evidence appeared to point to cell dissolution of the type described by Endo (1971), but nuclear division was also detected thereby supporting the theory of Huang and Maggenti (1969) and McClure *et al.* (1974) who contend that the multinucleate condition is produced by such nuclear division only.

11.3.3. Management

Because of large amounts of mulch and decomposed material is found underneath the canopy area, biological control could be a possible alternative in nematode control. The efficacy of *Purpureocillium lilacinum* gave good nematode control.

11.4. DAGGER NEMATODE, *XIPHINEMA BREVICOLLE*

11.4.1. Symptoms

The presence of nematodes was causing severe "stubby root" and darkening of the roots. Examination of roots from declining trees under the stereo dissecting microscope revealed both *X. brevicolle* attached to the roots, appearing to be partially embedded in the roots, thereby showing direct and positive evidence of the feeding of this species on litchi roots.

The head of *X. brevicolle* inserted fairly deeply into the cortex where some disruption of the cells was apparent. *X. brevicolle* feeds more superficially.

The penetration of *X. brevicolle* is associated with destruction of the cell walls of the cortex, evacuation of the cell contents and subsequent death of the affected tissue. This in turn leads to browning of the roots, sloughing of the root cortex and the development of many irregular, adventitious roots, resulting in the lumpy "stubby root" appearance of the roots (Milne *et al.*, 1971).

The development of these symptoms on the great mass of feeder roots, together with its consequent deleterious effect on nutrient and water uptake by the plant, is regarded as the primary cause of die-back in the above-ground parts of the litchi tree.

11.4.2. Management

INM includes application of selective chemical products such as cadusaphos or fenamiphos together with compost, manures and organic material as the latter amendments enhance root growth and soil health.

Because of large amounts of mulch and decomposed material found underneath the canopy area, biological control could be a possible alternative in nematode control. Several years ago, trials were conducted to test the efficacy of *Purpureocillium lilacinum* (previously known as *Paecilomyces lilacinus*) and the results showed good nematode control (Daneel and Steyn, 2022).

11.5. OTHER NEMATODES

Two species, e.g. *Hemicriconemoides litchi* and *Rotylenchulus reniformis* with high population density (654 and 267, respectively) and high frequency level in both litchi regions (84.2-95.2% and 38.1-78,9%, respectively) has determined as most significant species related with decline disease on litchi. Apart from these species, other one more or less important was determined as *Xiphinema radicicola*. 100 percentages of litchi death was related with full combinations of three species of nematodes and two species of fungi (*Pythium vexans, Rhizoctonia solani*) in the condition of cutting roots. The attack of *Trichodorus* sp. has also been recorded.

11.6. FUTURE THRUSTS AND CONCLUSION

More focus needs to be directed to alternative nematode control strategies. Breeding and subsequent screenings will always be a very important factor in nematode control. However, research is necessary on improving biodiversity in orchards to decrease nematode problems. Biodiversity can be enhanced by planting a variety of cultivars/ selections together in one orchard, by planting combinations of cover crops may be even as a permanent cover in orchards. This above-ground biodiversity will most likely influence and increase biodiversity below-ground. Additional methods for improving biodiversity below-ground include the addition of composts, manures and organic amendments. There is a wealth of organic materials that need to be evaluated and might provide excellent nematode control.

Currently, chemical control products will not be available for nematode control and that producers will use a completely new strategy in controlling nematodes. A major aspect will be the use of resistant or tolerant cultivars. In addition, a variety of commercial biological control agents and softer chemical control agents will hopefully be available for use on the majority of crops produced. Future developments will entail precision farming even in perennial crops such as fruit trees.

Chapter - 12

Avocado, *Persea americana*

12.1. INTRODUCTION

Avocado is the most nutritive among fruits and is regarded as the most important contribution of the New World to human diet. The fruit is relished by some people, but not by others. The pulp is rich in proteins (up to 4%) and fat (up to 30%), but low in carbohydrates. The fat is similar to olive oil in composition and is widely used in the preparation of cosmetics. Avocados have the highest energy value (245 cal/100 g) of any fruit besides being a reservoir of several vitamins and minerals.

12.1.1. Origin

Avocados are ancient fruits originated from Central America and Mexico.

12.1.2. Area and production

In 2021, global avocado production amounted to about 8.69 million metric tons, an increase from the previous year's production volume of 8.1 million metric tons. Mexico harvested about 2.4 million tons of avocados in 2021, making that country the top producer of avocados world-wide. Mexico is the largest avocado producer followed by Colombia, Peru, Indonesia, Dominican Republic, Kenya, Brazil, Haiti, Vietnam, and Chile.

Avocados are grown scattered in Indian southern tropical states like Tamil Nadu, Kerala, Karnataka, and Maharashtra. Also popular in the north-eastern Himalayan state of Sikkim on hill slopes at elevations of 800 - 1,600 meters.

12.1.3. Different types of avocados

- **Mexican race:** Fruit type is small (250 g) and ripens 6 to 8 months after blossoming. The fruit has a thin, smooth skin and a big seed that fits loosely in the centre chamber. The fruits can have up to 30% oil content. This species of tree is cold hardy.
- **Guatemalan race:** Fruits are enormous and borne on long stalks, weighing up to 600 g. 9-12 months after blossoming, the fruits mature. The skin of the fruit is thick and often warty. The seeds are little and tucked away in the hollow of the fruit. Fruits have an oil content that ranges from 8% to 15%.

» **West Indian race**: Fruit is medium in size, with smooth, leathery, glossy skin. Fruits are carried on long stalks and might take up to 9 months to ripen after flowering. Its seeds are big and fit into the cavity only loosely. The fruit has low oil content (3-10 per cent). This race is best suited to tropical climates.

12.1.4. Varieties

The avocado varieties grown in India include Green Type, Purple, TKD-1, Nabal, Linda, Puyevla, Gott-Froid, Furete, Pullock, and Waldin. The varieties Pollock, Waldin, Nabal, Linda, Puebla, Gottfreid, Collins, Booth 8 and Fuerte are recommended for commercial cultivation.

The Central Horticultural Experiment Station, Chethalli (Kodagu) (a regional Centre of The Indian Institute of Horticultural Research, Bangalore) has evolved an avocado variety "Arka Supreme" with an yield potential of 125-200 kg fruit per tree.

Due to increasing health consciousness among the educated population and the high nutritive value of avocado, it is expected to find its rightful place in the Indian market in due course. The successful introduction of avocado and its wide acceptance among the tribal population of the hilly state of Sikkim indicate that for household nutrition security, avocado should be a potential fruit crop in India. With some governmental support it can be popularized in the country and it can provide much needed household nutrition security to the Indian population.

12.1.5. Potential for avocado production development

The agro-climatic conditions prevailing in various parts of the country appear to be favourable for bringing more areas under avocado. Presently, quite a good number of improved varieties are now available with higher yield potential. Vegetative propagation techniques have also been standardized. Multiplication of a large number of high-quality nursery plants of selected varieties and their systematic planting, both in tropical southern India and humid semi-tropical areas of the north-eastern region of India, could help to place avocado properly on the fruit map of India. The research support for the crop is still very poor, but the available research information from Tamil Nadu and Sikkim amply demonstrate that size, color and quality of fruits attained in India are comparable with those of avocado fruits grown elsewhere. Avocado is grown successfully in neighbouring Sri Lanka, where good-quality fruits are harvested during May to August and December to January at different regions. Similar agro-climatic conditions to Sri Lanka are available in the Andamans and Nicobar Islands and in the tropical southern India. With proper varietal selection, it should be possible to exploit also the possibility of out-of-season production, thus enhancing the availability of fruits for a longer period during the year. Avocado fruits produced in the country can be marketed without much difficulty, particularly to meet the requirement of the growing tourist industry. The mainland India and the Andaman and Nicobar Islands are attracting foreign tourists in a large number of places, where avocado could find a good market access. Avocado has also a good export potential.

12.1.6. Nematode problems

Avocado, in comparison with other tree crops, appears to be relatively free of aggressive nematode pests, and it is difficult to determine the economic importance of the identified nematode parasites to avocado production. However, the lesion nematode, *Pratylenchus vulnus* and the burrowing nematode, *Radopholus citrophilus* are responsible for decline of avocados.

12.2. LESION NEMATODE, *PRATYLENCHUS VULNUS*

12.2.1. Symptoms

Sher (1955) attributed plant damage in California to *Pratylenchus vulnus*, and reduced tree growth was shown to be caused by this nematode, both in greenhouse inoculation experiments and in pre-plant fumigation trials with DD (Sher *et al.*, 1959). Root-lesion nematode disease of avocado has only been seen in Ventura where avocados have been planted after walnuts. *Pratylenchus vulnus* can cause extensive damage to young avocado trees. *P. vulnus* is more likely to cause problems in avocado orchards planted adjacent to walnut orchards, where infested trees were found to be stunted, had fewer feeder roots, the trees appeared more susceptible to frost injury (Westerdahl, 2003).

There was a marked reduction in size of all the plants infested with nematodes. Feeder roots were lacking in all the pots infested with the root-lesion nematode, *P. vulnus* was recovered in large numbers from all the soil and root samples in all the pots that were infested. No root-lesion nematodes were recovered from the healthy check plants (Table 12.1). The dry weights of the uninfested plants were significantly higher than the dry weight of the infested plants (Table 12.1).

Table 12.1. Pathogenicity of *Pratylenchus vulnus* to avocado

Treatment	Dry weight of plants (g)	*P. vulnus*/100 ml soil	*P. vulnus*/2 g roots
P. vulnus at 100/plant	54.8	183.9	63.4
P. vulnus at 1000/plant	38.4	459.6	47.3
Control (Uninoculated)	202.3	0	0

12.2.2. Host range

P. vulnus was found to reproduce well on horticultural crops such as walnut, almond, hazelnut, pistachio, grape, apple, cherry, plum, pear and loquat, as well as on avocado in Spain (Pinochet *et al.*, 1992).

12.2.3. Management

Avocado plants infected with the lesion nematode responded dramatically to pre-plant fumigation. However, with the exception of pre-plant fumigation, practical nematode control recommendations to growers are unavailable.

12.3. BURROWING NEMATODE, *RADOPHOLUS CITROPHILUS*

Work done in Florida during the mid1950s implicated *Radopholus citrophilus* (citrus race) in reduced performance of avocado trees (Young and Ruehle, 1955). However, it appears in retrospect that much of this and other work in Florida (McSorley, 1981) was related to surveys carried out in areas of citrus spreading decline, which at that time was a major economic disaster. No practical conclusions or recommendations regarding these nematode species in commercial avocado orchards have been developed since.

12.3.1. Symptoms

Ducharme and Suit (1953) demonstrated their capacity to create root lesions. An initial inoculum density of 10,000 burrowing nematodes per plant caused 10, 10, 17.46 and 62 per cent reduction over control in length and weight of shoot, number of leaves, weight and volume of root respectively over a period of 4 months. The pathogenic threshold level of *R. citrophilus* was 100 nematodes per plant or one nematode in 12 cm^3 of sandy loam soil for root growth characters. The study has established the potential of the burrowing nematode as a pathogen of avocado.

12.3.2. Management

Variability in the reaction of avocado varieties to different *R. citrophilus* populations is not well characterized. Milne and Keetch (1976) reported avocado to be a non-host for the banana race of *R. similis* in Natal, South Africa, whereas the nematode was shown to be a pathogen of avocado in pot trials in India (Jasy and Koshy, 1992).

12.4. DAGGER NEMATODE, *XIPHINEMA BREVICOLLE*

12.4.1. Symptoms

In Israel, populations of *X. brevicolle*, sometimes as high as 500/100 g of soil, are often recovered from around avocado roots, and reduced seedling growth in pots as a result of inoculation with this nematode has been demonstrated (Cohn, 1968).

12.4.2. Management

Benomyl suppressed dagger nematode population recovery completely from 7 days to at least 60 days after treatment.

12.5. OTHER NEMATODES

Interestingly, most of the economically important sedentary plant nematodes do not appear to be serious problems on avocado. *Rotylenchulus reniformis* has been observed on avocado roots in West Africa (Peacock, 1956), where Caveness (1967) found avocado to be a good host, and in Brazil (Sharma, 1978). *Meloidogyne hapla* was detected on avocado roots in New Zealand (Knight, 2001). There is no evidence that *R. reniformis* or *M. hapla* cause economic damage to avocado plants. The genera *Helicotylenchus* and *Meloidogyne* were encountered with some frequency at population densities between 150 and 350 nematodes/100 cm^3 of soil in a survey of avocado nurseries and orchards in Colombia (Saltaren *et al.*, 1999). Inoculation with *Helicotylenchus dihystera* reduced avocado seedling growth by 20–50% (Saltaren *et al.*, 1999). Similarly, granular nematicides reversed a decline of avocados infested with *H. dihystera*, *Mesocriconema xenoplax* and *Xiphinima elongatum* (Willers, 1999), suggesting a need for further study of the economic importance of ectoparasitic nematodes in avocado.

Possibly, the role of nematodes in damaging avocado roots has been overshadowed by the attention aroused by the severe avocado root disease caused by the soil fungus *Phytophthora cinnamomi* and, as suggested by Milne (1982a), it would be interesting to establish whether plant parasitic nematodes are capable of affecting the severity of this disease or the susceptibility to it of avocado trees.

12.6. CONCLUSION

In summing up, it can be said that nematodes are not considered a serious problem in on avocado. There are nematodes like lesion nematodes, that could become established here and cause the problem. Pratylenchus vulnus can cause extensive damage to young avocado trees. As the orchards are getting older, nematode problems will probably increase, and as we solve some of our more important disease problems nematodes will be a lot more conspicuous than they now are.

Possibly, the role of nematodes in damaging avocado roots has been over-shadowed by the attention aroused by the severe avocado root disease caused by the soil fungus *Phytophthora cinnamomi* and, as suggested by Milne (1982), it would be interesting to establish whether plant parasitic nematodes are capable of affecting the severity of this disease or the susceptibility to it of avocado trees. Investigations on avocado nematodes and their management ned to be pursued.

Chapter - 13

Mango, *Mangifera indica*

13.1. INTRODUCTION

13.1.1. Origin

Mango (*Mangifera indica*) is the most ancient among the tropical fruits and believed to have originated in the Indo – Burma region (South Asia).

13.1.2. Area, production and yield

The global production volume of mangos was a little over 57 million metric tons in 2021, a small increase from around 56.69 million metric tons in 2020. Area, production and yield of mango in different countries during 2016 is presented in Table 13.1.

Table 13.1. Major mango producing countries in the world (2016)

Country	Area (Million ha)	Production (Million tons)	Yield (Tons/ha)
India	2.24	18.78	8.39
China	0.59	4.77	8.14
Thailand	0.41	3.43	8.36
Mexico	0.21	2.20	10.64
Indonesia	0.17	2.18	13.02
Pakistan	0.17	1.61	9.57
Brazil	0.08	1.42	17.95
Egypt	0.11	1.28	11.30
Bangla Desh	0.15	1.16	7.59
Nigeria	0.13	0.92	6.87
Others	1.07	8.87	8.29
WORLD	5.43	46.51	8.57

Mango is commercially cultivated in Andhra Pradesh, West Bengal, Karnataka, Kerala, Bihar, Uttar Pradesh, Uttara Khand, Punjab and Haryana, Maharashtra, and Gujarat. State-wise area, production and yield of mango in India is presented in Table 13.2.

Table 13.2. State-wise area, production and yield of mango in India (2017-18)

State	Area ('000 ha)	Production ('000 MT)	Yield (tons/ha)
Andhra Pradesh	363.00	4373.61	12.048
Assam	4.68	48.44	10.350
Bihar	149.28	2443.47	16.368
Chhattisgarh	77.03	461.73	5.994
Gujarat	162.77	1207.78	7.420
Haryana	9.35	98.60	10.545
Himachal Pradesh	41.99	31.35	0.746
Jammu & Kashmir	12.96	30.35	2.341
Jharkhand	54.53	435.86	7.993
Karnataka	183.23	1760.60	9.608
Kerala	83.12	439.20	5.284
Madhya Pradesh	45.52	654.79	14.384
Maharashtra	166.76	791.36	4.745
Mizoram	0.91	4.19	4.604
Nagaland	0.64	4.24	6.625
Odisha	199.08	805.77	4.047
Punjab	6.90	116.52	16.886
Rajasthan	4.97	87.37	17.579
Tamil Nadu	152.57	1234.00	8.088
Telangana	115.99	1080.14	9.312
Tripura	10.33	54.93	5.317
Uttar Pradesh	265.62	4551.83	17.136
Uttarakhand	36.48	152.71	4.186
West Bengal	103.25	918.35	8.894
Others	7.17	35.14	4.901
TOTAL	2258.13	21822.32	9.664

Source: Hort. Statistics Divn., Dept. of Agri., Co-op. &Farmers Welfare

The major producing districts in different states in India include:

» Uttar Pradesh (Lucknow, Saharanpur, Unnao, Bulandshahr, Amroha etc.)

» Andhra Pradesh (Chittoor, Vizianagaram, Krishna, Anantapur, Kadapa etc.)

» Telangana (Mahbubnagar, Karimnagar, Khammam, Warangal, Adilabad etc.)

» Karnataka (Kolar, Ramanagara, Tumkur, Chikballapur, Dharwad etc.)
» Bihar (Darbhanga, Samastipur, Muzaffarpur, Purbi Champaran, Vaishali etc.)
» Gujarat (Valsad, Navsari, Gir Somnath, Kachchh, Surat etc.)
» Maharashtra(Ratnagiri, Sindhudurg, Raigarh, Satara, Sangli, Kolhapur, Latur, Nasik, Ahmednagar, Osmanabad etc.)

13.1.3. Major commercial varieties

The major commercial varieties of mango cultivated in different states of India are presented in Table 13.3.

Table 13.3. Major commercial varieties of mango in different states of India

States	Varieties
Uttar Pradesh	Bombay Green, Chausa, Dashehari, Mallika, Amrapali, Ambika 2000, Arunika 2008 and Langra
Andhra Pradesh	Banganapalli, Suvarnarekha, Neelum, Ratna, Sindhu, Ambika 2000, Arunika 2008 and Totapuri
Telangana	Banganapalli, Suvarnarekha, Neelum, Ratna, Sindhu, Ambika 2000, Arunika 2008 and Totapuri
Karnataka	Alphonso, Totapuri, Banganapalli, Pairi, Neelum, Ratna, Sindhu, Ambika 2000, Arunika 2008, Arka Puneet, Arka Anmol, Arka Neelkiran and Mulgoa
Bihar	Bombay Green, Chausa, Dashehari, Fazli, Mallika, Amrapali, Gulabkhas, Kishen Bhog, Himsagar, Zardalu, Ambika 2000, Arunika 2008 and Langra
Gujarat	Kesar, Alphonso, Rajapuri, Jamdar, Totapuri, Neelum, Dashehari and Langra
Maharashtra	Alphonso, Mankurad, Mulgoa, Pairi, Rajapuri, Kesar, Gulabi, Vanraj

13.1.4. Advance technology : High density planting.

13.1.5. Pre-cooling

» Forced air pre-cooling or hydro-cooling at -1 to 0 ^{0}C and RH – 90-95% are suitable for cooling up to 8 months.

13.1.6. Storage conditions

» Temperature: 13°C
» Relative humidity: 85-90%
» Storage period: 2-3 weeks
» Freezing point: -1°C

13.1.7. Sensibility

» Refrigeration, freezing and ethylene exposure

13.1.8. Peak harvesting season

» Uttar Pradesh: June
» Andhra Pradesh: April to May
» Telangana: April to May

» Karnataka: April to June
» Bihar: June to July
» Gujarat: May

13.1.9. Export from India (2017-18)

» Quantity: 49.18 Thousand MT
» Value: 38,234.02 Lakh Rupees

In 2020, India exported 128,026 metric tons of mango.

13.1.10. Nematodes problems

Mango appears to be relatively free from severe nematode damage, despite the fairly long list of nematode species associated with it (Ghorab *et al.*, 1987; Petit, 1990; Korayem and Koura, 1993; Yin, 1994; Anita and Chaubey, 2003). The plant parasitic nematodes reported from Brazil include, *Helicotylenchus dihystera, Pratylenchus brachyurus, Rotylenchulus reniformis, Meloidogyne* sp., *Macroposthonia* sp. and *Xiphinema* sp. The mango varieties grafted on Sinamaica rootstocks showed the presence of *Hoplolaimus, Trichodorus, Pratylenchus, Hemicriconemoides, Rotylenchus, Helicotylenchus, Xiphinema, Tylenchus* and *Meloidogyne*; while the varieties grafted on Bocao rootstock had *Hoplolaimus, Helicotylenchus, Rotylenchus, Xiphinema, Trichodorus* and *Meloidogyne*, with the most frequent genera being *Hoplolaimus, Hemicriconemoides* and *Rotylenchus*.

Economic responses to chemical treatment in mango were reported when DBCP was used to control *Hoplolaimus columbus* and *Xiphinema* spp. in Egypt (Shafiee and Osman, 1971), and fenamiphos applications were found effective in controlling *Pratylenchus brachyurus*, but not *R. reniformis* in Florida.

13.2. RING NEMATODE, *HEMICRICONEMOIDES MANGIFERAE*

Probably the most widely distributed nematode associated with mango is *Hemicriconemoides mangiferae* (Siddiqi, 1977; McSorley, 1992), which has been shown to be potentially damaging to mango seedlings at a population level of 6 nematodes/cm^3 of soil (Saeed, 1974). The nematode is widespread in mango orchards and on numerous other crops in India, particularly in sandier soils, when population density was strongly and positively related to soil moisture (Ashok Kumar *et al.*, 1991). Although the pathogenicity of *H. mangiferae* to mango has been demonstrated (McSorley, 1992), its economic importance in the field is not clear. McSorley *et al.* (1981) reported a wide distribution and strong association between *H. mangiferae* and declining mango in Florida, but the relationship is not always evident (Ashok Kumar *et al.*, 1991). The nematode was observed feeding on mango roots together with *Xiphinema brevicolle* in South Africa, but chemical treatment of existing trees, while reducing nematode populations, failed to induce a favorable tree response (Milne, 1982). Mango infested by *H. mangiferae* and *Rotylenchulus reniformis* responded strongly to treatment with systemic nematicides, however, control of the mite *Eriophyes mangiferae* that is involved in witches' broom syndrome may have caused at least part of the response (Noriega *et al.*, 1988).

13.2.1. History

Hemicriconemoides mangiferae was discovered in 1961 by Siddiqi. It is one of the smallest parasitic plant nematode. It was originally identified as a parasite of mango roots from India (MacGowen, 1984). A complete description of developmental stages of *H. mangiferae* was provided by Ashok Kumar and Vadivelu (1990) and Ashok Kumar (1988).

13.2.2. Distribution

H. mangiferae has been reported in the following countries: Australia, Brazil, Egypt, Fiji, Ghana, India, Israel, Korea, Mexico, Nigeria, Pakistan, Philippines, South Africa, Sudan, Thailand, United States of America, Vietnam, Venezuela, and the West Indies (MacGowen, 1984; Ashok Kumar *et al.*, 1991). McSorley *et al.* (1981) reported a wide distribution and strong association between *H. mangiferae* and declining mango in Florida.

13.2.3. Symptoms

The visible symptoms can include leaf chlorosis, leaf tip burn, and excess fruit drop. The latter is the most noticeable symptom of infected trees. Levels of 6 nematodes per 1 cubic centimetre are considered detrimental to the tree (MacGowen, 1984).

In South Africa, litchi trees may be killed by *H. mangiferae* and many exhibit bare twigs and branches, leaf chlorosis, leaf-tip burn, poor flowering, and excessive fruit drop. Young replanted trees in decline areas are slow to become established and begin to show decline symptoms after 7 or 8 years. In Florida, USA, *H. mangiferae* is significantly correlated with poor tree conditions in mango groves. The correlation was higher in samples with more than 100 *H. mangiferae* per 100 cubic centimetres of soil (McSorley *et al.*, 1981).

13.2.4. Host range

Besides mango, *H. mangiferae* infects grape (*Vitis vinifera*), pineapple (*Ananas sativus*), pomegranate (*Punica granatum*), jackfruit (*Artocarpus heterophyllus*), papaya (*Carica papaya*), lime (*Citrus aurantifolia*), lemon (*C. limon*), grapefruit (*C. paradisi*), mandarin (*C. re*ticulata), sweet orange (*C. sinensis*), longan (*Euphoria longan*), litchi (*Litchi chinensis*), sapota (*Manilkara zapota*), banana (*Musa paradisiaca*), date palm (*Phoenix dactylifera*), amla (*Phyllanthus emblica*), apricot (*Prunus armeniaca*), guava (*Psidium guajava*), tamarind (*Tamarindus indicus*), custard apple (*Annona reticulata*), okra (*Abelmoschus esculentus*), coconut (*Cocos nucifera*), cocoa (*Theobroma cacao*), black pepper (*Piper nigrum*), bamboo (*Bambusa vulgaris*), Indian hemp (*Cannabis sativa*), cotton (*Gossypium hirsutum*) and corn (*Zea mays*). Nematode is able to survive on many weed hosts.

In Florida and in other countries, damage is documented for litchi, mango (*H. mangiferae*), sugarcane, and tea (Inserra *et al.*, 2014).

13.2.5. Life cycle and reproduction

The juvenile is a J1 to J4, the J2 is that hatches out of the egg. J4 stage is where sexual differentiation occurs. Adult males do not feed or have a stylet. They are rare and not needed for reproduction. Parthenogenic reproduction may be the rule. *H. mangiferae* lays eggs in the soil once mating has occurred.

13.2.6. Host parasite relationship

H. mangiferae anterior becomes embedded in the cortex of the root. It is typically an ectoparasite of the root. It can become completely embedded in the root and become an endoparasite. Necrotic tissue is often found as a result of nematode feeding. The roots become stubby as an effect of the nematode feeding. Nutrient and water uptake is interrupted and this will cause decline of the tree (MacGowen, 1984).

13.2.7. Ecology

The population was larger at 10 cm depth followed by 20 and 40 cm depths. The largest nematode population was observed when the percent soil moisture was high. Both soil temperature and soil moisture are equally important. The soil pH also indirectly affects the nematode population densities (Ghorab *et al.*, 1987).

13.2.8. Management

13.2.8.1. Cultural methods: When orchards are established, it should be made sure that the planting stock is nematode free (Milne *et al.*, 1975). Soil fumigation is one way to make sure that this is achieved (MacGowen, 1984). High organic content in the rhizosphere can be used for achieving a temporary decline in the nematode population. The management of this nematode is difficult. It is almost impossible to get rid of the nematode once the infection has started. The nematodes affect fruit yield and losses can be easily overcome by having an efficient and good irrigation and fertilization program (Richard, 2009).

13.2.8.2. Host plant resistance: Haseeb *et al.* (1986) reported resistance to *H. mangiferae* in ten different fruit trees.

13.3. ROOT–KNOT NEMATODES, *MELOIDOGYNE* SPP.

Although detected in soil associated with mango (Maqbool, 1991; McSorely, 1992), there is only one report documenting infection of mango by *M. incognita* (Mani and Al Hinai, 1995; Mani *et al.*, 1995).

13.3.1. Symptoms and histopathology

Gross morphology of mango roots infected with root–knot nematodes showed blackened roots with galls. Examination of sections of mango roots showed that the root–knot nematode developed and reproduced normally within the root tissue. The juveniles of *M. javanica* penetrated the roots by puncturing action of the stylet. In the infected roots, great variation in the cell sizes was noticed, some cells were enlarged with broken walls while others were hypertrophied due to the presence of large number of egg– masses and larvae. The eggs were laid in gelatinous matrix and partial cell lysis was observed. A large number of giant cells can be seen, with walls that are in the process of liquefaction (Sayeed *et al.*, 2010).

The cells tended to be prolonged along the stele region occupying a considerable area and portrayed thick walls. Cells of cortex region were full of matrix causing disruption. The disruption of cortical cells due to migration of root–knot nematode larvae partly contributed to the development of knots in mango roots.

13.3.2. Management

Cover cropping with African marigold (*Tagetes* spp.) minimizes root-knot nematode damage. Intercrop with different mustards (e.g. *Brassica juncea* var. *integrifolia* or *Brassica juncea* var. *juncea*) on infested fields. As soon as mustards are flowering they are mulched and incorporated into the soil. While incorporated plant parts are decomposing in a moist soil, nematicidal compounds of this decomposing process do kill nematodes.

Maintain high levels of organic matter in the soil (manure or compost). Various botanicals such as neem cake, mahua cake, mustard cake *etc.* show nematicidal action when applied in the field.

13.4. DAGGER NEMATODES, *XIPHINEMA BREVICOLLE, X. BASIN*

Mango tree decline has been reported in association with *X. brevicolle* in South Africa.

Xiphinema basin was isolated from around the roots of diseased mango trees grown at a site at Wad Medani and at the GRS in numbers of up to 200 or more nematodes per 200 g soil.

13.4.1. Economic significance

These dagger nematodes are potentially very serious parasites and could severely limit production especially of tree crops.

13.4.2. Symptoms

When tree hosts are attacked, the typical above ground symptoms are the presence of several bare twigs and branches, leaf chlorosis, leaf–tip burn, poor flowering and excessive fruit drop. Trees can be killed. Generally the symptoms are of plants under stress due to impeded translocation of water and nutrients.

Below-ground symptoms are characteristic stubby dark–colored roots and loss of a large amount of feeder roots. The root ends become swollen and gall–like. Distinctly stunted roots were associated with high densities of *X. basin* in the rhizosphere of diseased plants. The stunted root systems could also be reproduced in potted seedlings grown in sterilized soil infested with *X. basin* (Yassin, 1974).

13.4.3. Life cycle

The dagger nematodes are migratory, ectoparasitic, obligate plant parasites. All juvenile and adult stages feed on plant roots with the aid of long massive spear. These nematodes are relatively long compared to other plant parasitic nematodes, being up to 5 mm. The length of life cycle varies throughout the world, ranging from 28 days in USA to 9 months in Israel and up to 3 years in Europe.

13.4.4. Epidemiology

Population numbers are generally highest in the summer/autumn. It prefers medium texture to heavy soils but is found in all soil types. The preferred range of pH is 6.5–7.5. It can survive in moist soils without food for 9 months and can survive for 4 years in orchards after plants (not roots) have been removed. It can survive freezing temperatures, though optimum range is 16–28 °C.

13.4.5. Management

- Only clean seedlings (nematode-free) obtained from clean nurseries should be used for planting.
- Infected plants with all roots should be removed and not replaced with the same species of plant for 4 years.
- Crop rotation with non–hosts should be followed.

13.5. RENIFORM NEMATODE, *ROTYLENCHULUS RENIFORMIS*

Anwar *et al.* (1991) considered *R. reniformis* to be damaging on mango, but McSorley *et al.* (1981) found no relationship between infestation by the nematode and tree decline symptoms.

R. reniformis appears to be the only sedentary nematode that commonly infects mango (McSorley and Parrado, 1983).

13.5.1. Symptoms

Above ground symptoms on host plants include dwarfing, shedding of leaves, formation of malformed fruit and seeds, and general symptoms of an impaired root system.

Below ground, roots are discolored and necrotic (dead) with areas of decay. Plant mortality is possible in heavy infestations.

13.5.2. Management

Interestingly, soil and root populations on seedlings were reduced effectively by application of the ethephon growth regulator (Badra and Khattab, 1982).

13.6. CONCLUSION

Mango appears to be relatively free from severe nematode damage, despite the fairly long list of nematode species associated with it (Ghorab *et al.*, 1987; Petit, 1990; Korayem and Koura, 1993; Anita and Chaubey, 2003). 26 species of nematodes have been reported to be attacking mango trees. Nematodes associated with mango include *Rotylenchulus reniformis, Criconemella sphaerocephala, Helicotylenchus dihystera*, and *Hemicriconemoides mangiferae*. *H. mangiferae* is known to be pathogenic to mango trees, as well as sapodilla and tamarind, and infestations of this nematode induce a slow decline of mango.

Populations of *H. mangiferae*, and *Rotylenchulus reniformis* were very high. Possible factors contributing to mango decline must include *H. mangiferae* and *R. reniformis*. Association of fungi with tissues symptomatic of mango decline in the presence of none to high population levels of plant parasitic nematodes is an area in need of further study to understand the relationship between nematodes and fungal pathogens in inducing decline syndrome in mango.

Chapter - 14

Peach, *Prunus persica*

14.1. INTRODUCTION

14.1.1. Origin

The cultivated form of peach has originated from China.

14.1.2. Area and production

Peach is basically a temperate zone plant and its commercial production is confined between the latitude of 30 and 40° N and S, although it is now grown almost all over the world. The major peach producing countries are Italy, China, Spain, Italy, Greece, Turkey, USA, Iran, Egypt, Chile, and Argentina (Table 14.1).

Table 14.1. Top 10 peach producing countries in the world – area, production and yield (2019)

Country	Area ('000 ha)	Production ('000 tons)	Yield (tons/ha)
China	838.878	15,825.757	188.654
Spain	77.700	1545.610	198.920
Italy	60.430	1244.940	202.704
Greece	41.410	926.620	223.767
Turkey	46.294	830.557	179.414
USA	36.380	730.900	203.381
Iran	32.155	591.412	183.925
Egypt	15.748	358.012	227.338
Chile	15.651	330.232	210.997
Argentina	12.835	210.000	163.615
WORLD	1,527.052	25,737.841	168.545

In India, peach is mainly cultivated in Jammu & Kashmir, Himachal Pradesh, Uttaranchal, Punjab and Delhi (Table 14.2). Low chilling peaches are grown in sub mountainous region and western part of the country i.e., Punjab, Haryana, Delhi and the Western Uttar Pradesh. Peach cultivation is being grown in the mid hill zone of the Himalayas extending from Jammu and Kashmir to Khasi hills. Peach is also grown on limited scale in the hills of south India and in the north-eastern region of the country.

Table 14.2. State-wise production and share of peach in India (2021-22).

State	Production ('000 tons)	Share (%)
Uttarakhand	52.86	48.88
Punjab	32.23	29.80
Jammu & Kashmir	7.35	6.80
Himachal Pradesh	7.11	6.57
Haryana	5.87	5.43
Nagaland	1.79	1.66
Tamil Nadu	0.36	0.33
Sikkim	0.34	0.31
Arunachal Pradesh	0.24	0.22
TOTAL	108.15	100

14.1.3. Varieties

Indigenous peach cultivars and their important characteristics are presented in Table 14.3.

Table 14.3. Indigenous peach cultivars and their important characteristics

Variety	Important characteristics
Saharanpur	An early ripening, white-fleshed peach hybrid.
Sharbati	Selection, chilling requirement of 30 to 40 hours.
Shan-e-Punjab	Most popular cv. due to potential larger size variety to grow under Ludhiana conditions which gave yield of 0.5 tons/ha.
Pant Peach-1	Chance seedling from cv. Sharbati.
Pratap	Selected for sub-tropical conditions in Punjab. Fruit matures one week earlier (third week of April) than that of Floridasun.

Peach varieties suitable for early-, mid-. and late season in different states of India are presented in Table 14.4.

Table 14.4. Peach varieties suitable for early-, mid-, and late season in different states of India

State	Early	Mid-season	Late
Himachal Pradesh	Alton, World's Earliest, Early White, Redhaven, Stark Red Gold,	July Elberta, Kanto 5, Shimizu Hakuto, Sunhaven,	J.H. Hale
Uttar Pradesh (mid-hills)	Early Candour, Redhaven, Sunhaven	July Elberta, Alexander, Crawford Early	Parrot Deluxe, J.H. Hale Peregrine

Jammu & Kashmir	Peshwari, Quetta, July Elberta, Saharanpur, Prabhat	J.H. Hale, Alexander, C.O. Smith	---
Punjab	Shan-e-Punjab, Floridasun, Prabhat, Sunred, Florida Red	Sharbati	---

14.1.4. Rootstocks

Rootstock for peaches is raised from seeds of desi peach trees. The seeds of commercial cultivars like Sharbati and Khurmani are also used. Peach plants can also be raised on peach-almond hybrid, apricot, almond seedling, plum and behmi (*Prunus mira*). Most commonly used rootstock are the seedlings of wild peach.

For peach, peach x almond hybrid GF677 clonal rootstock is commercially used. This rootstock is useful on alkaline soil due to resistance to chlorosis. The other important clonal rootstocks are Siberian C (cold hardy), St. Julien hybrid No.1 and 2 and Damas GF677 (resistant to water logging and alkaline soil), Headguard and Shalil (nematode resistant) (Table 14.5).

Table 14.5. Characteristics of different rootstocks

Rootstock	Characteristics
Bailey	Cold hardy rootstock with good overall performance; best in the northern states.
Guardian	Vigorous rootstock with resistance to peach tree short life syndrome; slightly more expensive. Used extensively in the eastern USA, and in areas where peaches were planted previously.
GF677 (Amandier)	A peach-almond hybrid for adaptation to high pH soils; highly vigorous; not well adapted for eastern US.
Halford	Good overall rootstock for northern states
Lovell	Moderately invigorating rootstock with fair resistance to peach tree short life.
Nemaguard	Invigorating rootstock with resistance to root knot nematode; susceptible to other nematodes and peach tree short life.
Siberian C	Cold hardy rootstock used only in the northern tier of states and Canada; short-lived and poor in the southern states.

14.1.5. Different aspects of peach cultivation

Different aspects of peach cultivation are presented in Table 14.6.

Table 14.6. Different aspects of peach cultivation

Variety	Rootstock	Training system	Spacing (m)	Density (trees/ha)	Yield (t/ha)	Reference
Royal Giant	Nemaguard	V-shape	2.0 x 5.5	909	353.0	USA
Usa	Nemaguard	V-shape	2.0 x 5.5	909	206.0	USA
Baby Gold-5	Nemaguard	V-shape	2.0 x 5.5	909	169.0	USA

Cresthaven	Nemaguard	Tatura trellis	2.5 x 2.5	1600	47.26 (6th yr)	CITH, 2013-14 (India)
Glohaven	Nemaguard	Tatura trellis	2.5 x 2.5	1600	42.99 (6th yr)	CITH, 2013-14 (India)
Red Globe	Nemaguard	Tatura trellis	2.5 x 2.5	1600	41.12	CITH, 2013-14 (India)
Glowhaven	Standard	Modified leader	3.0 x 3.0	1111	40.08 (5th yr)	CITH, 2010 (India)
Fantasia	Standard	Modified leader	2.5 x 2.5	1600	20.60 (3rd yr)	CITH, 2010 (India)
Red Globe	Standard	Modified leader	3.0 x 3.0	1111	16.86 (3rd yr)	CITH, 2010 (India)
Red Globe	Standard	Modified leader	2.5 x 2.5	1600	15.77 (3rd yr)	CITH, 2010 (India)
Cresthaven	Standard	Modified leader	2.5 x 2.5	1600	13.80 (3rd yr)	CITH, 2010 (India)
Glohaven	Standard	Modified leader	3.0 x 3.0	1111	13.22 (3rd yr)	CITH, 2010 (India)
Fantasia	Standard	Modified leader	3.0 x 3.0	1111	12.12 (3rd yr)	CITH, 2010 (India)
Cresthaven	Standard	Modified leader	3.0 x 3.0	1111	10.89 (3rd yr)	CITH, 2010 (India)

14.1.6. Export

The top 3 exporters of Peaches are China with 65,211 shipments followed by India with 28,129 and Turkey at the 3rd spot with 27,151 shipments. India exports most of its peaches to United States, United Arab Emirates and Bangladesh and is the second largest exporter of peaches in the World.

14.1.7. Nematode problems

The important nematodes of peach include *Meloidogyne* spp., *Paratylenchus* spp., *Pratylenchus* spp., *Mesocriconema* spp. and *Xiphinema* spp. Root-knot, lesion and ring nematodes have been recognized as pathogenic causing economic damage to peach crop.

14.2. ROOT-KNOT NEMATODES, *MELOIDOGYNE JAVANICA, M. INCOGNITA, M. ARENARIA*

Of the many nematodes associated with peach trees, the root-knot nematode was the first known to cause damage in the form of knotted, swollen roots, reduced root and top growth and leaf chlorosis.

14.2.1. Distribution

Infestation of *Meloidogyne* spp. has been reported from Delhi, Punjab, Haryana and Himachal Pradesh. *M. javanica, M. incognita,* and *M. arenaria* occur commonly in peach growing areas.

14.2.2. Symptoms

Above-ground symptoms on peach plants infected with *Meloidogyne* spp. include – poorly developed foliage, lack of vigor, die-back of twigs and yield reduction. In periods of draught, infected trees wilt more quickly than non- infected ones. The peach grove infested with root-knot nematode is usually composed of unevenly sized trees due to uneven distribution of nematodes and their population variation within one field.

The root system of nematode infested trees has poorly developed secondary and tertiary branches bearing galls of varying sizes. When many juveniles penetrate a root, root tip is devitalized and elongation of the root ceases resulting in the production of many fine tertiary branches forming brush-like appearance. The nematode prevents the availability of water and nutrients and desired energy for the production of fruit resulting into yield loss.

14.2.3. Histopathology

The larvae of *M. javanica* readily penetrate susceptible peach rootstock (Sharbati) and develop into mature females. The production of giant cells was detected as early as fourth day after nematode penetration. Giant cell formation takes place in pericycle area, a few cells away from nematode head. The cytoplasm of affected cells becomes dense, followed by enlargement of nuclei and nucleoli. The affected cells undergo fast multiplication of nuclei without cytoplasmic division, thus forming syncytium. In most cases, giant cells are located within vascular tissues.

14.2.4. Interaction with other pathogens

Crown gall of peach, caused by *Agrobacterium tumefaciens*, is increased by high populations of *M. javanica* (Nigh, 1966).

In an interaction study, Lovell seedlings were more susceptible to crown gall causing organism, *A. tumefaciens* in the presence of *M. javanica*. Not only there was significantly higher incidence of bacterial infection but also a larger number (27.5% increase) of nematode galls when the two organisms occurred together.

14.2.5. Management

14.2.5.1. Physical methods: Hot water treatment of peach seedling roots at 50-52 °C for 5-10 min eliminates the nematode infection.

14.2.5.2. Cultural methods: Rotations with cereals or grasses have been used to suppress nematode populations before establishing stone fruit orchards. Thus, coastal Bermuda grass (*Cynodon dactylon*) has been grown in sites infested with *Meloidogyne* before establishing peach orchards (Bertrand and Nyczepir, 1989).

Nematode-free planting stock is essential for transplanting in field. Planting of cover crops like tall fescue grass cultivar Jesup (Max-Q- a novel fungal non-toxic endophyte, *Neotyphodium coenophialum*) before taking up peach helps in reducing the root-knot nematode population in soil and enhance plant growth. Max-Q has been shown as a non-host/poor host to *M. incognita* and *M. hapla*, a poor host for *M. javanica* and *Pratylenchus vulnus*, and a good host for *M. arenaria* and *Mesocriconema xenoplax* (Nyczepir and Meyer, 2010; Nyzcepir, 2011).

14.2.5.3. Chemical methods: Treatment of nursery beds with Carbofuran 3G @ 3 kg a.i. /ha; root dip treatment of nursery plants in 0.1% solution of Fenamiphos for 6 hr. before planting in the main field; and soil application of Carbofuran 3G @ 100 and 150 g/plant twice or thrice a year in the main field proved beneficial.

The post-plant nematicides, Spirotetramat (Movento) at 0.42 kg/ha and Fluensulfone (MCW-2) at 0.014 kg/ha significantly reduced *M. incognita* numbers ($P < 0.05$) compared to the control in peach (Shirley, 2013).

14.2.5.4. Biological methods: *G. mosseae* at 400 g/pot (50 spores/g soil) colonized 67% peach roots and significantly enhanced plant growth as indicated by increased shoot and root weight and length and suppressed the multiplication of *M. javanica* on peach cv. Floridasun (Baghel and Bhatti, 1990). Strobel *et al.* (1982) reported that AM fungus, *Gigaspora margarita* decreased the number of *M. incognita* eggs produced per g of peach root growth at a low P level.

The egg parasite *Dactylella oviparasitica* suppresses the population of root-knot nematodes on peach trees (Stirling *et al.*, 1979).

14.2.5.5. Host resistance: Nemagaurd, Nemared, Flordaguard, Okinawa, Rancho, 'Cadaman' and 'Guardian' are the most popular root-stocks resistant to *M. incognita, M. javanica* and *M. arenaria* (Sharpe *et al.*, 1969; Fernández *et al.*, 1994; Marull *et al.*, 1994; McKenry, 1999; Simeone and Di Vito, 1992). Yunnan, Shalil and Bokhara have been found resistant to *M. incognita* but susceptible to *M. javanica*. Root-knot nematode resistant peach cultivars include Chinaflat, Bidbillos Early, Japanese Double and Elberta (Verma, 1987).

Tall fescue (*Schedonorus arundinaceus*) cv. Jesup (Max-Q - a novel fungal non-toxic endophyte, *Neotyphodium coenophialum*) served as a poor host to both *M. floridensis* and *M. incognita*, but was effective in only suppressing reproduction of *M. arenaria* (Shirley, 2013).

Hybrid rootstocks such as Atlas and Vikings are also reported to be root-knot resistant. Recent introductions having root-knot resistance include peach hybrids Barrier 1 and Cadaman (both *P. persica* x *P. davidiana*), Penta (Empyrean 2) and Tetra (Empyrean 3) (both *P. domestica*), Adesoto 101 (*P. insititia*), plum hybrids Myran (*P. cerasifera x P. salicina*), Ishtara [*P. cerasifera* x (*P. persica x P. cerasifera*)], Julior (*P. insititia x P. domestica*), Hiawatha (*P. besseyi x P. salicina*), peach-almond (*P. dulcis x P. persica*) hybrids Garnem (GxN15) and Felinem (G x N22), and Pumiselect (*P. pumila*) (Esmenjaud *et al.*, 1997; Fernandez *et al.*, 1994; Pinochet *et al.*, 2002; Nicotra and Moser, 1997). Most of these rootstocks are adequately compatible with peach.

14.3. ROOT-KNOT NEMATODE, *MELOIDOGYNE FLORIDENSIS*

M. floridensis is endemic to Florida. In 1966, *M. floridensis* was first detected infecting peach rootstocks ‹Okinawa› and ‹Nemaguard› in Gainesville, Florida. This Florida-endemic species overcomes the resistances in 'Okinawa', 'Nemaguard', 'Nemared', and 'Guardian' peach rootstocks (Esmenjaud *et al.*, 2009; Nyczepir and Beckman 2000); therefore, these rootstocks commonly used in major peach production regions in California, Georgia, and the Carolinas are not suitable for commercial peach production in Florida.

14.3.1. Distribution

Its distribution in Florida extends from Miami-Dade County northward to Alachua County (Brito *et al.*, 2008; Brito *et al.*, 2010). A recent survey conducted by nematologists at the University of Florida and the Division of Plant Industry of 25 Florida peach orchards shows the incidence of peach infected by *M. floridensis* at 36%.

14.3.2. Symptoms

Root-knot disease on a peach tree is often overlooked because the infective juvenile is microscopic and dwells underground. Aboveground symptoms of diseased trees generally develop within a year after transplanting seedlings. The severity of the disease manifestation is dependent on the population density of juveniles in the soil. The aboveground symptoms include stunting, yellowing, premature leaf drop, and reduced vigor. The symptoms mock those caused by other factors such as nutrient deficiencies, water stress, and root injuries from other root-damaging pathogens and pests. These symptoms become most apparent near the end of the growing season (late summer or fall) and may be most pronounced when infected trees are under stress from lack of moisture. The leaves turn pale-green to yellowish in color, and older leaves may drop prematurely due to the reduced level of photosynthates to support plant growth. With the reduced foliage and lower carbohydrate reserves in the roots, the following season's fruit yield and quality are reduced on bearing trees grafted on susceptible rootstocks. The belowground symptoms include severe root galling.

14.3.3. Host range

Meloidogyne floridensis is a endoparasitic nematode that feeds and reproduces on the roots of several ornamentals (dracaena, hibiscus, impatiens, and snapdragon), vegetables (cucumber, eggplant, squash, watermelon, and tomato), herbs (verbena, basil, and dill), weeds (morning glory, velvetleaf, amaranth, redroot pigweed, red tassel flower, dichondra, American pokeweed, lion›s ear, wild cucumber, and spurge nettle), and fruits (peach).

14.3.4. Life cycle

The development time from egg to egg-laying adult is dependent on temperature and soil moisture. The cycle is slower (greater than 5 weeks) during the cooler months and more rapid (3–4 weeks) during warmer months.

14.3.5. Management

14.3.5.1. Chemical methods: Vapam (metam sodium) formulation currently is the only pre-plant soil fumigant labeled for peach growers in the Southeastern United States.

14.3.5.2. Host resistance: Standard rootstocks in the southeastern United States, such as 'Nemaguard', 'Nemared', and 'Guardian', are believed to be susceptible to the endemic *M. floridensis*. Currently, only 'Flordaguard' peach rootstock (Sherman *et al.*, 1991) has been suggested for commercial, low-chill peach production in root-knot nematode infested, non-alkaline soils in Florida. This rootstock has the combined nematode resistance and good horticultural features from 'Okinawa' peach and Chinese wild peach (*Prunus davidiana*).

An interspecific peach-plum hybrid rootstock 'MP-29' offers attributes lacking in 'Flordaguard'. It has additional resistance to predominant pathogens in the Southeast such as Armillaria root rot and peach tree short life disease.

14.4. RING NEMATODE, *MESOCRICONEMA XENOPLAX*

The ectoparasitic ring nematode causes economic damage to peach and is considered as one of the important components in peach tree short life (PTSL) disease complex. *Mesocriconema xenoplax* predisposes peach trees to PTSL. Under field conditions in Byron, Georgia, trees on 'MP-29' showed higher yield efficiency (higher yields with smaller trunk cross-sectional area) than those on 'Guardian' and 'Sharpe' rootstocks (Beckman *et al.*, 2012).

14.4.1. Economic importance

The estimated losses that occurred in peach orchards suffering from PTSL in Georgia were 30-70% after five years (Wehunt *et al.*, 1980). *M. xenoplax* was responsible for 33% loss in fruit yield of peach (Anon, 1990).

14.4.2. Distribution

M. xenoplax was first observed on peach grown in Northern India. It is distributed in Himachal Pradesh and Haryana.

14.4.3. Symptoms

The infected plant roots are darkened and often contain longitudinal cracks. The most serious infection appears in the form of the death of finer roots due to direct feeding by nematodes. Due to excessive killing of feeder roots, plants are stunted, showing mineral deficiency syndrome, and more susceptible to water stress. Parasitism of *M. xenoplax* causes chlorosis, leaf drop, stunting and loss of vigor on top of peach trees. Peach roots showed lesions, pits and fewer feeder roots. The roots are predisposed to bacterial canker caused by secondary infection of *Pseudomonas syringae*.

14.4.4. Life cycle

According to Seshadri (1964), eggs are deposited singly, 8-15 in number per female in 2-3 days egg laying period. Total life cycle was completed in 25-34 days at 22-26 °C in sandy soil at pH 7.0 and 15.5% moisture. It takes 11-13 days for egg stage, 3-5 days for second stage, 4-7 days for third stage and 5-6 days for fourth stage development and pre-oviposition period of adult. The nematode is multivoltine and completes several generations in a year.

14.4.5. Interaction with other pathogens

Parasitization by *M. xenoplax* predisposes peach trees to bacterial canker caused by *Pseudomonas syringae* (Lownsberry *et al.*, 1977). *M. xenoplax* has been strongly implicated in the peach tree short life (PTSL) condition in some parts of USA. PTSL disease syndrome is characterized by a sudden collapse of the new growth above soil line and death of trees in late winter or early spring shortly after foliation (Taylor *et al.*, 1970). In addition, bacterial spot damage caused by *Xanthomonas campestris* pv. *pruni* becomes more severe on peach trees if the soil is infested with *M. xenoplax* (Shepard *et al.*, 1999).

14.4.6. Management

14.4.6.1. Physical methods: Hot water treatment of peach seedlings at 50-52 °C for 5-10 min. eliminates the nematode infection.

A 37-100% reduction in population of *M. xenoplax* and other nematodes has been achieved by soil solarization in South Africa (Barberecheck and Broembsen, 1986). Soil solarization used to suppress populations of *M. xenoplax* in peach orchards was found effective for up to 19 months, and increased peach tree growth (Nyczepir and Kluepfel, 2007).

14.4.6.2. Cultural methods: Sorghum as a green manure with and without plastic did suppress the population of *M. xenoplax* in the early stages (Nyczepir and Rodriguez-Kabana, 2004).

Pre-plant crop rotation with wheat has been shown to be comparable to methyl bromide fumigation in suppressing the ring nematode (*M. xenoplax*) population prior to establishing a peach orchard. Pre-planting bahia grass (*Paspalum notatum*) or wheat (*Triticum aestivum* 'stacy') increased peach tree growth and survival from PTSL in soil infested with *M. xenoplax* (Nyczepir and Bertrand, 2000).

Coastal Bermuda grass and small grains, especially wheat, are poor to non-hosts for ring nematode. Several consecutive years of growing coastal Bermuda grass or wheat on an anticipated peach orchard site may possibly reduce the necessity for using a preplant fumigant.

14.4.6.3. Chemical methods: The post-plant nematicides, Spirotetramat (Movento) at 0.42 kg/ha and Fluensulfone (MCW-2) at 0.014 kg/ha significantly reduced *M. xenoplax* numbers ($P < 0.05$) compared to the control in peach (Shirley, 2013).

14.4.6.4. Biological methods: *Hirsutella rhossiliensis* reduced nematode population by 50%. Rate of nematode decay was much faster for parasitized second or third-stage juveniles (Jaffee *et al.*, 1988).

Treatment of peach seedlings with a strain of *Pseudomonas aureofaciens* isolated from suppressive soil inhibited infection by the ring nematode in soil (McInnis *et al.*, 1990).

14.4.6.5. Host resistance: Lovell rootstock of peach has been recorded as most tolerant to *M. xenoplax* (Rom *et al.*, 1985).

14.5. PEACH TREE SHORT LIFE (PTSL) DISEASE COMPLEX

The PTSL complex is caused by the susceptibility of young peach trees (approximately 3-5 years in age) to cold injury and/or bacterial canker caused by (*Pseudomonas syringae* pv. *syringae*). Both of these are initiated by parasitism from the ring nematode, *Mesocriconema xenoplax* (Beckman and Nyczepir, 2011; Brittian and Miller, 1978; Nyczepir, 1989; Ritchie and Clayton, 1981). Ring nematode presence appears to be the most important predisposing factor to consider (Nyczepir, 1990).

14.5.1. Symptoms

One common symptom is the unexpected collapse of growth and eventual death of a young tree (Brittain and Miller, 1978; Beckman and Nyczepir, 2011; Nyczepir, 1989; Ritchie and Clayton, 1981), which are similar to the symptoms observed on any plant deprived of water. Removal of bark in the affected areas will show a transition between healthy and brown necrotic vascular tissue. At this division, the characteristic brown tissue will extend and follow down the tree trunk to the soil line, where it ceases. This brown tissue will give off a distinct "sour sap" smell which is characteristic of PTSL. This odor is caused by the fermentation of carbohydrates released by the plant's damaged cambial tissue. This symptom may also occur along with water-soaking of the bark and leakage or ooze from lenticels within the bark. The primary root system below the soil line remains alive and appears healthy, but upon further examination of the feeder roots one will detect unhealthy, necrotic roots and a reduced number of tertiary roots caused by parasitism from *M. xenoplax*. In mid- to late summer, suckers may form at the base of the trunk from the surviving root system (Brittian and Miller, 1978; Nyczepir, 1989). Symptoms associated with bacterial canker usually coincide with a delay in flowering and emergence of new foliage in the spring, with decline and death of the affected limbs or tree by late summer. Cracking of damaged bark may occur, and it can easily be peeled away from the limbs and trunk of the tree (Beckman and Nyczepir, 2011).

14.5.2. Management

14.5.2.1. Cultural methods: Lime should be applied to bring the pH levels between 6.0-6.5 and nutrients should be applied during liming. Subsoiling the land to remove any hardpan which may exist must be done during this process. Subsoiling the new site will improve water infiltration and drainage, encourage root growth and development, increase tree survival during weather extremes, and enhance nutrient uptake (Beckman and Nyczepir, 2011; Brittian and Miller, 1978).

The rotation of land with wheat for a period of three years before planting a new peach orchard has been shown to be as successful as a pre-plant fumigation with methyl bromide in suppressing ring nematode and increasing tree survival in an old peach site with a history of PTSL (Nyczepir, 2011).

Pruning at the proper time will have an adverse effect on the development of PTSL within an orchard. The proper time to prune is in the spring from February to June (Horton *et al.*, 2013).

14.5.2.2. Host resistance s: The rootstock Lovell has no resistance when it comes to nematodes which parasitize peach, but has some tolerance to PTSL as compared to Nemaguard. Trees grown on Lovell live longer and are more productive than trees on Nemaguard in a PTSL site (Nyczepir, 2011). The rootstock Guardian also has resistance to both common species of root-knot (*M. incognita* and *M. javanica*), and is generally more resistant to PTSL than either Nemaguard or Lovell rootstocks.

14.5.2.3. Integrated methods: the use of pre-plant fumigation, resistant rootstocks, and rotation will not totally eliminate nematodes and over time the population will build back up to damaging levels if no post-plant management practice is utilized.

In conclusion, management of PTSL currently includes; proper site management, correct cultural practices, proper rootstock selection, use of a viable cover crops, along with the application of a soil fumigant.

14.6. LESION NEMATODES, *PRATYLENCHUS VULNUS, P. PENETRANS*

In Canada, the lesion nematode causes 'peach replant problem'.

14.6.1. Symptoms

Above-ground symptoms include tree decline with general nutrition deficiency symptoms and reduced growth. Under-ground symptoms are root lesions or discoloration of roots usually reddish brown later turning dark and black. The track of lesions on root becomes visible as the nematodes feed and migrate from one place to another. Rapid deterioration of feeder roots and reduced vigor are symptoms caused by *P. vulnus*. The nematode poses threat to replanting of peach.

14.6.2. Life cycle

The nematode enters roots and moves inter and intracellularly and feed on cortical cells, the nematode larvae and adults both have capacity to enter or leave the roots. Eggs are laid by the adult female nematode in the soil or in roots. Hatching begins to take place immediately after egg laying. After hatching, different stages of larvae feed on roots and develop into adults. Under unfavorable conditions, nematodes leave the roots and migrate to new favorable roots for penetration and feeding.

14.6.3. Interaction with other pathogens

Nematodes of the genus *Pratylenchus* along with *Pythium* fungus has been associated with a large percentage of peach trees suffering from peach decline. *Criconemoides* spp. and *Tylenchorhynchus* spp. have also been associated with this condition (Hendrix *et al.*, 1965).

14.6.4. Management

14.6.4.1. Physical methods: Hot water treatment of peach seedlings at 52 °C for 15 min. eliminates *P. penetrans* infection.

14.6.4.2. Chemical methods: Combining pre-planting treatment with Carbofuran and post-planting treatment with Phorate annually to standing crop proved beneficial.

14.6.4.3. Biological methods: Pinochet *et al.* (1995) found that pre-colonization of peach roots with *Glomus mosseae* was effective in reducing the *P. vulnus* population.

14.6.4.4. Host resistance: Rubira, GF 305, Penta and Tetra are listed as having tolerance to *P. vulnus* in Europe (Alcaniz *et al.*, 1996). Pinochet *et al.* (2002) also reported that Krymsk 86 has resistance to *P. vulnus* and to a lesser extent 'Krymsk 1'.

14.7. DAGGER NEMATODE, *XIPHINEMA AMERICANUM*

The major damage to peach trees from dagger nematode feeding is that it serves as the vector for tomato ringspot virus (TmRSV), which causes stem pitting.

Prunus stem pitting (PSP), which is caused by TmRSV. This virus is vectored by *Xiphinema rivesi* and *X. americanum*. In addition, many broad leaf weeds serve as hosts to both the virus and nematode vectors.

14.7.1. Symptoms

Affected trees appear weak and show a general decline. Leaves may have upward cupping, turning prematurely yellow or reddish purple, droop, and then prematurely drop. The bark is abnormally thick and spongy and the wood underneath has a severely pitted, indented texture. Symptoms are most severe in the wood just above and below the soil line.

Above-ground symptoms of nematode damage can be hard to detect and easily confused with other plant stresses resulting from water or nutrient deficiencies. The most commonly recognized symptoms that can be attributed to this pest are those that result from transmission of the associated viruses.

Plant symptoms resulting from viral infection commonly include: chlorotic mottling and defoliation; bright yellow discoloration of foliage; yellowing along main leaf veins; spots, blotches, and crinkling of leaves; and stunting.

Dagger nematode is a migratory ectoparasite which means it attacks roots from the outside. It is a relatively large nematode (1.3 to 2.2 mm) (Akinbade *et al.*, 2014). Dagger nematode has a long, hollow, needle-like stylet called an odontostyle which it uses to feed on cells at root tips. Feeding can cause root tip cells to enlarge, disrupting function and resulting in roots that have a knobby, stunted appearance, with swellings at the tips.

14.7.2. Host range

The virus has a wide host range affecting all *Prunus* species (Cherries and peaches) hop and rose.

14.7.3. Life cycle

The life cycle may be up to one year in duration with individuals living three to five years.

14.7.4. Virus transmission

Dagger nematodes (*Xiphinema rivesi* and *X. americanum*) can harbor virus particles in the esophageal lumen near the stylet extension and in the glandular region of the esophagus. As the nematode feeds, virus particles can be transmitted to the plant within 24 hours of exposure (Ferris, 2017).

14.7.5. Management

14.7.5.1. Cultural methods: Since many weed species such as dandelions are hosts for this virus, dagger nematode resistance in rootstocks is the only way to prevent infection. Thus, broadleaf weed control is important in the mid-Atlantic region to prevent reinfestation of orchards with TmRSV and to reduce natural reservoirs of the virus.

TmRSV can survive in a number of weed hosts. Thus, good weed control and pre-planting site preparation can help to reduce the incidence of this disease. Certified virus-free planting stock should be purchased.

The prunus ringspot disease can be managed through strategies such as the use of virus-free planting material produced by regular serological and molecular indexing of mother trees by ELISA and RT-PCR. Additionally, use of biotechnological approaches such as *in vitro* micropropagation using axillary and terminal buds as explants.

Crop rotation with tolerant/resistant cover crops like tall fescue, crotalaria and perennial ryegrass was found effective in reducing the dagger nematode population.

14.7.5.2. Host resistance: Marianna 2624 (incompatible with peach) was reported in Michigan to have some field resistance (Kommineni *et al*, 1998). Also, some cherry plum (*P. cerasifera*) genotypes appear to be less sensitive to tomato ringspot virus (Hoy and Mircetich, 1984; Halbrendt *et al.*, 1994). Therefore, rootstocks like Mr. S. 2/5 (*P. cerasifera*), Krymsk 1, and Krymsk 2 (*P. incana* x *P. cerasifera*) may offer some tolerance.

14.8. PIN NEMATODE, *PARATYLENCHUS* SPP.

These are the smallest representatives among the plant parasitic forms which attack herbaceous as well as woody plants. At least seven out of total of more than 100 species are associated with peach. Of these, *P. prunii, P. nannianus* and *P. curvitatus* are more common.

14.8.1. Symptoms

General plant decline may be difficult to diagnose. The nematode produces non-specific symptoms like stunting of the plant, yellowing of foliage and under-sized fruits. Pin nematode is a small nematode which feeds either singly or in groups on the outside of plant roots. Feeding causes small lesions and death of lateral roots. In general, it is found randomly distributed in the orchard across tree rows and alleyways (Howland *et al.*, 2014). Small root lesions may be noticeable. The extent of damage depends upon the age and vigor of the plant at the time of infestation.

Roots of peach seedlings inoculated with surface-sterilized P. *neoamblycephalus* were smaller, darker, and had fewer feeder roots than those of non-inoculated controls. Nematodes were observed feeding ectoparasitically, but with heads embedded in roots as deep as the cortex. They were associated with small lesions and dead lateral roots. Clusters of nematodes were common at ruptures in the epidermis, and where lateral roots emerged (Braun and Lownsbery, 1975).

14.8.2. Host range

Rose, apricot, peach, and all selections and hybrids of *Prunus cerasffera* tested were hosts for *P. neoamblrcephalus* (Braun and Lownsbery, 1975).

14.8.3. Ecology

Limitation of plant growth by *P. neoamblvcephalus* was greater at 20 and 27 ^{0}C than at 30 ^{0}C, and was not affected by pH over the range 4.5 to 6.5 (Braun and Lownsbery, 1975).

14.8.4. Management

Nursery should be raised in nematode free soil. Only healthy, nematode free planting stock should be transplanted in the field. Weed removal, nutritional supply in form of organic amendments and use of resistant rootstocks help in keeping the nematode population under control. Root dip treatment in oxamyl @ 0.024 g/L kills the nematode population.

14.9. CONCLUSION

With the uncertainty of the effects of global climate change, increased demand for "higher quality" fruit, the threat of the introduction of exotic nematode pests, and a demand for alternatives to chemical nematicides, there is a need for new management practices. Due to a lack of pre and post-plant nematicidal options afforded to peach producers and the increased cost of those that remain, reduced rates of soil fumigants, alternatives to chemical controls, nematode non-host groundcovers/rotation crops, biorational nematicides, and improvements to our cultural practices are becoming ever more important. There is a growing interest in the use of suppressive groundcovers as a nematode management strategy in peach producing areas. One groundcover that has shown to be a promising option is the tall fescue grass cultivar Jesup (Max-Q) which is a non-toxic fungal endophyte (*Neotyphodium coenophialum*) infested tall fescue developed as a viable forage crop for cattle production. It also imparts drought tolerance to peach (Phillips *et al.*, 2009). Max-Q has been shown as a non-host/poor host to a number of nematodes. Further, there is a need for a sustainable system of nematode management practices and IPM strategies. A portion of these challenges will be met with the goal of researching new cultural practices (in terms of planting, rotations, cover crops, *etc.*), reductions in chemical inputs, and alternatives to chemicals for the suppression of plant-parasitic nematodes.

Chapter - 15

Strawberry, *Fragaria* × *Ananassa*

15.1. INTRODUCTION

15.1.1. Origin

The garden strawberry was first cross bred in Brittany, France in the 1750s via cross of *Fragaria virginiana* from eastern north America and *Fragaria chiloensis*, which was of Chile and later run commercially and the first strawberry named woodland strawberry which was first cultivated in the early 17th century was replaced by this hybrid.

15.1.2. Area and production

Strawberry is a native of temperate regions, but varieties are available which can be grown in subtropical areas The cultivated strawberry (*Fragaria* x *ananassa*) was first tried in Europe in the early 18th century and represents the accidental cross of *F. virginiana* from eastern North America, which was noted for its fine flavor and *F. chiloensis* from Chile and noted for its large size. Strawberry industry in America was started in the beginning of 18th century, and today nearly 50% of worlds production is in Europe. The strawberry is commercially grown on large scale in Europe and North America. Among the European countries, France, Italy, Poland and Spain are the major producers. The other countries having a sizable production of strawberries are Russia, Japan and Turkey (Table 15.1).

Table 15.1. Top 10 strawberry producing countries in the world – area, production and yield (2019)

Country	Area ('000 ha)	Production ('000 tons)	Yield (tons/ha)
China	125.637	3212.814	255.722
USA	18.130	1021.490	563.425
Mexico	16.429	861.337	524.278
Turkey	16.090	486.705	302.489
Egypt	11.772	460.245	390.966
Spain	7.260	351.960	484.793

Russia	31.122	208.800	67.091
South Korea	6.212	192.971	310.642
Poland	49.900	185.400	37.154
Morocco	3.455	167.827	485.751
WORLD	396.401	8885.028	189.003

In India strawberry is commercially grown in plains mainly for fruits and in hills for raising planting material production. The major strawberry growing states in India are Haryana, Himachal Pradesh, Jammu and Kashmir, Kerala, Madhya Pradesh, Maharashtra (Mahabaleshwar), Meghalaya, Mizoram and Tamil Nadu (Table 15.2).

Table 15.2. State-wise area, production and productivity of strawberry in India (2017-18)

State	Area ('000 ha)	Production ('000 tons)	Productivity (tons/ha)
Haryana	0.19	3.11	16.368
Himachal Pradesh	0.05	0.05	1.000
Jammu & Kashmir	0.09	0.14	1.555
Kerala	0.05	0.08	1.600
Madhya Pradesh	0.10	0.85	8.500
Maharashtra	0.17	1.08	6.353
Meghalaya	0.10	0.85	8.500
Mizoram	0.17	1.08	6.353
Tamil Nadu	0.01	0.05	5.000
TOTAL	0.65	5.36	8.246

Source: Horticulture Statistics Division, Dept. of Agri. Co-op. & Farmers' Welfare.

15.1.3. Varieties

Several varieties have been bred in different countries and evaluated under local conditions. The varieties recommended for Indian conditions include Chandler, Camarosa, Sweet Charlie, Ofran, Oso Grande, Gorella, Selva, Sanga Sangana, Belrubi, Shasta, Fern, Tioga and Pajaro. Other varieties include Premier, Red Cost, Local Jeolikot, Dilpasand, Bangalore, Florida 90, Katrain sweet, Pusa Early Dwarf, and Blakemor.

15.1.4. Export

The total volume of Strawberry export around the world in year 2018 was 26722920. Netherlands is the largest exporter of Strawberry. The figures show the great potential for Indian exporters of Strawberry to increase their participation in global trading and improve their numbers. In YTD 2018, India exported 26722920 million metric tons of Strawberry showing the fall of -3.42% compared to YTD 2017.

Strawberry is exported to over 100 countries. In the year 2020-2021 (Apr-Nov), India has exported Strawberry worth of 32.24 USD million. The total volume of export in 2020-2021 (Apr-Nov) was around Rs. 28,053,799.

Below are the top countries which imported the highest dollar value worth of Strawberry from India in 2020-21 (Apr-Nov) (Table 15.3).

Table 15.3. Export of strawberry from India to other countries (2020-21).

Country	Value (Million USD)	Share (%)
Netherlands	11.61	36.01
China	3.14	9.74
USA	3.04	9.43
UK	2.66	8.25
Saudi Arabia	1.83	5.68
Canada	1.19	3.69
Japan	1.17	3.63
Italy	1.14	3.54
France	0.99	3.07
Australia	0.70	2.17

The top 5 trading partners of India are Netherlands (11.61 USD Million), China (3.14 USD Million), USA (3.04 USD Million), United Kingdom (2.66 USD Million), Saudi Arabia (1.83 USD Million) . The total export value of Strawberry in these countries is 22.28 USD million. These top 5 countries account for over 69.11% of the total Strawberry export from India.

15.1.5. Nematode problems

The economically important nematodes infecting strawberry include bud and leaf, lesion, and root-knot nematodes. *Aphelenchoides* spp. (*A. fragariae*, *A. ritzemabosi*), which are causal organisms for disease 'Dwarf Crimp', 'Spring dwarf of strawberry' and 'Red plant of strawberry'. *Aphelenchoides besseyi* causes 'Summer crimp of Strawberry'.

15.2. BUD AND LEAF NEMATODE, *APHELENCHOIDES* SPP.

Strawberry dwarf is caused by at least three species of *Aphelenchoides* (*A. besseyi, A. fragariae*, and *A. ritzemabosi*). The nematode is responsible for a cool weather disease of strawberry called 'spring dwarf' caused by *A. fragariae*. The summer dwarf is caused later in the season by *A. besseyi.* Susceptible varieties showed 60 to 82% reduction in yield after 2 years of infestation.

15.2.1. Symptoms

Foliar nematode symptoms on strawberry include deformation of buds, leaves, and flowers; undersized leaves with crinkled edges; tight crowns; reddened and stunted petioles; and few flowers and fruits. Spring dwarf (caused by *A. fragariae*) symptoms appear in the early spring as soon as plant growth starts. Bud and leaf development is retarded and plants fail to develop normal foliage. Leaves and petioles in the crown are modified into a compact mass of crinkled, twisted, and distorted structure. When the apical bud is killed, numerous adventitious buds develop and give rise to multiple-crowned plants. Generally fruit buds are killed. The petioles of certain cultivars develop a prominent red coloration, and hence the name 'red plant'.

15.2.2. Life cycle

They have short life cycles (egg to adult in 14 days) and high reproductive potentials, so they can increase in number very rapidly (Warner, 2018). The life cycle is completed in 10-11 days at 18°C. The eggs hatch in 4 days and the larvae mature in 6-7 days. About 32 eggs are laid by a female. Occasionally the mother is killed when eggs hatch in the uterus.

15.2.3. Interaction with other pathogens

A. fragariae and/or *A. ritzemabosi* occasionally interact with a bacterium, *Rhodococcus fascians*, causing "cauliflower disease". Both nematode and bacterium must be present to produce the disease. Diseased plants are stunted with deformed stems, leaves and flowers which in advanced stages resemble tiny cauliflowers.

15.2.4. Favorable conditions

Wet weather during the strawberry growing season is most favorable for spread of foliar nematodes.

15.2.5. Survival and spread

- Nematodes are found to over-winter as juveniles and adults in soil, on dead leaves and in dormant buds.
- Spread by contact between plants in the presence of water.
- They migrate in the film of water on plant leaves.

15.2.6. Management

15.2.6.1. Regulatory methods: The bud and leaf nematode can be managed by selecting nematode-free propagation stock. Whenever possible, certified stocks should be used.

15.2.6.2. Physical methods: Hot water treatment of strawberry runners at 50 °C for 5 min or 48 °C for 15 min or 46 °C for 20 min eliminates the nematode infection.

Tissue culture technology can be employed to regenerate nematode free strawberry plants from clean, dormant, excised axillary buds.

15.2.6.3. Cultural methods: *Aphelenchoides* spp. control includes removal and destruction of infected strawberry leaves. Avoiding periods of relatively long plant wetness should be considered since this nematode relies on moisture to move up the plant and between plants. Following this line of thinking, drip irrigation is preferable over overhead spray irrigation.

Uprooting and destroying infected plants and propagation through healthy planting stock are recommended for the control of nematodes.

Crop rotation with grain crops such as barley and rye proved beneficial.

15.2.6.4. Chemical methods: Foliar sprays with Fenamiphos at 4-6 ml/3.8 liters of water gave good control of *A. fragariae.* Chemicals used for disinfestation of strawberry plants include Parathion as a foliar spray and soil application of Carbofuran in granular form.

15.2.6.5. Host resistance: In Poland, the strawberry varieties "George Soltwedel", "Regina" and "Talizman" are comparatively resistant as are "Saksonka" and "Festival'naya" in USSR.

15.3. LESION NEMATODE, *PRATYLENCHUS PENETRANS*

15.3.1. Economic importance

Initial lesion nematode densities of 100, 150, and 200/g root reduced yield by 20.5%, 30.8%, and 41.0% respectively but reduced accumulated profit by 63.7%, 95.3%, and > 100%, respectively. The decline in yield experienced in these scenarios and resulting economic losses would likely result in the early removal and replanting of the crop (LaMondia, 1999).

15.3.2. Symptoms

Damage is frequently seen as patches in the field. The above-ground symptoms include stunting of plants, pale green leaves, reduced yields and eventually death of plants. The below-ground symptoms include distinct brown necrotic lesions on roots in early or light infestation. Shortened crop cycle, reduction in crop yield, and decreased production of runners are also noticed. Fewer fine feeder roots and a brushy appearance of the plant can be observed. Roots have individual reddish-brown lesions that can be so numerous on the feeder roots. The root is girdled. The entire root system eventually becomes black and necrotic as populations increase and secondary organisms invade.

15.3.3. Interaction with other pathogens

The presence of the lesion nematode makes the plant susceptible to soil-borne disease pathogens like species of *Verticillium*, *Fusarium*, *Cylindrocarpon*, *Rhizoctonia*, and *Pythium*.

15.3.4. Management

15.3.4.1. Regulatory methods: Use of certified nematode-free plants.

15.3.4.2. Physical methods: Hot water treatment of dormant strawberry plants at 54.4 °C for 1 min or 52.8 °C for 3 min or 51 °C for 7.5 min or 49.4 °C for 17.5 min eliminates the nematode infection.

Soil biosolarization with chicken manure reached efficacies of 67% reduction for *P. penetrans*, which seems to provide sufficient nematode control in most situations. Soil biosolarization is proposed as an effective alternative to chemical soil fumigation for the management of nematodes in strawberry cultivation (Talavera *et al.*, 2019).

15.3.4.3. Cultural methods: Crop rotation with potato (moderately susceptible and causing moderate nematode reproduction), oats or barley (not susceptible but causing nematode reproduction), and beet (not susceptible and suppressing nematode reproduction) was found effective. Specifically, rotating pearl millet with susceptible crops such as strawberry has been used successfully in Egypt and in Michigan, USA (Warner, 2018) to reduce population densities of *P. penetrans*.

Cover crops such as switch grass has been shown to reduce populations of lesion nematodes (Pritts and Nonnecke, 2004). Use of cover crops like Yellow mustard (*Brassica juncea*) which contains allyl isothiocyanate which is effective against nematodes. For effective nematode management, *Brassica carinata* also can be used.

Fallow periods can disrupt the nematode life cycle. Practice good weed control using herbicides or tillage.

15.3.4.4. Biological methods: MeloCon WG (*Purpureocillium lilacinum* strain 251) at 2 to 4 kg/ha plus a soil wetting agent can be applied to established plants, although it might be better used when applied to plants just before planting.

15.3.4.5. Integrated methods: Integration of cultural (crop rotation) and chemical (nematicides) methods can be employed to manage the lesion nematode:

» Soil application of nematicide reduces nematode population to a very low level.
» Rotation of strawberry with potato (which is moderately susceptible and causing moderate nematode reproduction), beet root (which is not susceptible and suppressing nematode reproduction), and oats or barley (which are not susceptible but causing nematode reproduction)

15.4. ROOT-KNOT NEMATODES, *MELOIDOGYNE* SPP.

15.4.1. Symptoms

Symptoms of RKN infection tend to occur in more or less definite areas where transplants fail to develop normally. Infested patchy areas are usually fairly well defined from those with healthy plants. Unless a proper nematode control measure is followed, a field infested with RKNs, which has only a few such patchy areas at the beginning of the growing season may then increase in size and number until almost the whole field will be infested. The general impact of RKNs on strawberries is to cause both stunting and decline in plant growth and subsequent yield, the intensity of which is related to initial nematode population level and the rate at which population increase in reaction to the infested plants during the growing season. Plantings which are stunted by nematodes often have worse weed problems than areas without nematode injury because the crop is less able to compete with weeds or even any other stresses than it should be. Most symptoms associated with them are the result of inadequate water supply or mineral nutrition to the tops.

Feeding by RKNs induces distinguished knot-like swellings (called galls) on the roots. When plants are severely infected by *Meloidogyne*, the normal root system is reduced to a limited number of severely RKN-galled roots with a completely disorganized vascular system. The RKN-infected roots are seriously hampered in their main functions of uptake and transport of water and nutrients. Consequently, the plants do not flower properly and therefore produce fruits of poor quality, and it is very easy to get drought damage. Rootlets are almost completely absent at severe infestation which may render plant death.

Leaf yellowing, stunting of plant growth, wilting of plants during hot sun, and yield reduction are some of the symptoms that appear on the above-ground plant parts. The main below-ground symptoms produced by the root-knot nematodes include galling at the tips of roots and abundant branching of roots above the root galls. Shortened crop cycle, reduction in crop yield, and decreased production of runners are also noticed.

In warm sandy soils, attack by *M. hapla* can cause severe stunting. When the nematode interacts with other soil-borne plant pathogens, death of plants is the possibility. In the presence of nematodes, starvation of plants for water is very severe during drought season.

15.4.2. Management

15.4.2.1. Physical methods: Soil solarization with transparent polythene mulch gave the highest reduction percentage of total nematode population and reproduction rates of RKN.

Biosolarization is a modified form of solarization that combines organic soil amendments, with passive solar heating under a transparent plastic mulch, which creates multiple pest and pathogen inactivation mechanisms in the soil (Domínguez *et al.*, 2016). Biosolarization was applied in mid-July each season. Dried chicken manure from nearby chicken farms was uniformly distributed onto the surface of the soil and then incorporated into the top 20-cm layer by crosswise plowing using a rotary cultivator. Plots were then drip irrigated until the soil reached field capacity. Solarization was carried out under a low-density transparent polyethylene film (0.03 mm thick) during July and August for about 6 weeks. Biosolarization with chicken manure was more effective in reducing *M. hapla* soil densities when applied at a rate of 25 (86%) than at 20 tons ha^{-1} (67%) ($p < 0.05$), but the dosage of chicken manure did not affect the efficacy of biosolarization in reducing *P. penetrans* soil densities. Soil biosolarization reached efficacies of 73% reduction for *M. hapla* and 67% for *P. penetrans*, which seems to provide sufficient nematode control in most situations. Soil biosolarization is proposed as an effective alternative to chemical soil fumigation for the management of nematodes in strawberry cultivation (Talavera *et al.*, 2019).

15.4.2.2. Cultural methods: Planting certified nematode-free runners is imperative. Crop rotation with non-host or poor crops can be effectively employed to reduce the northern root-knot nematode population. Typically, two years out of strawberries with corn or small grains is recommended. Cover crops like Yellow mustard (*Brassica juncea*) which contains allyl isothiocyanate) and *Brassica carinata* are an excellent way to disrupt the nematode life cycle.

Cover crops, such as velvet bean (*Mucuna pruriens*), sun hemp (*Crotalaria juncea*), and sorghum-Sudan grass, are all poor hosts to root-knot nematodes and good rotational crops for soil improvement and root-knot nematode management.

Many growers tend to intercrop garlic with strawberry. They grow one or two rows of garlic plants per one-meter width bed which usually has four strawberry rows. Both oil (El-Habashy, 2010) and aqueous extract (Tibugari *et al.*, 2012) of garlic caused significant decrease in number of RKN galls and egg masses on roots and J2 number/ 250 g soil as well as improved root and shoot growth of strawberry cv. Festival compared with the untreated control (El-Habashy, 2010).

Strawberry is usually rotated with traditional summer crops such as cotton (*Gossypium hirsutum*), soybean (*Glycine max*), maize (*Zea mays*), or peanut (*Arachis hypogaea*). Such summer crops have been useful in rotation sequences for suppressing certain species and races of RKNs (McSorley and Gallaher, 1991; Rodriguez-Kabana *et al.*, 1991; Chen and Tsay, 2006; Abd-Elgawad and Askary, 2015). Rotation with corn could suppress RKN populations and resulted in an increase in strawberry yield compared to bare fallow. Based on economic analysis, it is concluded that corn and bare fallow were the most appropriate rotation strategies for nematode management in strawberry.

Cover crop such as marigold, has been shown to reduce populations of root-knot nematodes (Osman *et al.*, 2008).

15.4.2.3. Biological methods: BCAs gave encouraging results concerning both reducing nematode development parameters and increasing strawberry fruit yield, including beneficial entomopathogenic nematodes (EPNs) (Hammam *et al.*, 2019). BCAs can be included in integrated pest management programs in trends that make them able to contribute synergistically or complimentary to other chemicals or BCAs.

Application of *Purpureocillium lilacinum* at 2 to 4 kg a.i./ha at the time of planting is beneficial.

15.4.2.4. Host resistance: Strawberry cvs. Tiago, Chandler and Torrey were found resistant to root-knot nematodes (Laqman Khan, 2001).

Jack Pinkerton and Finn (2005) reported that strawberry cvs Camarosa, Chandler, Allstar, Firecracter and Dimante were highly resistant to *M. hapla*.

15.4.2.5. Integrated methods: Application of Neem cake in combination with Carbofuran applied at the time of planting, increased the yield by 35.2% over control (Laqman Khan, 2001).

15.5. CAULIFLOWER DISEASE COMPLEX

15.5.1. Symptoms

Both nematode (*Aphelenchoides fragariae, A. ritzemabosi*) and bacterium (*Rhodococcus fascians*) must be present to produce the disease. The disease manifests in the form of malformation of the inflorescence, shortening and thickening of the aerial parts of strawberry plants.

The disease manifests in the form of malformation of the inflorescence, shortening and thickening of the aerial parts of strawberry plants. *A. fragariae* and *A. ritzemabosi* act as vectors of *R. fascians* and play an obligatory role in the cauliflower disease complex on strawberry. Strawberry plants infected with both *A. ritzemabosi* and certain strains of *R. fascians* resemble small cauliflowers. In extreme cases, plants are stunted and flowers are deformed. The petals fail to develop or become small and greenish. The sepal becomes much enlarged. Stamens and receptacles are malformed so that fruits do not develop. Auxiliary buds are continually produced in the crowns. Therefore, both the organisms actively contribute to the cauliflower syndrome. The bacterium attaches to the cuticle of the nematodes and is transported to the crown tissue of strawberry plants.

Cauliflower disease complex on strawberry is vectored by the nematodes like *A. fragariae* and *A. ritzemabosi* and the bacterium, *R. fascians* which play an obligatory role. Strawberry plants infected with both *A. ritzemabosi* and certain strains of *R. fascians* resemble small cauliflowers.

In extreme cases plants are stunted, petals are small and greenish in color or fail to develop, and flowers are deformed. The sepal becomes much enlarged. Development of fruits is prevented due to the malformation of stamens and receptacles. Auxiliary buds are continually produced in the crowns. Therefore, both the organisms actively contribute to the cauliflower syndrome. The bacterium attaches to the cuticle of the nematodes and is transported to the crown tissue of strawberry plants.

15.5.2. Management

- » Use certified nematode-free propagation material.
- » Root-dip treatment of strawberry runners in hot water maintained at 50 °C for 5 min or 48 °C for 15 min or 46 °C for 20 min.
- » Regenerating plants from clean, dormant, excised axillary buds.
- » Uprooting and destroying infected plants.
- » Foliar sprays with Fenamiphos at 4-6 ml/3.8 liters of water.
- » Strawberry varieties "George Soltwedel", "Regina", "Talizman", "Saksonka" and "Festival'naya" are comparatively resistant.

15.6. STRAWBERRY LATENT RINGSPOT VIRUS (SLRV)

Strawberry latent ringspot virus (SLRV) is a plant pathogenic virus vectored by dagger nematodes, *Xiphinema diversicaudatum* and *X. coxi*.

15.6.1. Symptoms

SLRV causes a reduction in size of shoots and leaves, a leathery texture and distortion of leaf shape; leaves also show yellow angular flecks. There is a high frequency of scion bud death on plants infected with this virus. SLRV causes a marked reduction in the number and diameter of shoots produced at the bud union, reduced number of flowers per bush, increased proportion of deformed flowers and strongly decreased vigor of diseased bushes. The size and vigor of infected rootstocks are also reduced (Murant and Lister, 1987).

15.6.2. Host range

SLRV is reported to infect many different crop species including, apricot celery, cherry, currants (*Ribes* spp.), delphinium, gladiolus, grapevine, hop, lily, narcissus, oleander, olive, parsley, parsnip, peach, raspberry, rhubarb, strawberry and rose.

15.6.3. Transmission

SLRV is transmitted by vegetative propagation and by nematode vectors *Xiphinema diversicaudatum* and *X. coxi*. Rapid infection during the 2-year production cycle can occur when viruliferous nematodes are present.

15.6.4. Management

In the absence of resistant material, healthy planting material should be used in areas free from known infection with SLRV in the crop species planted and in soil free from viruliferous *X. diversicaudatum*.

Control of SLRV is best achieved by allowing a crop-free interval of 2 years, either by bare-fallowing or by planting virus-immune hosts. During this period the vector (*X. diversicaudatum*) gradually loses its infectivity, provided the virus is not inadvertently re-introduced into the field. If a 2-year crop-free interval is not practicable, the interval should be as long as possible and followed by soil fumigation to kill nematodes. Whilst this treatment is unlikely to be fully successful, it may delay serious re-infection with SLRV for several years (Harrison and Murant, 1996).

Good weed control is also necessary to minimize the risk of SLRV-infected germinating seedlings acting as a source for the acquisition of virus by surviving *X. diversicaudatum* (Harrison and Murant, 1996).

The virus has been eliminated from developing meristem tip cultures of infected plants of several species, including Narcissus (Stone *et al.*, 1975).

15.7. CONCLUSION

No one nematode management strategy or tactic, with the exception of illegal MB soil fumigation, will effectively control all PPNs on strawberry. Therefore, adoption of integrated procedures for nematode control boosted by plasticulture technologies, multi-irrigation system, fertilization, and improved strawberry cultivars usually growing in 3–4 rows on bedded soil should be exercised. On the other hand, the need to back nematode quarantine and certification programs is apparent. Growers should use only the best quality transplants available and should if possible make special effort to inspect sources of strawberry transplants prior to planting in the field. Adequate cultural practices such as solarization, plastic mulching, intercropping, crop rotation, flooding, cover cropping, and/or plant resistance/tolerance should be selected in integrated pest management programs. soil biosolarization is proposed as an effective alternative to chemical soil fumigation for strawberry cultivation. Such programs may better include BCAs (Eissa and Abd-Elgawad, 2015) in trends that make them able to contribute synergistically or complimentary to other chemicals/nematicides or BCAs.

Chapter - 16

Raspberry, *Rubus idaeus*

16.1. INTRODUCTION

16.1.1. Origin

Raspberries originate from Asia.

16.1.2. Area and production

The three major raspberry production regions during 2020 are Russia, USA, and Poland. Worldwide 778,405 tons of raspberry is produced in 2020. Russian Federation is the largest raspberry producer in the world with 164,602 tons production volume per year. United States of America comes second with 137,829 tons yearly production (Table 16.1).

Table 16.1. Country-wise production of raspberry (2020)

Country	Area ('000 ha)	Production ('000 tons)	Yield (tons/ha)
Russian Federation	21.025	164.602	7.829
USA	8.765	137.829	15.725
Poland	29.282	129.063	4.407
Mexico	6.208	112.661	18.147
Serbia	11.041	61.875	5.604
Ukraine	4.000	31.920	6.939
Bosnia & Herzegovina	2.647	22.1000	8.372
Chile	4.783	19.132	4.000
Spain	1.801	17.808	9.886
UK	1.552	14.908	9.606

16.1.3. Varieties

Basically there are 'summer-bearers' and 'ever-bearer' types of raspberries. Usually, summer-bearers bear one crop per season and that too in summer time where as ever-bearers bear 2 crops yearly, one in summer and one in fall. Some of the varieties grown across the globe are; summer red raspberries, Black raspberries, Ever-bearing red raspberries, Purple raspberries and Golden raspberries are main varieties of raspberries. Mysore Raspberries are mostly cultivated in India.

The Indian ones are more likely to be yellow like the Himalayan raspberry (*Rubus ellipticus*) or red-black like the Mysore raspberry (*Rubus neveus*) and the taste when you try them is much more sour than sweet.

16.1.4. Export

Value of raspberries export increased for 56% in 2019 compared to 2012, and in value (USD) export increases for 61%. The top 10 world export countries with their export value (in USD) in 2019 are presented in Table 16.2. Spain is on the first place with export value of 520.592.404 USD and South Africa is on the tenth place of top 10 world export countries with export value of 17.445.023 USD (Table 16.2).

Table 16.2. Top 10 world export countries by value (USD) in 2019

Country	Rank	Value (Million USD)
Spain	1	520.592
Mexico	2	402.268
United States	3	354.789
Morocco	4	250.923
Portugal	5	237.331
Netherlands	6	202.249
Germany	7	42.355
Belgium	8	41.873
France	9	29.725
South Africa	10	17.445

16.1.5. Nematode problems

Some species of plant parasitic nematodes can be troublesome in raspberry plantings. Chief among these are the lesion nematode, which damages roots, and dagger and needle nematodes, which transmit viruses.

16.2. LESION NEMATODES, *PRATYLENCHUS* SPP.

Lesion nematodes (*Pratylenchus* species, most importantly *P. penetrans*) feeds on roots and contributes to plant decline. Decline caused by nematode infestation may occur gradually over a number of years.

16.2.1. Symptoms

Root-lesion nematodes are migratory endoparasites; they tunnel into the root to feed and return back to the soil. Lesion nematodes enter and destroy the cortex of feeder roots, creating the lesions that they are named for. These wounds interfere with proper root function and can serve as an ingress for soil-borne fungal pathogens. Above ground, raspberry plants may appear stunted, wilted, and yellowed.

Plants often have reduced vigor, produce fewer canes, and are less resilient when stressed by factors such as winter injury, drought, and insect feeding.

Root lesion nematodes are migratory endoparasite that moves between soil and roots, but feeds on and migrates in root cortical cells. These activities kill tissues in the root cortex, which appear as necrotic lesions or spots on the roots. Feeder roots are destroyed over time, leaving only woody roots. Infested plants often decline over a 3- to 4-year period, produce fewer canes and fruit, and are more susceptible to winter injury (Vrain and Copeman, 1987; Vrain and Pepin, 1989).

Plants infected with root-lesion nematodes are dwarfed, off-color and grow poorly. Symptoms are most often observed on fruiting canes. Dead roots or lesions may be on roots with injury being more severe on small feeder roots. Yields may be reduced when nematode populations reach high levels.

The root-lesion nematodes *Pratylenchus penetrans* and *P. crenatus* cause lesions, especially on feeder roots, that result in stunted plants with small and yellowish leaves. The affected plants are spindly with fewer and smaller canes, small fruit and poor yields.

Small, elongated, discolored lesions appear on new roots of infected plants. The fine-feeder roots die as the damage increases. This damage stimulates the formation of many fine roots and a witches' broom appearance. Without feeder roots, the larger diameter roots are unable to take up nutrients and water and often are invaded by secondary fungi. Above-ground symptoms are most often observed on fruiting canes and include reduced cane number and diameter, stunting, and off-color leaves.

16.2.2. Thresholds

The damage potential of *P. penetrans* was greater on sandy loam versus silt loam soils. Populations greater than 500 nematodes/500 cc soil (100/100 g soil) at planting will affect stand establishment, first-year growth, and yield. As a rule of thumb, 100 individuals per 100 cm^3 soil is considered the economic threshold for lesion nematodes in raspberry. The soil nematode population densities that might warrant treatment are presented in Table 16.3.

Table 16.3. The soil nematode population densities that might warrant treatment

Nematode	Planting age	Threshold level
Lesion nematode	New planting	500 per pint soil
Lesion nematode	Established planting	pint soil

16.2.3. Management

16.2.3.1. Cultural methods: If lesion nematodes are a concern, plants in the field should be killed and plowed under first, as this forces the nematodes out of roots and into the soil where they are more vulnerable. The use of soil solarization (using clear plastic over soil during the summer to heat the soil to a high temperature) has been used with some efficacy in the season before planting. Solarization can be effective when combined with raised beds.

Other cultural methods include:

- Use of certified planting stock.
- Planting in soil free of high populations of root-lesion nematodes.
- Work machinery in fumigated and non-infested fields first, before moving to nematode infested fields. Thoroughly clean machinery after leaving infested fields.

Population densities of *P. penetrans* in soil were lower and root biomass was greater under the composts (broiler manure) mulch relative to the non-mulched treatment. Fine root biomass was negatively correlated with *P. penetrans* per gram root. The abundance of omnivorous and predacious nematodes was negatively correlated with *P. penetrans* population densities and positively correlated with root biomass. The organic mulches increase soil food web structure and the abundance of nematode antagonists (e.g., predacious nematodes), resulting in reduced crop damage by *P. penetrans*. The improved raspberry root growth and primocane productivity was the result of beneficial changes in soil chemical properties, such as increases in exchangeable calcium and pH (Forge and Kempler, 2009).

Lesion nematode multiplication was suppressed in roots of cover crops such as 'Saia' oat (*Avena sativa*), creeping red fescue (*Festuca rubra*), and redtop (*Agrostis alba*) compared to barley or white clover (Vrain *et al.*, 1996).

16.2.3.2. Chemical methods: Velum Prime at 6.5 to 6.84 fl. oz./Acre can be used after planting using drip, trickle of micro-sprinklers.

Fall spray-applied fenamiphos, suppressed lesion nematode briefly (less than 6 months) (Walters *et al.*, 2009).

Fosthiazate and oxamyl significantly reduced lesion nematodes per gram dry root population densities by 97% and 87%, respectively, compared with the infested, non-treated control (Zasada *et al.*, 2010).

16.2.3.3. Biological methods: MeloCon WG (*Purpureocillium lilacinum* strain 251) at 2 to 4 lb/Acre plus a soil wetting agent can be applied to established plants, although it might be better used when applied to plants just before planting.

Fall and spring spray-applied DiTera (a fermentation product of the fungus *Myrothecium verrucaria*), and spring shallow-incorporated abyssinian mustard (*Brassica carinata*) seed meal suppressed root lesion nematodes (Walters *et al.*, 2009).

16.2.3.4. Host resistance: *Rubus niveus* productivity is not impacted by root lesion nematode.

16.3. DAGGER (*XIPHINEMA* SPP.) AND NEEDLE, (*LONGIDORUS* SPP.) NEMATODES

Dagger nematodes (*Xiphinema* spp.) are most important in raspberries as vectors of viruses. The presence of dagger nematodes in a raspberry planting is a cause for concern because of the potential for virus transmission. Some species of dagger nematode also cause direct damage to raspberry by feeding on root tips; however, the most common species in the US, *X. americanum*, generally does not. Dagger nematodes vector tomato ringspot, tobacco ringspot, cherry leaf rasp, grapevine fanleaf, and peach rosette viruses. Strawberry latent ringspot virus is transmitted by *X. diversicaudatum*, which exists in northwestern America but is more common in Europe. Cherry rasp leaf and strawberry latent ringspot viruses have not been reported on *Rubus* species in the US, but both have been detected in other host species.

Needle (*Longidorus* spp.) nematodes are vectors of cherry rasp leaf and tomato black ringspot virus. All of the above named viruses have a large host range that includes common weeds.

16.3.1. Symptoms

Nematode damage can occur without above-ground symptoms, though a gradual decline in yield over a period of years often indicates a problem. Dagger nematodes are widespread in most agricultural soils. They can act indirectly as vectors of viruses or directly as parasites on raspberry roots. *Xiphinema bakeri* causes considerable damage by feeding mainly on root tips where it causes swelling and curling, reducing root and top growth.

Dagger nematode species are ectoparasites; they do not enter the root, but feed on the growing root tips (McElroy, 1991). *X. bakeri* feeding causes direct damage to roots, observed as diagnostic "fishhook" swelling of the root tips (McElroy, 1991). Continued feeding stunts root and cane growth and reduces fruit yields within several years. Planting raspberry into soil infested with high population densities of *X. bakeri* may result in reduced establishment and high mortality of young plants (McElroy, 1992).

X. americanum does not cause direct damage to raspberry, but is a vector of tomato ringspot virus (ToRSV) (McElroy, 1991, 1992). ToRSV infected plants are stunted, produce crumbly, non-marketable fruit, and may die within 3 to 4 years after being infected. Even low population densities of *X. americanum* are problematic because they can transmit ToRSV. Virus transmission by dagger nematodes produces symptoms on leaves such as yellowing of veins, mosaic, and malformation of the plant. In the absence of ToRSV, dagger nematodes can cause stubby roots and dwarfing.

16.3.2. Thresholds

The soil nematode population densities that might warrant treatment are presented in Table 16.4.

Table 16.4. The soil nematode population densities that might warrant treatment

Nematode	Planting age	Threshold level
Dagger nematode (*X. bakeri*)	New planting	100 per pint soil
Dagger nematode (*X. americanum*)	Any	Zero tolerance when tomato ringspot virus is present

16.3.3. Life history

Dagger nematodes have the life stages of egg, juvenile and adult. Juvenile stages are smaller than the adult stage. They reproduce once per year, live four or five years and reach highest population levels in undisturbed areas and in zones within soil where oxygen is most plentiful. Dagger nematodes are migratory ectoparasites; they do not enter plant tissue and instead feed on root cells from the outside, piercing cells to feed on the interior.

16.3.4. Management

Solarization can be effective when combined with raised beds in SW Washington and Oregon. Anti-condensing plastic films may be more effective.

Biofumigation with selected *Brassica* species may be useful for reduction of nematode populations. Mustard cover crops can help at low densities with *Brassica juncea* more effective than *Sinapis alba* or *Brassica napus*. Plants growing in fertile soils are able to withstand nematode infestation better than those grown in poor soils. Adding compost to the soil boosts plant nutrition and increases microbial competition and populations of natural enemies in the soil. Control weeds, as many species can harbor important viruses. Do not move infested soil to an uninfested area, clean tools, tractors, and boots before moving from one area to another.

Anaerobic disinfestation has shown some efficacy. Organic acid and volatiles produced during the anerobic phase contributes to suppression.

Spring spray applications of oxamyl or fosthiazate reduced dagger nematode population densities for up to 2 years. Two spring applications of oxamyl at 2 lb/acre provided more than 2 years of suppression, while two spring applications of 0.8 lb/acre suppressed nematodes for only 1 year.

16.4. RASPBERRY RINGSPOT NEPOVIRUS (RRSV) (COMOVIRIDAE)

16.4.1. Geographical distribution

EPPO region: Found in Luxembourg, Poland, Yugoslavia. Widespread in France, Latvia and Norway; locally established in Belgium, Bulgaria, Czech Republic, Finland, Germany, Greece, Hungary (unconfirmed), Ireland, Netherlands, Norway, Slovenia, Spain, Switzerland, Russia (European, Far East), Turkey and UK (excluding Northern Ireland).

Asia: Kazakhstan, Russia (Far East), Turkey.

EU: Present.

16.4.2. Economic importance

RRSV causes a severe disease, reducing both growth and fruit yield and killing plants. It is of great economic importance in Germany and Russia, and of some economic importance in other countries where it is established.

16.4.3. Symptoms

On raspberries, infected plants are usually patchily distributed in crops, reflecting vector distribution. In the very susceptible cultivar Norfolk Giant, part or all of the canes may die in winter. Those that survive produce stunted and brittle shoots which have rolled leaves and show dieback progressively from the tip, accompanied by necrosis. In less susceptible cultivars, such as Malling Jewel, Malling Enterprise and Malling Notable, symptoms appear on the leaves of new shoots in spring, but are mild or absent on the leaves of fruiting canes. Young, expanding leaves develop a conspicuous, yellowish-green, ringspot or oak-leaf pattern; chlorotic blotches and a net-like chlorosis along the smaller veins may also be present. Some necrosis and dieback may occur. Malling Promise develops few symptoms and does not dieback; Lloyd George and Malling Landmark are immune to the common strain but are susceptible to the Lloyd George yellow blotch strain (Murant, 1970b).

16.4.4. Host range

The main host is raspberries (*Rubus idaeus*). Cultivars differ considerably in susceptibility; thus, Norfolk Giant is highly susceptible to the type strain, while Lloyd George is immune to it.

Other species of *Rubus* can also be infected by the virus. Other important hosts are strawberries and cherries.

16.4.5. Biology

The virus is mechanically transmissible to a number of herbaceous plants and is also transmitted through seed; in some hosts, more than 50% of progeny seedlings are infected. Besides aiding dissemination of the virus, infection of the seed, especially in weeds, provides an important means of survival of RRSV in soils. Virus can be transmitted to seed of raspberry and strawberry.

16.4.6. Transmission

RRSV is transmitted by two species of the nematode genus Longidorus. These free-living, soil-inhabiting vectors are specific for serologically distinctive forms of RRSV (Harrison, 1964); thus, the Scottish and Dutch strains are transmitted most efficiently by *L. elongatus* (Taylor, 1962), while the English form is transmitted by *L. macrosoma* (Debrot, 1964). Other nematode species (including *Xiphinema diversicaudatum* and other species of *Longidorus*) have been suspected of transmitting strains of RRSV, but the methods used to demonstrate the transmissions were not thought to be sufficiently rigorous and these transmissions are not, therefore, considered acceptable (Trudgill *et al.*, 1983).

Both larvae and adults of *L. elongatus* transmit the virus, but the adult does not pass the virus to its progeny, nor is it retained when the nematode molts. Starved *L. elongatus* retain infectivity for up to about 9 weeks. *L. elongatus* also transmits tomato black ring nepovirus (EPPO/CABI, 1996), which produces identical symptoms and is often found together with RRSV.

16.4.7. Spread and dispersal

Natural dispersal of RRSV is accomplished over long range by transport of infected seeds by animals or wind. Short-range dispersal occurs by means of transmission by viruliferous nematodes from plant to plant and, to a more limited extent, by the movement of the nematodes themselves.

In international trade, the disease can be carried in infected Rubus, Fragaria and cherry plants or parts of plants, including seeds. RRSV might also be present in viruliferous nematodes or infected seeds in the soil accompanying consignments of plants. The nematodes can survive the physical disturbance of international transport but not if the soil becomes dry.

16.4.8. Management

Fumigant such as dazomet, directed at the nematode vectors, give good control of virus transmission in raspberry plantations.

The consignments of plants for planting of Rubus must come from a field found free from RRSV during the last growing season. The use of certified planting material can ensure the absence of virus diseases.

Chapter - 17

Cherry, *Prunus* spp.

17.1. INTRODUCTION

17.1.1. Origin

Cherries are prehistoric fruits, originating throughout West and North Africa, Europe and parts of Turkey.

17.1.2. Area and production

Worldwide 2,317,958 tons of cherry is produced in 2020. Turkey is the largest cherry producer in the world with 599,650 tons production volume per year. United States of America comes second with 288,480 tons yearly production (Table 17.1). India is with 10,936 is ranked at 26.

Table 17.1. Country-wise area, production and yield of Cherry (2020)

Country	Area ('000 ha)	Production ('000 tons)	Yield (tons/ha)
Turkey	84.746	599.650	7.076
USA	37.110	288.480	1.773
Iran	28.397	220.293	7.761
Chile	24.496	123.224	5.030
Uzbekistan	10.808	95.267	8.814
Italy	29.970	94.888	3.166
Spain	25.252	94.138	3.728
Romania	6.127	73.834	12.050
Greece	13.678	71.858	5.253
Syria	35.004	69.153	1.975
Ukraine	10.100	63.320	6.269
India	3.553	10.936	3.077

In India, Cherry is mainly grown in the North-Western Himalayan region of Jammu and Kashmir (J&K), Himachal Pradesh (H.P.) and Uttarakhand hills. India ranked 26th and 31st in terms of area harvested (3,599 ha) and production (11,107 tons) of cherry during 2019; but ranked very low in terms of the yield i.e. 49th (30.896 tons/ha). Jammu and Kashmir harvested 15555 tons of cherries from 2700 ha area in 2019-20 (Anonymous, 2022) while Himachal Pradesh contributed only 12 per cent of national cherry production and 26 per cent of harvested area with -1.0% and 2.6% CGR respectively (Diksha, 2018).

The tree takes 3-4 years from the planting to start bearing fruits that are red to peach in color with a tarty and sweet flavor.

17.1.3. Varieties

There are many varieties of Cheery fruit that are cultivated throughout India. Some of the most popular commercial varieties are Lambert, White Heart, Pink Early, Black Tararian, Black Heart, Guigne Noir Hative. The recommended sweet cherry varieties for different states of India are presented in Table 17.2.

Table 17.2. Recommended sweet cherry varieties for different states in India

State	Varieties
Jammu & Kashmir	Black Heart, Early Purple Black Heart, Guigne Noir Gross Lucenta, Guigne Noir Hative, Guigne Pour ova Precece, Bigarreau Napolean, and Bigarreau Noir Gross.
Himachal Pradesh	Black Tartarian, Bing, Napolean White, Sam, Sue, Stella, Van, Lambert, Black Republican, Pink Early, Black Heart, Early Rivers, Sunburst, Dero-nero II, and Merchant.
Uttar Pradesh	Bedford Prolific, Black Heart, and Governor's Wood

Important exotic collections of sweet cherry suitable for different altitude zones are presented in Table 17.3.

Table 17.3. Sweet cherry varieties suitable for different altitude zones

Altitude zone	Suitable varieties
True temperate zone	Black Heart, Bigarreau Noir Gross (Misri), Guigne Pour Pora Pecoce (Awal), Guigne Pourpeara Prece, Bigarreau Noir Grosse, Stella, Compact Stella, Van, and Bing.
Mid to high altitude	Black Tartarian, Napolean Lambert, and Bing.
Mid hill to moderate	Black Heart and Bedford Prolific.

17.1.4. Clonal rootstocks

» Clonal rootstocks are commercially recommended for raising its plants as trees on paja show symptoms of delayed incompatibility.

» Clonal rootstocks are Colt, Giesela, Charger, SL64 and Mazzard F 12/1.

» Colt is semi-dwarf, compatible with almost all varieties of sweet cherry, has good anchorage, and is tolerant to gummosis, crown-rot, moderately resistant to stem pitting virus and bacterial canker but susceptible to oak-root fungus.

» Mazzard F 12/1 rootstock is semi-vigorous and difficult-to root.

17.1.5. Cultivation aspects

Different aspects of sweet cherry cultivation are presented in Table 17.4.

Table 17.4. Different aspects of sweet cherry cultivation

Variety	Rootstock	Training system	Spacing (m)	Density (trees/ha)	Yield (t/ha)	Reference
Bing	Gisela-6	Palmette	2.6 x 4.9	864	27.1 (9th yr)	Whiting *et al.*, 2005 (USA)
USBing	Gisela-6	Central leader	2.6 x 4.9	864	25.8 (9th yr)	Whiting *et al.*, 2005 (USA)
Bing	Gisela-6	Y- trellis	2.6 x 4.9	864	23.9 (9th yr)	Whiting *et al.*, 2005 (USA)
Van	Damil	Y- trellis	0.5 x 4.0	5000	14.5	MeLand, 1998 (USA)
Summit	Tabel Edabriz	V- shape	0.7 x 3.8	3759	19.3 (6th yr)	Radunic *et al.*, 2011 (USA)
CITH-Cherry-07	Colt	Modified centre	3.0 x 3.0	1111	14 (12th yr)	CITH, 2014-15 (India)
CITH-Cherry-07	Colt	Modified centre	3.0 x 3.0	1111	13 (12th yr)	CITH, 2014-15 (India)
CITH-Cherry-07	Colt	Modified centre	3.0 x 3.0	1111	12.5 (12th yr)	CITH, 2010 (India)
CITH-Cherry-07	Colt	Modified centre	3.0 x 3.0	1111	11.0 (12th yr)	CITH, 2010 (India)

17.1.6. Export

In 2019, Chile was the largest exporter of sweet cherries worldwide with most of its shipments to China, contributing 42.5% of world's export ($1.58B), followed by Hong Kong 20.6% ($765M), United States 13.5% ($501M), Turkey 5.26% ($196M), and Spain 2.92% ($109M).

During 2020-21, India exported fruits worth Rs. 4,971.22 crores/ 674.53 USD Millions (APEDA, 2021). As shown in Table 17.5, countries like Nepal, Australia, Sri Lanka and UAE were the top cherry importing countries for India in 2020-21; among which Australia and Sri Lanka were new importers

Exports of fresh cherries in India have more than doubled since 2011-12, rising above 350,000 tons to 622,000 tons in 2020-21.

Table 17.5. Country wise-export of cherry products from India in 2020-21

Country	Quantity (Metric Tons)	Value (Rs. '000)
Nepal	3.49	72
Australia	0.05	65
Sri Lanka	0.22	36

UAE	0.20	17
Nigeria	0.03	6
Saudi Arabia	0.06	6
Oman	0.08	4
Bahrain Islands	0.03	3
Maldives	0.01	2
Trinidad	0.01	1
TOTAL	4.18	212

17.1.7. Nematode problems

Root-knot, lesion, ring and dagger nematodes are the major limiting factors in the successful production of cherry.

17.2. ROOT-KNOT NEMATODES, *MELOIDOGYNE* SPP.

17.2.1. Symptoms

Feeding by root-knot nematodes causes swellings of the entire root, which impairs normal root functions. Root-knot nematode produces characteristic galls (swellings) on roots. Known for the gall formation it induces on plant roots, root-knot nematode can cause plant nutrient and water stress by disrupting plant root function. Plants may show signs of water stress or nutrient deficiency due to compromised root systems.

17.2.2. Management

17.2.2.1. Physical methods: Hot water treatment of cherry seedlings at 50-52°C for 5-10 min eliminates the nematode infection.

17.2.2.2. Biological methods: The highest reduction percentage in root-knot nematode (*M. javanica*) population (94.7%) was achieved by the application of (*G. mosseae* + *P. lilacinum* + *Hirsutella rhossiliensis*) compared to the nematicide Carbofuran (93.5%). In addition, by using this combination, gall index and egg masses were 1.0 and 5.0 compared to 2.0 and 12.0 for Carbofuran, respectively. Moreover, the combination showed better enhancement in various growth parameters of the Ground cherry plant. However, inoculation of Ground cherry with *G. mosseae* (mychorrhiza) alone, 15 days before nematode inoculation, significantly suppressed number of galls and egg-masses (-33.3 to 45.3% of control values) (Abo-Korah, 2017).

17.2.2.3. Host resistance: Cherry rootstocks – Mazzard, Stockton Morello, English Morello and Montmorency were entirely free of root-knot galls.

Mazzard is immune to *M. incognita* and resistant to *M. javanica.* (Immune rootstocks are not attacked by nematodes, whereas resistant or non-host rootstocks may be invaded by the nematodes and show damage, but do not allow population increases). Mahaleb is resistant to *M. incognita* and susceptible to *M. javanica.* Stockton Morello is immune to *M. incognita* and susceptible to *P. vulnus.*

17.3. THE LESION NEMATODE, *PRATYLENCHUS VULNUS, P. PENETRANS*

17.3.1. Symptoms

Poor growth and off-color, usually in patches in the orchard. Lesion nematodes penetrate into the roots and cause damage by feeding and tunneling through the cortical root tissues. Lesion nematodes may cause reddish brown lesions on roots that later turn dark and ultimately black. Roots are brown or black and decayed or have dark lesions. Large roots can become necrotic while the tips of fine root hairs may be swollen. In heavy infestations, yields are poor and growth stunted.

17.3.2. Management

17.3..2.1. Cultural methods: Avoid sites with high soil populations of root-lesion nematodes. Long rotations may help, but control weeds since dandelion can be a virus host.

17.3.2.2. Chemical methods: Velum Prime at 6.5 to 6.84 fl oz/Acre can be used after planting using drip, trickle of micro-sprinklers.

17.3.2.3. Biological methods: MeloCon WG (*Purpureocillium lilacinum* strain 251) at 2 to 4 lb/Acre plus a soil wetting agent can be applied to established plants, although it might be better used when applied to plants just before planting.

17.3.2.4. Host resistance: Cherry replants on Mahaleb rootstock grew well and *P. vulnus* and *P. penetrans* caused no root lesions.

17.4. CHERRY RASP LEAF NEPOVIRUS (CRLV)

17.4.1. Geographical distribution

CRLV is native to western North America: Canada (British Columbia, Ontario, Quebec), USA (California, Colorado, Idaho, Montana, New Mexico, Oregon, Utah, Washington).

17.4.2. Economic importance

In North America, CRLV causes serious stunting in peaches, and fruit yield and quality reductions in cherries and apples. Young trees and seedling rootstocks are sometimes killed. Because of its slow spread, the disease is mainly a nuisance in nuclear stock propagation. However, it can reach high levels of infection in older orchards. Trees planted on previously infected sites often become infected.

17.4.3. Symptoms

Prominent projections (enations) are found on undersides of leaves. These take the shape of leafy outgrowths or raised protuberances between lateral veins and along the midrib. Leaves are more or less deformed; many are extremely narrow, folded, and distorted. Upper surfaces have a rough, pebbly texture with depressions corresponding to projections on the lower surface. Not all parts of a tree show symptoms at the same time. Early infections usually are limited to low branches. Spread is slow within a tree and between trees. Fruit production is reduced. Severely affected branches and young trees may die.

Large enations (leafy outgrowths or protuberances) are found on the lower surface of narrow deformed leaves from affected parts of trees. The enations differ from those caused by other diseases in cherry in that they are formed between the lateral veins. The symptoms are frequently restricted to the lower part of the tree and are often, in recent infections, restricted to one or two limbs.

17.4.4. Host range

The principal hosts are cherries and peaches, and also apples. The rootstock species Prunus mahaleb is also susceptible. Raspberries are also attacked (Jones *et al.*, 1985).

17.4.5. Biology

CRLV is a nepovirus, transmitted by the nematode vector *Xiphinema americanum* (EPPO/CABI, 1996). Since the form of *X. americanum sensu lato* found in western USA is now known as *X. californicum*, this is the vector in practice. CRLV is readily transmitted by sap inoculation. Seed transmission has been shown to occur in some herbaceous hosts. Virus has been detected in pollen from infected cherry trees, but transmission by pollen has not been confirmed. Spread of the virus in the field is generally slow due to the slow movement of the nematode vector.

17.4.6. Virus transmission

CRLV is spread only slowly by its nematode vector, and the most likely means of international movement is in infected propagating material. It could possibly be carried by the nematode vector in soil accompanying plants. The virus has been intercepted several times in imported plant material from North America (Jones *et al.*, 1985).

The main importance of the dagger nematode is it acts as a vector for virus diseases such as Tomato ringspot, *Tobacco ringspot virus* (found in the Grand Ronde area of Oregon) and *Cherry rasp leaf virus.* However, *X. americanum* is more important on stone fruit trees as a vector of the *Cherry rasp leaf virus,* which causes rasp leaf disease, and strains of *Tomato ringspot virus,* which cause yellow bud mosaic, cherry mottle leaf, and Prunus stem pitting diseases. These virus diseases can reduce the productivity of the trees significantly, and infected trees may eventually die

The dagger nematode, *X. americanum* transmits the cherry rasp leaf virus disease. The cherry rasp leaf virus disease occurs in the western United States and in British Columbia, and Canada (Bodine *et al.*, 1951). It has been shown that the disease is caused by the sap-transmissible CRLV (Mansen *et al.*, 1974) and therefore differs from the English rasp leaf virus disease which is induced by the synergistic action of two viruses, prune dwarf virus and raspberry ringspot virus or arabis mosaic virus (Cropley, 1964). Enation symptoms caused by CRLV are very severe and typically restricted to the lower part of the tree. *Xiphinema americanum* transmits the virus. *X. americanum* was recovered from 43 of 64 samples collected and CLRV from 12 of those containing this species. Viruliferous nematodes were present from the soil surface layers to a depth of at least 60 cm. Several common orchard weeds such as dandelion (*Taraxacum officinale*), plantain (*Plantago major*), and balsamroot (*Balsamorhizer sagittata*) are virus hosts and may serve as reservoirs from which nematodes can transmit the virus to cherry (Mansen *et al.*, 1974).

17.4.7. Management

Cultural control measures include:

- Use nursery stock that has been tested and found to be free of all known viruses.
- Remove infected trees as soon as symptoms are seen.
- Control the nematode vector.
- Thermo therapy has been found effective.

Perhaps the most important step in prevention is to always plant stock that is free of viruses. Use healthy planting material. It should originate from a certification scheme giving appropriate guarantees.

Colt rootstock appears to slow the development of symptoms in cherry scions.

17.5. RING NEMATODE, *MESOCRICONEMA XENOPLAX*

Ring nematodes, *Mesocriconema* spp. are ectoparasites and may feed for several days at a single site. They predispose trees to other pathogens such as bacterial canker and to environmental stresses.

Stress resulting from nematode feeding, particularly by ring nematode *Mesocriconema (=Criconemoides) xenoplax,* can increase tree susceptibility to bacterial canker, *Pseudomonas syringae.*

17.5.1. Symptoms

Heavy feeding results in death of the feeder roots. Fine roots that remain are necrotic or grouped in short clusters. The shoot or root growth of trees may be reduced.

17.5.2. Management

17.5.2.1. Cultural methods: Avoid sites with high soil populations of ring nematodes. Pre-planting the future orchard site with winter wheat followed by fallow have helped to reduce nematode populations.

17.5.2.2. Chemical methods: Velum Prime at 6.5 to 6.84 fl oz/Acre can be used after planting using drip, trickle of micro-sprinklers.

17.5.2.3. Biological methods: MeloCon WG (*Purpureocillium lilacinum* strain 251) at 2 to 4 lb/Acre plus a soil wetting agent can be applied to established plants, although it might be better used when applied to plants just before planting.

Chapter - 18

Walnut, *Juglans regia*

18.1. INTRODUCTION

18.1.1. Origin

Early history indicates that English walnuts came from ancient Persia, where they were reserved for royalty. Thus, the walnut is often known as the "Persian Walnut." Walnuts were traded along the Silk Road route between Asia and the Middle East.

18.1.2. Area and production

It is grown extensively in USA, China, France, Italy, Turkey, Poland, Yugoslavia, Rumania, Iran.

Worldwide 3,758,565 tons of walnut is produced per year. China is the largest walnut producer in the world with 1,785,879 tons production volume per year. United States of America comes second with 607,814 tons yearly production. China and United States of America produce together more than 60% of World's total. India is with 33,000 tons production is ranked at 11 (Table 18.1).

Table 18.1. Area, production and productivity of walnut (2020).

Country	Area ('000 ha)	Production ('000 tons)	Yield (tons/ha)
China	487.007	1785.879	3.667
USA	127.476	607.814	4.768
Iran	153.642	405.281	2.637
Turkey	86.853	195.000	2.245
Mexico	83.513	141.818	1.698
Ukraine	13.200	107.900	8.181
Chile	30.964	73.529	2.375
Uzbekistan	5.629	53.116	9.436
France	20.340	39.410	1.937

Romania	1.673	34.095	20.379
India	31.000	23.000	1.064
WORLD		3,758.565	

In India, it is grown in the states of Jammu and Kashmir, HP and Uttarakhand (Table 18.2).

Table 18.2. State-wise area, production and productivity of walnut in India (2017-18)

State	Area ('000 ha)	Production ('000 tons)	Productivity (tons/ha)
Arunachal Pradesh	1.00	0.63	0.630
Himachal Pradesh	4.44	2.46	0.554
Jammu & Kashmir	85.62	275.45	3.217
Uttarakhand	17.56	21.17	1.205
TOTAL	108.63	299.71	2.759

Source: Hort. Statistics Divn., Dept. of Agri, Co-op. & Farmers' Welfare.

The major walnut growing districts in different states in India are detailed in Table 18.3.

Table 18.3. The major walnut growing districts in different states in India.

State	Districts
Jammu and Kashmir	Anantnag, Pulwana, Budgam, Baramulla, Srinagar, Doda, Punch, Udampur, Rajouri and Kathua.
Himachal Pradesh	Kullu, Mandi, Shimla, Kinnur, Srimour, and Chamba
Uttarakhand	Nainital, Dehradun, Pauri, Tehri, Chamoali, Almora and Pithoragarh

18.1.3. Varieties

18.1.3.1. Exotic varieties: Hartley, Payne, Franquette, Serr, Sunland, Chico, Vina, Howard, Chandler, Tehana, Tulari, Lara.

18.1.3.2. Local selections: Gobind, Pratap selection, Solding selection, Kotkhai Selection, Kashmir budded, Wilson Wonder, Chakrata Selection, Sulaman and Hamdam.

Walnut varieties grown in different states of India ae presented in Table 18.4.

Table 18.4. Walnut varieties grown in different states of India

State	Varieties grown
Jammu and Kashmir	Lake English, Drainovsky and Opex Caulchry
Himachal Pradesh	Gobind, Eureka, Placentia, Wilson, Franquetfe and Kashmir Budded
Uttarakhand	Chakrata Selections

Important walnut varieties grown at different altitudes are presented in Table 18.5.

Table 18.5. Important walnut varieties grown at different altitudes.

True temperate zone	Mid to high altitude	Mid-hill to moderate
Suliman, Hamdan. CITH-W-1, CITH-W-2, CITH-W-3, CITH-W-4, CITH-W-5, CITH-W-6, CITH-W-7, CITH-W-8, CITH-W-9, CITH-W-10	Hartley, Howard, Chandler, Blackmore	Local Selection

18.1.4. Export

The country has exported 2,482.59 MT of Walnuts to the world for the worth of Rs. 73.98 crores/ 9.88 Million USD during the year 2021-22.

Major export destinations (2021-22) include United Arab Emirates, France, U.K., Djibouti, Germany and New Zealand (Table 18.6).

Table 18.6. Walnut exports to different countries from India (2021-22)

Country	Quantity (Lakh tons)	Value (Rs. in million)
UAE	654.56	2,066.07
France	430.75	1,286.83
UK	361.45	1,032.93
Djibouti	177.78	701.77
Germany	183.05	619.30
Algeria	170.99	263.57
New Zealand	48.05	262.08
Netherland	70.50	228.64
Iran	140.00	214.02
Spain	40.00	171.00
TOTAL	2,277.13	6,846.21

18.1.5. Nematode problems

Three groups of parasitic nematode are common in walnut; lesion nematode (*Pratylenchus vulnus*), ring nematode (*Mesocriconema xenoplax*), and root knot nematode (*Meliodogyne incognita*). Lesion and ring nematodes cause the greatest damage to walnut production (Westerdahl *et al.,* 2003). Nematode infestation resulted in yield loss and poor plant growth.

18.2. LESION NEMATODE, *PRATYLENCHUS VULNUS*

18.2.1. Symptoms

Pratylenchus spp. are migratory endoparasite that feed within root cortical tissue and are also found in the surrounding soil. Infected plants have roots with black lesions and fewer feeder roots resulting in stunted root growth. Top growth may exhibit general symptoms of an impaired root system including lack of vigor, dieback, chlorotic and small leaves, and reduction of yield. In general, root lesion infection results in plants exhibiting symptoms of chlorosis, wilting, and stunting. Infected roots show initial symptoms of small, water-soaked lesions that soon turn brown to black. Lesions are formed along the root axis and may coalesce laterally to girdle the roots, which are killed later. Affected root tissue may slough off leaving a severely reduced root system. Secondary infection by fungi and bacteria may further destroy the root system by causing sloughing off the root tissues and rot. Plant yield is reduced and in severe infections, plants may be killed (Chitambar *et al.*, 2018).

In California, as in many regions worldwide, this nematode is the primary cause of tree decline and replant problems in orchards (McKenry 1999). *Pratylenchus vulnus* is the root lesion species most found in walnut orchards in California. Walnut tree vigor and yields are reduced by the feeding activity of *P. vulnus*, which places infected trees under stress (Westerdahl *et al.*, 2017). Growth of young walnut trees can be arrested by *P. vulnus* and even at the density of one nematode/250 cm^3 and established walnut orchards in California are able to support 500 *P. vulnus*/250 cm^3 soil (Buzo *et al.*, 2009).

18.2.2. Host range

Root systems of young grapevines may be restricted in growth with absence of major roots and dead feeder roots while root lesions at feeding sites may not be present (Raski *et al.*, 1973). *Pratylenchus vulnus* reduces plum yields (McKenry, 1989). During the 1970s, *P. vulnus* was also found to affect rose production in California (Lear *et al.*, 1970).

18.2.3. Spread

On its own, *Pratylenchus* species can move 1-2 m per season from an infected root. The main mode of long and short distance spread is mainly through infected roots, bare root propagative plant materials, soil debris, cultivation tools, equipment and human activity that can move soils from infested to non-infested sites. It can also be spread by drainage, irrigation, or flood water (Corbett, 1974).

18.2.4. Management

18.2.4.1. Cultural methods

» If possible rotate plantings to new land.

» Use certified trees.

18.2.4.2. Chemical methods: Pre-plant soil fumigation with one of the following methods:

» Basamid G. Do not harvest within one year of application. 5-day re-entry.

» Dominus at 10 to 40 gal/Acre. Apply as a shank injection to an entire field or raised beds when soil temperatures are above 60 °F but not greater than 90 °F. 5-day re-entry.

» Nema-Q was ineffective alone but reduced nematodes when combined with DiTera in California. 24-hr re-entry.

18.2.4.3. Biological methods

» MeloCon WG (*Purpureocillium lilacinum* strain 251) at 2 to 4 lb/Acre plus a soil wetting agent can be applied to established plants, although it might be better used when applied to plants just before planting. Stable for only days at room temperature, weeks in the refrigerator or for a year if frozen. 4-hr re-entry.

» DiTera DF (*Myrothecium verrucaria* strain AARC-0255 fermentation solids and solubles) is registered for many nematodes on walnuts. Was ineffective alone but reduced nematodes when combined with Nema-Q in California. 4-hr re-entry.

18.2.4.4. Host resistance: Walnut rootstock VX211 is a highly vigorous. Paradox hybrid clone that outgrew other walnut seedlings in the presence of lesion nematodes. Three years after inoculation, soil population levels of *P. vulnus* on VX211 were significantly reduced compared to that of the moderately vigorous AX1. Growth of VX211 was 35% greater than that of AX1 regardless of the Pi. VX211 has high vigor, exhibits tolerance to low population levels of *P. vulnus,* reduces nematode feeding and reproduction within the root terminus, and is currently available to California growers (Buzo *et al.*, 2009).

18.3. RING NEMATODE, *MESOCRICONEMA XENOPLAX*

Mesocriconema xenoplax was frequently found associated with the rhizosphere of walnut in Italy (Ciancio *et al.*, 1996). The damage caused by this nematode on walnuts is extremely severe.

18.3.1. Distribution

The ring nematode is considered as an economically important nematode pest on walnuts and is distributed worldwide.

18.3.2. Symptoms

Artificially infested walnut seedlings showed reduced development and growth, with extensive root necrotic areas and phloem lesions (Lownsberry *et al.*, 1978). Ring nematode infestation will result in stunted roots, which sometimes proliferate and form dense mats.

18.3.3. Host range

Some of the important crop plants known to be hosts of *Mesocriconema xenoplax* are: peach, grape, chrysanthemum, walnut, clover, fescue, vetch, ryegrass, curly dock, and cowpea.

18.3.4. Histopathology

Roots of walnut (cv. Bleggiana) seedlings showed lesions produced by the nematode penetration through the first epidermis layers and the cortex, with associated cell disruption. In the lesions were seen epidermis and cortex fragments tat had been removed or detached by nematodes during their migration between the two root layers. Specimens of *Mesocriconema xenoplax* were frequently observed deeply inserted inside roots and an extensive area of epidermis erosion was also visible surrounding the lesions. Eggs were observed in the roots, where they were deposited beneath the epidermis. Root penetration by nematodes and their migration under the epidermis were considered to be the primary cause of root damage (Ciancio and Grasso, 1998).

18.3.5. Management

Seedlings planted into soil-solarized field plots remained at or below the economic threshold for nematicide treatment for nearly 18 months. Soil solarization alone induced a shift toward a microbial community that was suppressive to nematode multiplication (Kluepfel *et al.*, 2002).

Paradox, with its hybrid vigor, is more tolerant to infestation by nematodes. However, it can be damaged when nematode numbers are very high.

18.4. ROOT-KNOT NEMATODES, *MELOIDOGYNE* SPP.

The root-knot nematodes (*Meloidogyne* spp.) cause problems on Paradox and English walnut rootstocks.

18.4.1. Symptoms

Root-knot nematodes cause swelling of roots, called galls, mainly on Paradox and English walnut rootstocks.

18.4.2. Histopathology

Histopathology of the infected walnut roots with root-knot nematode, *Meloidogyne incognita* showed that roots were completely destroyed because of infestation. *Meloidogyne* species are potential pest of associated plants and trees. The second stage juveniles (J2s) established themselves in the roots very successfully and resulted in the formation of giant cells. Presence of feeding sites, giant cells, dense cytoplasm, destroyed cortex, second stage juveniles and females in the infected roots clearly indicate that *Meloidogyne incognita* is not only potential pest of walnut trees in studied area but also resulted in great economic loss. Feeding sites were developed and necrosis of root along with hypertrophy was seen. Cortex was completely destroyed which is a major component of ground tissue. Transverse section of the root cells showed presence of female sections and egg masses. The environment is suitable for the growth of *Meloidogyne incognita*, therefore damaging roots of the host trees in the studied area which go unnoticed because of the lack of the awareness to the local farmers (Saeed *et al.*, 2022).

18.4.3. Management

Use clonal Paradox VX211 rootstock where lesion or root-knot nematodes are suspected or present. This rootstock has the ability to tolerate the presence of these nematodes, even though the nematodes can still reproduce.

18.5. CONCLUSION

Clearly, more sustainable methods of nematode management in walnut orchards are urgently needed. Three main avenues include: (i) development of walnut rootstocks with elevated resistance or tolerance to root lesion and other plant-parasitic nematodes, (ii) implementation of chemical nematode suppression using novel nematicidal chemistries, and (iii) biorational methods that rely on the use of organic amendments and cover crops for nematode suppression. Lastly, novel organic amendments and cover crops for nematode suppression are tested. Some of the by-products of anaerobically digested food waste and animal manure through fermentation process, called digestates, can suppress plant-parasitic nematodes, while also serving as a nutrient source, is being explored in walnuts and other nut crops.

In summary, a multi-pronged approach should be investigated to alleviate the devastating impacts of plant-parasitic nematodes in walnuts. The goal is to increase profitability and sustainability of walnut production while reducing its environmental impacts.

Chapter - 19

Kiwi, *Actinidia deliciosa*

19.1. INTRODUCTION

19.1.1. Origin

The Kiwi fruit (*Actinidia deliciosa* also known as Chinese Gooseberry) is native to China and centre of origin is in the mountain ranges of South Western China where it occurs naturally as a deciduous fruiting vine. The fruit is primarily known as Chinese gooseberry until 1962, when New Zealand growers began to market the fruit as kiwifruit.

19.1.2. Area and production

Kiwifruit is mainly grown in China, Italy, New Zealand, Iran, Chile, Greece, France, Turkey and USA. Worldwide 4,274,870 tonnes of kiwi is produced per year. China is the largest kiwi producer in the world with 2,390,287 tonnes production volume per year. Italy comes second with 523,595 tonnes yearly production (Table 19.1).

Table 19.1. Area, production and productivity of kiwi fruit (2020).

Country	Area ('000 ha)	Production ('000 tons)	Yield (tons/ha)
China	197.048	2390.287	12.130
Italy	26.510	523.595	19.750
New Zealand	12.574	434.048	34.519
Iran	10.227	294.413	28.787
Chile	8.866	225.797	25.467
Greece	7.710	182.589	23.682

France	3.795	65.036	17.137
Turkey	2.487	43.950	17.672
USA	1.497	28.300	18.905
Japan	1.875	25.600	13.656
Spain	1.554	22.965	14.874
WORLD		4,274,870	

In India, it is grown in mid hills of Himachal Pradesh, Jammu & Kashmir, Uttarakhand, Sikkim, Arunachal Pradesh, Meghalaya, Nagaland and Nilgiri hills of south India. Presently, India produces around 16,620 MT of Kiwi fruit in an area of 4,000 Ha in states like Arunachal Pradesh, Manipur, Sikkim, Nagaland, Himachal Pradesh, Mizoram, and Jammu and Kashmir (Data of 2021-22) (Table 19.2).

Table 19.2. State-wise production of kiwi fruit in India (2021-22)

State	**Production ('000 tons)**	**Share (%)**
Arunachal Pradesh	7.43	44.71
Manipur	3.01	18.11
Sikkim	2.16	13.00
Nagaland	1.80	10.83
Himachal Pradesh	1.16	6.98
Mizoram	1.03	6.20
Jammu and Kashmir	0.03	0.18
TOTAL	16.62	100.00

19.1.3. Varieties

There are two types of kiwi fruit varieties, namely

» Pistillate varieties: Hayward, Allison, Abbott, Monty, and Bruno

» Staminate varieties: Allison, Tomari, and Matua.

19.1.3.1. Cultivars of *A. deliciosa*: The most widely planted kiwifruit cultivar is *A. deliciosa* 'Hayward' selected in New Zealand in 1925. 'Hayward' fruit represents about 90-95% of the international kiwifruit trade. The second most widely planted *A. deliciosa* cultivar is 'Qinmei' which is selected in the Qinling Mountains in China; it is only commercially grown in this country. The third most widely planted *A. deliciosa* cultivar is 'Jinkui', also grown in China as well. Other significant cultivars are Abbott, Allison, Bruno, Monty (Montgomery), Koryoku and Greensill.

19.1.3.2. Cultivars of *A. chinensis* (Chinese gooseberry): The most widely planted cultivar of *A. chinensis* is 'Hort16A', marketed as ZESPRI Gold. The second most widely planted cultivar of *A. chinensis* is the Chinese cultivar 'Jintao', marketed in Italy as Jingold or Kiwigold. Some other cultivars grown outside China are 'Jiangxi 79-1' (syn. 'Koushin', 'Kosuei 79-1', 'Lushan 79-1 ', 'Red Princess'), and more recently (registered in 2005) Sanuki Gold.

Maturity duration and harvesting time of various kiwi cultivars is presented in Table. 19.3.

Table. 19.3. Maturity duration and harvesting time of various kiwi cultivars

Cultivars	Days from full bloom	Harvest time
Allison	188-195	4-6 November
Abbott	192-200	7-12 November
Bruno	190-199	1-3 November
Monty	189-197	6-10 November
Hayward	194-202	15-20 November

19.1.4. Rootstocks

Seedlings of some cultivated varieties like Bruno and Abbott are commonly used as a rootstock in kiwifruit. Seeds of these two cultivars are preferred because of good germination and strong seedling vigor.

19.1.5. Strengths and sustenance/improvement

» Orchards are easy to establish and farmers can readily learn what is needed for kiwi cultivation

» The benefits of the technology are easy to observe; farmers generate cash income from selling kiwifruit, juice, and jam.

» Soil erosion is decreased due to increased groundcover.

» Kiwi cultivation provides on-farm employment opportunities.

» Awareness and training programs can help farmers quickly learn what is needed for kiwi cultivation.

19.1.6. Conclusions

» Increase kiwi growing area and production each following year.

» There is a need to increase number of the cultivars otherwise price will go down dramatically

» Early ripening cultivars and yellow cultivars are good alternatives

» There is a need for new cold storage units around kiwi growing areas

» We should grow all cultivars in suitable ecology

» All kiwifruit growers should join same union to have big bargain potential

» Kiwifruit processing industry should be developed

19.1.7. Nematode problems

The major plant parasitic nematodes associated with kiwi fruit include root-knot nematodes (*Meloidogyne* spp.).

19.2. ROOT-KNOT NEMATODES, *MELOIDOGYNE* SPP.

Major species of RKN infecting Kiwi plants are *Meloidogyne hapla, M. incognita, M. arenaria* and *M. ethiopica*.

19.2.1. Economic importance

Meloidogyne spp. caused significant yield losses to the extent of 36 to 49% in kiwifruit orchards (Aydinli and Mennan, 2022).

19.2.2. Symptoms

The only widespread nematode damage reported on kiwi fruit is caused by *Meloidogyne hapla* which induces small, discrete root galls associated with unthrifty plants.

- » Vines become slightly weaker, especially in sandy areas, and fruit size on infected vines is reduced.
- » Formation of galls/knots on host root system is the primary symptom.
- » In severely infected plants, the root system is reduced and the rootlets are almost completely absent.
- » The roots are seriously hampered in their function of uptake and transport of water and nutrients showing nutrient deficiency symptoms like chlorosis.
- » Plants wilt during the hot part of day, especially under dry conditions and are often stunted.
- » The nematode causes gall on the root systems and impairs the absorption of water and minerals, which results in wilting, growth suppression, reduction in yield, and eventually death (Xia *et al.*, 2019).
- » Nematode infection predisposes plant roots to opportunistic fungal and bacterial soil-borne pathogens breaking primary defence barrier causing disease complexes.
- » Di Vito *et al.* (1988) found that *M. incognita* race 1 caused serious growth suppression in kiwi cv. Howard.

19.2.3. Biology

- » RKN have a relatively simple life cycle consisting of the egg, four larval stages and the adult male and female.
- » Development of the first stage larvae (J1) occurs within the egg where the first molt occurs. Stylet develop in the second stage larvae (J2) which helps them to hatch out from eggs and find feeder roots of host plants for infection.
- » Under suitable environmental conditions and availability of host, the eggs hatch and new larvae emerge to complete the life cycle within 4 to 6 weeks depending on temperature.
- » Mode of reproduction is usually through parthenogenesis or amphimixis in the genus *Meloidogyne*.
- » Nematode development is generally most rapid within an optimal soil temperature range of 70 to 80 °F.
- » Eggs are deposited in the gelatinous matrix by the females to protect them from unfavourable condition and longer survival which are 200-250 in number.

19.2.4. Survival and spread

- » **Primary infection:** Egg masses in infected root debris and soil or in collateral and other hosts of Solanaceae, Malvaceae, Gramineae and Leguminaceae plants which act as sources of inoculums.
- » **Secondary infection:** Autonomous, second stage juveniles (J2) released after completing one cycle within 30-40 days, that can also be dispersed through water.

- **Favorable conditions:** Moist light sandy loam soils and temperature ranging between 25- 34° C.

19.2.5. Management

Reduced nematode population and improved plant status can be achieved with various organic amendments, mulches and biological control agents.

19.2.5.1. Physical methods: Root-knot nematodes in the root tissue of kiwi fruit can be killed without damaging root tissues by the hot water dipping of seedlings at 48 °C for 10 min. This could be useful for organic production and distribution of kiwi fruit seedlings by avoiding the synthetic nematicides which are not easily decomposed in soil once applied (Ma *et al.*, 2011).

Soil solarization in hot summers before planting so that the inoculum desiccates. Solarization is an option in individual vine replants

19.2.5.2. Cultural methods

- Ensure the inoculum free field by soil testing before planting, and testing should be repeated periodically every year.
- Use infection free plants from a certified nursery.
- Irrigate with greater frequency to avoid plant stress.
- Avoid planting cover crops susceptible to RKN.
- Crop rotation with grass crops such as barley, wheat, or Sudan grass.

19.2.5.3. Chemical methods

- Soil drench up to 15-20 cm depth with Valium Prime (Bayer) at 0.625 l/ha mixed in 200-300 litres of water before planting or after fruit harvesting.
- Broadcast Nimitz (Adama) 7-8 kg/ha for effective management once a year after fruit harvesting.
- The most effective control of *M. hapla* was obtained with 0.1% a.i. phenamiphos without any phytotoxic symptoms. To control *M. hapla* in nursery stock, a root dip in aqueous solution of 0.1% a.i. phenamiphos for 1 hr. can be recommended (Grandison, 1983).

19.2.5.4. Biological methods

- Soil amendment of field/ plants with any one of *Trichoderma harzianum*, *Purpureocillium lilacinum*, *Pseudomonas fluorescens*, *Pasteuria penetrans* or *Pochonia chlamydosporia* enriched compost @ 20 kg/ha can be done before and after planting.
- The endophytic *Bacillus altitudinis* 123EN strain was found effective by induction of resistance resulting in lower sensitivity of kiwifruit seedlings to *Meloidogyne incognita*.
- The antagonistic ability of two isolates *Pantoea ananatis* 121.en and *Pseudomonas chlororaphis* 54.en against *M. incognita* resulted in a reduction in the number of gall and egg masses in the roots, a decrease in the J2 population in the soil, and improvement in plant growth parameters. Thus, these endophytic bacteria can be both plant growth stimulators and biocontrol agents against root-knot nematodes (Banihashemian *et al.*, 2022).
- Soil application of castor oil at 30 L/ha and *Purpureocillium lilacinum* strain PL1 (Bionema

SL) at 20 kg/ha found effective in significantly reducing root gall index of *M. incognita* (Table 19.4) (Kepenekci *et al.*, 2017).

Table 19.4. Biological control of root-knot nematodes on kiwifruit

Treatment	Root galling/(% Reduction over control)		Reproduction rate	
	2013	2014	2013	2014
Inoculated control	7.70 ± 0.26	7.20 ± 0.13	5.092	5.120
Uninoculated control	3.60 ± 0.16 (53.24)	3.40 ± 0.16 (52.77)	2.010	5.120
Castor oil	1.70 ± 0.26 (77.92)	1.90 ± 0.10 (73.61)	0.596	0.620
Sesame oil	4.30 ± 0.15 (44.15)	4.00 ± 0.14 (44.44)	2.060	2.180
M. verrucaria	5.10 ± 0.17 (33.76)	5.30 ± 0.15 (26.38)	3.490	3.800
Glomus spp.	4.50 ± 0.16 (41.55)	4.90 ± 0.23 (31.94)	2.120	2.280
P. lilacinum	3.00 ± 0.21 (61.03)	3.20 ± 0.13 (55.55)	1.254	1.280

- Mycorrhizal kiwi were shown to be more tolerant of *M. javanica* than non-mycorrhizal seedlings (Verdejo *et al.*, 1990).
- Biopesticides like castor oil and neem oil cakes have been found effective in managing RKN.
- The positive correlations between *Pasteuria penetrans* and *Meloidogyne incognita* found in kiwi fruit orchards suggest the capacity of the bioagent in regulating the nematode populations but only when the density of the bioagent has reached certain levels (Verdejo-Lucas, 1992).
- Clandosan (chitin amendment from Igene Biotechnology, Inc.) at 5,000 and 7,500 kg/ha gave more effective control of *Meloidogyne incognita* and *Pratylenchus* sp. than the nematicide fenamiphos, at the rate of 25 liters/ha commercial product at the 40%. The saprophytic nematode fauna was also higher in terms of soil populations in the Clandosan treatments, thus indicating a high soil microbiological activity (Gonzalez, 1993).

19.2.5.5. Host resistance: Kiwi cv. Hayward seedlings were relatively tolerant of *M. hapla* in pot trials in Chile (Philippi and Budge, 1992).

Chapter - 20

Apple, *Malus domestica*

20.1. INTRODUCTION

Apple cultivation in India started during Eighteenth century and now it is an important crop of temperate region. Apple ranks 6th among fruit production in India. China is largest apple producer in the world with 43% share, however, India has only 3% share.

20.1.1. Origin

Apple originated in Eastern Europe and Western Asia

20.1.2. Area and production

Worldwide, 87,236,221 tons of apple is produced per year. China is the largest apple producer in the world with 42,426,578 tons production per year. United States of America comes second with 4,997,680 tons yearly production. China and United States of America produce together more than 50 % of world's total apple. With 3,618,752 tons of production per year, Turkey is the third largest producer of apple. India, with 2,316,000 tons of production per year is ranked at fifth (Table 20.1).

Table 20.1. Area, production and productivity of apple fruit (2020).

Country	Area ('000 ha)	Production ('000 tons)	Yield (tons/ha)
China	2,041.197	42,426.578	20.785
USA	118.860	4,997.680	42.047
Turkey	174.439	3,618.752	20.745
Poland	166.440	3,080.600	18.509
India	308.000	2,316.000	7.520
Italy	55.000	2,303.690	41.885
Iran	100.759	2,241.124	22.242

Russian Federation	210.822	1,950.800	9.253
France	50.370	1,753.500	34.812
Chile	32.371	1,621.321	50.085
Brazil	32.405	1222.979	37.740
WORLD		87,236.221	

Table 20.1 indicates that apple productivity in India is much lower than other apple producing countries.

State-wise area, production and yield of major apple growing states in India during 2019-20 is presented in Table 20.2.

Table 20.2. State-wise area, production and yield of apple in India (2019-20)

States	Area ('000 ha)/ Share (%)	Production ('000 tons)/ Share (%)	Productivity (tons/ha)
Jammu and Kashmir	136.54 (49.24)	1672.72 (74.62)	12.25
Himachal Pradesh	110.68 (39.92)	492.10 (21.95)	4.45
Uttarakhand	25.00 (9.02)	67.48 (3.01)	2.70
Arunachal Pradesh	4.73 (1.71)	7.35 (0.33)	1.55
Others	0.32 (0.12)	2.05 (0.09)	6.41
TOTAL	311.5 (100.00)	1914 (100.00)	8.08

Source: Horticulture Statistics Division, Department of Agri. Co-op. & Farmers Welfare.

Figures within parentheses indicate percentages Source: National Horticulture Board (NHB)

Major apple producing states/districts in India include:

» Jammu & Kashmir (Baramulla, Kupwara, Shopian, Kulgam, Badgam *etc.*)

» Himachal Pradesh (Shimla, Kullu, Kinnaur, Mandi, Chamba *etc.*)

» Uttrakhand (Uttar Kashi, Almora, Nainital, Dehradun, Chamoli *etc.*)

» Arunachal Pradesh (West Kameng, Tawang *etc.*)

Among Indian states productivity of Jammu and Kashmir is higher than the other two states. It is mainly due to the best climate available in this state.

20.1.3. Major commercial varieties and rootstocks (Table 20.3)

Table 20.3. Superior apple cultivars and rootstocks

Type	Varieties
Spur and semi spur type	Red Chief, Red Spur, Ruby Red, Hardi Spur, Sturdee Spur, Well Spur, Super Red, Stark Spur Red, Spur Type Red Delicious, Bright-n-Early, Oregon Spur, Starkrimson
Color strains	Top Red, Hi Red, Vance Delicious, Hardeman

Low chilling	Tropical Beauty, Schlomith Michal, Mayaan, Vered, Tamar, Anna, Naomi
Scab resistant	Priscilla, Sir Prize, Macfree, Freedom, Coop 12, Coop 13. Firdous and Shireen are indigenously developed scab resistant varieties
Varieties of promise	Fuji, Red Fuji, Gala, Scarlet Gala, Gala Mast, Granny Smith, Breaburn, Jonagold, Empire, Criterion
Processing cultivars	Delicious, Granny Smith, Liberty, Rome Beauty, York Imperial, Stayman Winesap, Northern Spy, Prima, Pricilla, Sir Prise, Freedom, Red Free, Summer Red, McIntosh, Maharaji, Prima
Indigenous varieties	Lal Ambri, Sunhari, Firdous, Shireen, Akbar, Chaubatia Princess, Chaubatia Anupam, Chaubatia Agrim, Chaubatia Swarniam, Chaubatia Alankar, Chaubatia Anurag, Ambred, Ambrich, Ambroyal, Ambstarking
Rootstocks	M-7, M-9, M-26, M-27, MM-106, MM-109, MM-111, EMLA-106

Table 20.4. Productivity potential of some indigenous apple cultivars

Cultivar	Year of release	Av. yield (kgs/tree)	Yield (tons/ha)
Lal Ambri	1973-74	150-160	35-40
Sunheri	1973-74	64-72	16-18
Akbar	2000-01	160-170	40-45
Firdous	1995	50-60	12-15
Shireen	1995	50-60	12-15
Gulshan	2001-02	119-120	29-30

20.1.4. Cultivation aspects

Table 20.5. Different aspects of apple cultivation

Variety	Rootstock	Training system	Spacing (m)	Density (trees/ha)	Yield (t/ha)	Reference
Empire	M-26	Y-trellis	2.1x3.7	1283	275.00	Robinson *et al.*, 1991
COE Red Fuji	M-26	Vertical axis	0.75x1.5	8889	135.3 (5th yr.)	CITH, 2013-14 (India)
COE Red Fuji	M-9	Expalier	1.5x3.0	2222	95.36 (5th yr.)	CITH, 2012-13 (India)
COE Red Fuji	M-9	Vertical axis	0.75x1.5	8889	93.82 (5th yr.)	CITH, 2012-13 (India)
Mollies Delicious	M-9	Head & spread	0.75x1.5	8889	126.33	CITH, 2012-13 (India)
Mollies Delicious	M-9	Modified leader	0.75x1.5	8889	93.82 (5th yr.)	CITH, 2012-13 (India)
Mollies Delicious	M-9	Central leader	2.5x3.5	1143	39.5 (8th yr.)	CITH, 2010 (India)
Red Delicious	M-26	Y-trellis	2.4x4.3	961	122.00	Robinson *et al.*, 1991
Royal Delicious	MM-106	Modified leader	2.5x3.5	1143	67.33 (10th yr.)	CITH, 2012-13 (India)
Marshal McIntosh	M-9	Slender spindle	1.7x4.0	1429	60.00	Wunsche & Lasko, 2000 (USA)
Golden Delicious	M-9	Central leader	1.5x3.0	2222	47.40 (8th yr.)	CITH, 2010 (India)
Golden Delicious	MM-106	Central leader	2.5x3.5	1143	39.5 (8th yr.)	CITH, 2010 (India)
Red Spur	Standard	Modified leader	4.0x4.0	625	39.5 (8th yr.)	CITH, 2012-13 (India)

Vista Bella	M-9	Central leader	1.5x3.0	2222	39.5 (8th yr.)	CITH, 2010 (India)
Red Fuji	EMLA-111	Modified leader	3.0X3.0	1111	16.9	Bhatia & Kumar. 2009 (India)
Scarlet Gala	EMLA-111	Modified leader	3.0X3.0	1111	15.6 (9th yr.)	Bhatia & Kumar. 2009 (India)
Oregon Spur	MM-106	Modified leader	2.5x2.5	1600	13.2 (7th yr.)	Verma, 2009 (India)
Red Chief & Silver Spur	MM-106	Modified leader	2.5x2.5	1600	12.6 (7th yr.)	Verma, 2009 (India)

20.1.5. Advance technology

High density planting on Mailing root stock

20.1.6. Pre-cooling

Forced air pre-cooling or hydro-cooling at -1 to 0 ^{0}C, RH – 90-95%, freezing temperature around -1.7 ^{0}C suitable for cooling up to 8 months.

20.1.7. Storage conditions

» Temperature: -1.1^0 to 0 ^{0}C
» Relative humidity: 85-90%
» Storage period: 4-8 Months

20.1.8. Handicaps

The main handicaps include non-availability of

» Irrigation
» Suitable cultivar which can produce sustainable yields under changing environment
» Nutrients for their efficient use during the growth stage when an apple tree demands
» Training and pruning
» Disease and pest management
» Protection of fruits from hailstorms.

20.1.9. Productivity enhancing technologies

» Looking to the topography of apple growing areas, water can be stored in tanks and can be used whenever necessary through drip system for irrigation.
» Cultivars have been developed which do well under fluctuating temperatures. These are to be popularized on priority basis.
» Fertigation can be used for application of nutrients along with irrigation water. The nutrient can thus be made available to the trees when they require them most.
» Although the system of training and pruning is recommended since long, however, it has not been used as an essential practice. Proper training and pruning gives strong framework to a tree and also enable the tree to yield a normal crop on a regular basis.
» The first precaution to be taken up for management of diseases and pests is to maintain

proper sanitary conditions in the orchard, provide conditions for aeration and sunlight up to the inner parts of the tree. Use of appropriate fungicides and insecticides at proper timings are very important for their control and to check their spread.

- In some pockets, during summer, the hailstorms occur regularly which adversely affect the quality and production. Planting of apple orchards in these areas should be avoided and be put under temperate nuts like walnut, pecan nut *etc.*

20.1.10. Apple export from India (2017-18)

- Quantity : 12,530 million tons
- Value : Rs. 391.659 million

The top countries which imported the highest dollar value worth of apple fruit from India in 2020-2021 (Apr-Nov) are presented in Table 20.6.

Table 20.6. Top countries which imported the highest dollar value worth of apple fruit from India in 2020-2021 (Apr-Nov).

Country	Value ('000 USD)	Share (%)
China	370	16.09
UK	320	13.91
Nepal	260	11.30
Thailand	240	10.43
Vietnam	220	9.57
Malaysia	160	6.96
Saudi Arabia	160	6.96
Qatar	150	6.52
Australia	110	4.78
Germany	60	2.61
TOTAL	1,410	100.00

20.1.11. Nematode problems

Of the many genera of plant parasitic nematodes detected in soils from apple orchards, two species of root lesion nematode, four species of root-knot and the dagger nematode are believed to be the most economically important ones.

20.2. LESION NEMATODES, *PRATYLENCHUS PENETRANS, P. VULNUS*

The root-lesion nematode *Pratylenchus penetrans* is the most important nematode affecting apple production. It causes 'soil sickness' of apple nurseries and orchards. The nematode is responsible for decline in apple tree fruit orchards. They reproduce better and cause more damage in sandy soils. Extensive feeding on fine roots results in symptom development. Nematode populations are highest in fall, at the end of the growing season. Many weeds are also act as hosts.

20.2.1. Symptoms

Root lesion nematodes penetrate into roots and cause damage by feeding and migrating through the cortical tissues. Interaction of root lesion nematodes with other soil-borne organisms can increase injury to roots. The lesion nematodes are sometimes responsible for the apple replant disease (characterized by young tree decline after transplanting). Apple trees (especially younger trees) may exhibit poor growth, stunting, lack of vigor, twig dieback, and a gradual decline in yields. Severely infected root systems may lack feeder roots. The economic threshold of *P. penetrans* is 15 nematodes per 100 g soil which cause significant growth reduction in apple trees (Crossa Raynand and Audergon, 1987).

Young apple trees infected with lesion nematodes exhibit poor growth. Yields of infected trees may be significantly lower than non-infected trees. Infected roots generally have discolored and necrotic (black) areas where nematode feeding has caused cell death and where additional pathogens have likely invaded. Severely affected root systems may lack feeder roots.

20.2.2. Life cycle

The juveniles of *Pratylenchus* sp. enter the roots wherever the tissue is immature so that the penetration may become easy. Walia and Bajaj (2003) observed that the larvae feed on root cortex and multiply by moving inter and intracellularly. Migration through host tissues and feeding activities result in destruction of host cells which leads to the formation of necrotic lesions. Sexual reproduction is common in *P. penetrans*. The adult female laid eggs in soil or in roots. Among four moults in nematode life cycle; the first one is inside the egg and subsequent three molts are outside the roots. All the life stages except J1 are parasitic. Entire life cycle is completed in 30–45 days. The population of the lesion nematode is much higher within the roots as compared to the surrounding soil, since it is an endoparasite.

20.2.3. Interaction with other pathogens

The addition of *Phytophthora cactorum* to apple replant disease soil containing the nematode significantly reduced plant growth compared with the corresponding individual treatments. *Phytophthora parasitica* did not by itself reduce plant height but in the presence of nematode it reduced plant height to a greater extent than did the nematode alone (Utkhede *et al.*, 1992).

20.2.4. Management

20.2.4.1. Physical methods: In order to eliminates the nematode infection, the bare roots of apple rootstocks should be dipped in hot water maintained at 50 °C for 15 min or at 47 °C for 30 min. The material should first be stored in a cold place (at 1 to 3.3 °C), the treatment conducted in January, and the plants subsequently restored in a cold place until planting. Immersion in hot water should be followed immediately by a cold bath.

20.2.4.2. Cultural methods: Plant ground cover with crops that are not suitable hosts such as tall fescue, red fescue or perennial ryegrass. Yellow mustard (*Brassica juncea*) contains allyl isothiocyanate, which is active against nematodes. *Brassica carinata* also has been reported to be effective against nematodes.

Use certified (nematode-free) planting material. Remove old roots and plant green manure cover crops (tall fescue, red fescue, oat cultivar Saia, or perennial ryegrass as orchard ground cover) resistant to root lesion nematodes (*Pratylenchus vulnus*) for 1 to 2 years or fallow the site for 4 years. If *Pratylenchus vulnus* is present, maintain a poor host for this nematode, such as tall fescue, red fescue, or perennial ryegrass as orchard ground cover. These crops are hosts for *Pratylenchus penetrans* and should not be used if this species of root lesion nematode is present.

Rape seed meal (RSM) amendment reduced recovery of *Pratylenchus* spp. from apple roots. Suppression of *Pratylenchus* was attained to an equivalent extent by amending soils with either RSM or soybean meal (SM) when applied to provide a similar N content. RSM amendment elevated populations of *Pythium* spp. and of ammonia-oxidizing bacteria that release nitric oxide but suppressed fluorescent pseudomonad numbers. *Streptomyces* spp. soil populations increased significantly in response to RSM but not SM amendment. The vast majority of *Streptomyces* spp. recovered from the apple rhizosphere produced nitric oxide and possessed a nitric oxide synthase homolog. The transformations in the bacterial community structure are associated with the observed control of *Pratylenchus* spp., with nitric oxide production by soil bacteria potentially having a role in the induction of plant systemic resistance (Cohen *et al.*, 2005).

Preplant soil fumigation and biofumigation with seed meal (SM) formulation consisting of either *Brassica juncea–Sinapis alba* or *B. juncea–B. napuseach* provided similar levels of disease control during the initial growing season. Although tree growth was similar in fumigated and SM-amended soil during the initial growing season, tree performance in terms of growth and yield was commonly superior in *B. juncea–S. alba* SM-amended soil relative to that in fumigated soil at the end of four growing seasons. SM-amended soils were resistant to reinfestation by *Pratylenchus penetrans* and *Pythium* spp. relative to fumigated soils and corresponded with enhanced tree performance. After 2 years, the rhizosphere microbiome in fumigated soils had reverted to one that was indistinguishable from the no-treatment control. In contrast, rhizosphere soils from the SM treatment possessed unique bacterial and fungal profiles, including specific microbial elements previously associated with suppression of plant-pathogenic fungi, oomycetes, and nematodes. Overall diversity of the microbiome was reduced in the SM treatment rhizosphere, suggesting that enhanced "biodiversity" was not instrumental in achieving system resistance or pathogen suppression (Mazzola *et al.*, 2015).

20.2.4.3. Chemical methods: Pre-plant soil fumigation in fall with Metam Sodium is highly recommended. Velum Prime at 6.5 to 6.84 fl. oz. /Acre can be used after planting using drip, trickle of micro-sprinklers.

20.2.4.4. Biological methods: MeloCon WG (*Purpureocillium lilacinum* strain 251) at 2 to 4 lb/Acre plus a soil wetting agent can be applied to established plants, although it might be better used when applied to plants just before planting.

20.2.4.5. Host resistance: The Geneva series rootstocks G11 and G30 support lower populations than Malling or Malling-Merton rootstocks in eastern Washington.

20.3. APPLE REPLANT DISEASE

20.3.1. Symptoms

Symptoms of apple replant disease (ARD) include uneven growth, stunting and shortened internodes aboveground and discolored roots, root tip necrosis and reduced root biomass belowground. The bearing of fruit is delayed by 2–3 years and yields are depressed by up to nearly 60% for the duration of the tree's commercial life (Willett *et al.*, 1994).

20.3.2. Causal agents

Predominantly biotic factors have been identified as the underlying cause (Tewoldemedhin *et al.*, 2011a, b, c). No single organism has been identified as the universal causal agent, but there is increasing evidence that multiple soil-borne plant pathogenic fungi (e.g. *Cylindrocarpon* spp., *Rhizoctonia* spp. and *Fusarium* spp.) and oomycetes (e.g. *Pythium* spp. and *Phytophthora* spp.) form a disease complex of variable species composition (Mazzola and Manici, 2012). Root damage caused by root lesion nematodes (e.g. *Pratylenchus penetrans*) has also been implicated in facilitating the entry of pathogens into root tissues resulting in the development of severe ARD (Galkovskyi *et al.*, 2012; Tewoldemedhin *et al.*, 2011a, b, c). The relative importance of these putative causal agents of ARD appears to be site-specific (Manici *et al.*, 2013),

20.3.3. Management

Current research efforts are focused on more selective approaches for the modification of indigenous soil microbial communities e.g. the addition of antagonistic and beneficial microorganisms (Utkhede and Smith, 2000; Hoitink and Boehm, 1999), incorporation of composted organic amendments and their indigenous suppressive microbial communities (Yao *et al.*, 2006) and specific orchard groundcover management regimes (St. Laurent *et al.*, 2008).

20.3.3.1. Cultural methods

1. **Plant nematode-free trees.** If the nursery site where trees were propagated had significant replant disease pressure young trees likely carry inocula in the roots which may serve to infest a grower's site.
2. **Crop rotation.** Many other crops can host the pathogens in this complex, thus crop rotation is not a guaranteed strategy to lower potential disease pressure. However, a five year plus rotation to a non-woody crop can help reduce pressure. For small-scale, soil in the planting hole can be replaced with good quality soil from a non-orchard site. In order to effectively replace the soil, the planting hole should be 7 feet (length) X 7 feet (width) X 2 ½ feet (depth) (Benson 1976); (Smith 1994).
3. **Water management**

a) Limit periods of soil saturation: Oomycetes in the apple replant disease complex can disperse between trees if the soil is saturated, conditions that will also aid the root infection process. Dispersal of the pathogen can be reduced by minimizing periods of soil saturation.

b) Choose well drained soils: Choose orchard sites with good drainage.

c) Maintain and improve soil structure: Add organic matter to improve aggregate stability and water infiltration and reduce compaction.

iv) Biofumigation: Biofumigants, especially brassica seed meals (Mazzola *et al.*, 2009; Mazzola and Manici, 2012), appear to offer reproducible reductions in the populations of ARD causal agents and ARD disease severity. This is particularly true against *P. penetrans*: application of *B. juncea* seed meal resulted in more than 98% reduction in populations of *P. penetrans*.

20.3.3.2. Host resistance: The planting of resistant apple rootstock genotypes has proven to be an effective means of reducing the severity of *Phytophthora* crown and root rots and in principle is an attractive option for managing ARD. In general, Geneva series rootstocks are tolerant to the disease. Some of the Cornell-Geneva rootstocks e.g. G.41, G935, G30. G65, G210, G214 have been reported to show greater tolerance of ARD than other rootstocks (Isutsa and Merwin, 2000; Leinfelder and Merwin, 2006; Mazzola *et al.*, 2009), notably the current commercial standard for dessert apple M.9. M.116 (clone 86.1.25) is slightly less vigorous than MM.106, but is reportedly resistant to ARD and other diseases (Webster *et al.*, 1986).

20.4. DAGGER NEMATODE, *XIPHINEMA* SPP.

Dagger nematode (*Xiphinema americanum*) is migratory ectoparasite, that attack roots from the outside, feeding on epidermal cells with their long stylets. They commonly act as vectors of viruses, for example the *Tomato ringspot virus*, which causes apple union necrosis and decline (especially for Red Delicious on M106 rootstocks). The dagger nematode (*Xiphinema* spp.) transmits the *Tomato ringspot virus* to apples causing Union Necrosis and Decline. These nematodes also transmit *Cherry rasp leaf virus* causing the Flat Apple disease. Both the nematode and virus are required for either disease to occur. Damage caused by the virus affects fruit production. The nematode is an ectoparasite found only in soil. Nematode populations are highest in fall, at the end of the growing season. *Xiphinema* was overall the third-most prominent genus of plant-parasitic nematode found in the Okanagan of BC, being found in 59% of apple blocks.

20.4.1. Symptoms

Dagger nematode is a migratory ectoparasite which means it attacks roots from the outside. It is a relatively large nematode (1.3 to 2.2 mm) (Akinbade *et al.*, 2014). Dagger nematode has a long, hollow, needle-like stylet called an odontostyle which it uses to feed on cells at root tips. Feeding can cause root tip cells to enlarge, disrupting function and resulting in roots that have a knobby stunted appearance, with swellings at the tips.

No known symptoms occur from nematode feeding alone. Union necrosis results a dark necrotic line that forms at the graft union flanked by spongy orange tissue. Fruits on Flat-Apple-infected trees are smaller, flatter, and the calyx end tends to be open.

Plant symptoms resulting from viral infection commonly include: chlorotic mottling and defoliation; bright yellow discoloration of foliage; yellowing along main leaf veins; spots, blotches and crinkling of leaves; and stunting.

20.4.2. Life cycle

While its life cycle has not been experimentally confirmed, researchers suggest that it may be up to one year in duration with individuals living 3 to 5 years. Dagger nematode can harbor virus particles in the esophageal lumen near the stylet extension and in the glandular region of the esophagus. As the nematode feeds, virus particles can be transmitted to the plant within 24 hours of exposure (Ferris 2016). Since the stylet, the stylet extension, and the lining of the esophageal lumen are shed when the nematode molts, virus particles are lost as the nematode passes from one life stage to another. However, changes in life stage occur when the nematode is at a feeding site and taking in food. Consequently, virus particles are rapidly reacquired from the plant on which the nematode is feeding.

20.4.3. Management

20.4.3.1. Cultural methods: Good weed management as many weeds are reservoirs of virus.

20.4.3.2. Chemical methods: Velum Prime at 6.5 to 6.84 fl. oz./acre can be used after planting using drip, trickle of micro-sprinklers.

20.5. APPLE UNION NECROSIS DECLINE (AUND)

AUND is due to an incompatibility at the graft union where a resistant scion is grafted onto a susceptible, but tolerant rootstock, most commonly MM.106.

20.5.1. Distribution

Common to all fruit-growing regions in eastern North America.

20.5.2. Symptoms

Symptoms appear approximately 4–6 years after planting, usually once the tree is capable of bearing a full fruit load. Affected trees show a general decline beginning with delayed budbreak. The canopy tends to be sparse, bearing small, pale green leaves, and premature defoliation is possible. A distinct black, sunken line at the union is apparent underneath the bark. The graft union may be weakened to the point where the scion and rootstock separate partially or completely or crack under stressful conditions.

20.5.3. Causal agent

Apple union necrosis decline is caused by Tomato Ringspot Virus (TmRSV).

20.5.4. Survival and transmission

TmRSV can survive in a number of weed hosts, is seed-transmitted in dandelion, and is vectored by the dagger nematodes *Xiphinema americanum* and *X. rivesi.*

20.5.5. Management

To avoid introducing the virus into new plantings, purchase only certified virus-free trees from a reputable nursery. Good weed control and pre-planting site preparation can help to reduce the incidence of this disease. Selecting resistant rootstock or, in some instances, using a resistant inter-stem will reduce the incidence of disease. Rootstocks considered tolerant to AUND (i.e., those in which the development of symptoms is possible) include MM.106 (particularly on Delicious), M.26, and P.2. Those considered resistant include M.4, M.7, Ottawa 3, and Novole. Several other rootstocks, including M.9, MM.111, G.16, G.30, and B.9, have given mixed results. Fruiting varieties resistant to TmRSV include 'Delicious', 'Quinte', 'Tydeman's Red', 'Jerseymac', and 'Jonathan'. 'Golden Delicious', 'Empire', and 'York Imperial' are tolerant.

20.6. ROOT-KNOT NEMATODE, *MELOIDOGYNE* SPP.

The root-knot nematode, *Meloidogyne incognita* has adverse effect on young apple trees (Utkhede *et al.*, 1992). The root-knot nematode damage root system and hamper uptake of water and nutrients. *Meloidogyne incognita* infection can cause noticeable damage to apple roots and consequently can affect fruit production. Root-knot nematodes are well known for their interaction and establishing relationship with fungi in disease complexes (Singh *et al.*, 2007). *Meloidogyne mali* was originally described in Japan on roots of an apple rootstock (*Malus prunifolia*) (Itoh *et al.*, 1969). It is a native of Japan.

20.6.1. Geographical distribution

To date, *M. mali* has been reported in France, Italy, Japan, the Netherlands and the USA (EPPO, 2017b).

20.6.2. Symptoms

Slight swellings were observed on the roots six weeks after inoculation and the number of root galls induced by the nematodes became high after tenth week. *M. mali* is a damaging nematode which can produce large root galls on its host plants, interfering with their water and nutrient uptake from the soil and thus reducing their growth. Stunting and severe decline of infected trees are reported in Japan in orchards infested by this nematode which penetrate host roots (Nyczepir and Halbrendt, 1993).

20.6.3. Host range

The host range of *M. mali* includes many species of economic importance in fruit trees (e.g., apple, chestnut, fig, and mulberry), forestry trees (e.g., elm, maple, oak, and yew), and vegetable crops (e.g., cabbage, carrot, cucumber, eggplant, soybean, and watermelon) (Kang *et al.*, 2022).

20.6.4. Biology

This root-knot nematode has sedentary endoparasitic habits. Second-stage juveniles (J2) penetrate host roots where they establish a specialized feeding site (giant cells) in the stele. As J2 develop, they cause root swellings and become swollen females. Females rupture root cortex and sometime protrude with the egg masses from the root surface. J2 emerge from the egg masses and migrate in the soil (Itoh *et al.*, 1969). It was found that *M. mali* needed 18 to 22 weeks for completing one full generation and that only one generation took place in a year.

20.6.5. Histopathology

The second-stage juveniles of *Meloidogyne incognita* after penetration entered into cortex and moved along the cortical layer of the cells. Disruption of cortical tissue was observed and a number of cells were present with broken walls. Numerous giant cells were also found, which were feeding sites for nematodes (Khan *et al.*, 2010). It is known that all the *Meloidogyne* species irrespective of their host range induce giant cells of same shape and size (Blok *et al.*, 1997). Feeding sites act as sinks for plant photosynthates and therefore impair plant growth and development (Singh *et al.*, 2007). Deformation and blockage of vascular tissue at feeding sites limit translocation of different nutrients and water resulting in suppression of plant growth and adversely affecting yield (Hussey and Williamsons, 1997). It is also reported that the infection weakens plants and predisposes them to bacterial infection (Singh *et al.*, 2007). Several sections of females and egg masses were present in the superficial layer of the medullary region. Prominent nuclei were present in cells near feeding sites as reported by Pasha *et al.* (1987) in eggplant roots infected by *M. incognita*.

20.6.6. Spread

The nematode spread through root material, soil debris and infected bare root propagative plant material. Natural spread in the soil is slow but *M. mali* can easily be transported with soil and plants for planting over long distances.

20.6.7. Management

The best method of controlling apple root-knot nematode is to plant nematode-free seedlings in nematode-free soil. Use certified rootstocks or seedlings to establish new orchards. Improve soil tilth and drainage and control other pests. Proper irrigation and fertilizer application will also reduce stress on trees.

Chapter - 21

Plum, *Prunus cerasifera*

21.1. INTRODUCTION

21.1.1. Origin

Five centers of origin for plums had been identified by Watkins (1976). These include Europe for *Prunus domestica* (European plum), Western Asia for *Prunus insititia* (Damson plum), Western and Central Asia for *Prunus cerasifera* (*P. cerasifera*) (cherry plum), China for *Prunus salicina* (Japanese plum), and North America for *Prunus americana* (American plum).

21.1.2. Area and production

Plums are grown in the temperate zone of all the countries of the world. Worldwide 12,063,776 tons of plum is produced during 2020. China is the largest plum producer in the world with 6,676,142 tons production volume per year. Romania comes second with 512,975 tons yearly production. India is ranked at 8th with production volume of 261,903 per year (Table. 21.1).

Table 21.1. Country-wise area, production and productivity of apple fruit (2020).

Country	Area ('000 ha)	Production ('000 tons)	Yield (tons/ha)
China	1,961.428	6,676.142	3.404
Romania	65.114	512.975	7.878
Serbia	77.949	463.115	5.941
USA	28.875	392.537	13.594
Turkey	20.811	297.589	14.299
Chile	17.278	294.873	17.066
Iran	33.881	269.113	7.943
India	31.410	261.903	8.338
Spain	16.344	222.020	13.584

Italy	12.720	220.729	17.353
France	14.741	211.269	14.332
WORLD		12,063.776	

In India, plum is mainly grown in Uttarakhand, Himachal Pradesh, and Jammu and Kashmir, but on limited scale in north-eastern states. In Himachal Pradesh, plums are mainly grown in Solan, Sirmour, Shimla. Mandi districts. Low chill varieties are also grown in Punjab, Haryana and Eastern U.P (Table 21.2).

Table 21.2. State-wise production and share (%) of plums in India (2021-22 First Advance Estimate)

States	Production ('000 tons)	Share (%)
Uttarakhand	34.84	44.40
Himachal Pradesh	16.70	21.28
Jammu and Kashmir	13.53	17.24
Punjab	7.04	8.97
Nagaland	2.72	3.47
Tamil Nadu	2.09	2.66
Haryana	1.29	1.64
Arunachal Pradesh	0.19	0.24
Maharashtra	0.07	0.09
TOTAL	78.47	100.00

Source: National Horticulture Board (NHB)

21.1.3. Varieties

21.1.3.1. Exotic cultivars

Table 21.3. Important exotic collections of plum for different altitudes.

True temperate zone	Mid to high altitude	Mid-hill to moderate
Satsuma, Santa Rosa, Burbank, Grand Duke, President, Early Italian, Frontier, Diurret, Friar.	Kelsey, Santa Rosa, Titron, Satsuma, Mariposa.	Jamuni, Kelsey, Santa Rosa, Titron

i) Lombard plum: The color of fruit is purplish red. Varieties are Lombard, Victoria.

ii) European plum: cultivars are California Blue, Washington, French Prunes, Early Italian, Stanley, Grand Duke, Victoria, Damson.

iii) Japanese plum: Cultivars include Beauty, Methley, Santa Rosa, Kelsey, Mariposa, Satsuma, Burbank Red Beauty, Fronteir

21.1.3.2. Indigenous cultivars

i) Alu Bokhara: Under Amritsar (Punjab) conditions, the selection showed promise for higher production.

ii) Kala Amritsari: Most popular cultivar in Punjab. This is self-fruitful but yield improves if pollinated with cv. Titron.

iii) Pant Plum-1: Selection from the seedling population. It can be dwarfing rootstock for plum cultivars.

iv) Satluj Purple: Tree is self-incompatible and must be planted with cv. Kala Amritsari as pollinator.

21.1.4. Rootstocks

21.1.4.1. Seedling rootstocks: The plums grown mostly on plum rootstocks though seedling of peach, apricot and almond can be used. In India, plum is propagated mainly on seedlings of wild apricot (Chuli).

21.1.4.2. Clonal rootstocks: Myrobalan B, Myrobalan 29C, Mariana GF 8/1, Mariana 2624, Mariana 8-6, Mariana 9-25, Pixy, St. Julien A, GF-667.

21.1.5. Cultivation aspects

Table 21.4. Different aspects of plum cultivation

Variety	Rootstock	Training system	Spacing (m)	Density (trees/ha)	Yield (t/ha)	Reference
Simca	Nemaguard	Central leader	2.0 x 5.5	909	169.8	---
Mallard	St. Julian	Y - trellis	0.5 x 4.0	5000	95.1	Meland, 2005 (Norway)
Opal	St. Julian	Y – trellis	0.5 x 4.0	5000	77.6	Meland, 2005 (Norwey)
Elda	St. Julian	Y - trellis	0.5 x 4.0	5000	53.00	Meland, 2005 (Norwey)
Mariposa	Standard	Modified leader	5.0 x 5.0	625	27.26 (9th leaf)	CITH, 2014-15
AU - Cherry	Standard	Modified leader	5.0 x 5.0	625	23.99 (9th leaf)	CITH, 2014-15
Tarrol	Standard	Modified leader	5.0 x 5.0	625	23.40 (9th leaf)	CITH, 2014-15

21.1.6. Export

Plum is exported to over 133 countries. In the year 2020-2021 (Apr-Nov), India has exported plum worth of 96.59 USD million. The total volume of export in 2020-2021 (Apr-Nov) was around 174.345 million tons. The total volume of plum export around the world in year 2018 was 175.068 million tons. The figures show the great potential for Indian exporters of plum to increase their participation in global trading and improve their numbers.

Below are the top countries which imported the highest dollar value worth of plum from India in 2020-2021 (Apr-Nov) (Table 21.5)

Table 21.5. Export of plum from India to other countries during 2021-22 (April-November)

Country	Value (Million US$)	Share (%)
Netherlands	22.52	23.32
Saudi Arabia	9.73	10.07
China	8.16	8.45
USA	7.55	7.82
UK	7.28	7.54
Russia	4.48	4.64
UAE	3.84	3.98
Japan	3.49	3.61
France	2.34	2.42
Belgium	1.85	1.92

The top 5 trading partners of India are Netherlands (22.52 USD Million), Saudi Arabia (9.73 USD Million), China (8.16 USD Million), USA (7.55 USD Million), and United Kingdom (7.28 USD Million) . The total export value of plum in these countries is 55.24 USD million. These top 5 countries account for over 57.19% of the total plum export from India.

21.1.7. Nematode problems

The major plant parasitic nematodes associated with plums include root lesion (*Pratylenchus vulnus*), ring (*Mesocriconema xenoplax*), dagger (*Xiphinema americanum*) and root-knot (*Meloidogyne* spp.) nematodes, which causes detrimental loss (McKenry, 2004; Nyczepir *et al.*, 2009; Nyczepir, 2011). Pin nematodes (*Paratylenchus* sp.) are the least infectious nematodes that are often found in plum orchards. Mature trees may withstand high population of nematodes before showing signs of damage, but young trees that are transplanted in nematode-infested areas develop poorly.

Nematodes on plum roots can cause serious damage by feeding on tree roots. Some are more damaging than others and infestations can be spotty throughout an orchard. But overall, the nematodes can cause loss of vigor, reduction in fruit yield, and ultimately death of branches or entire trees.

21.2. ROOT LESION NEMATODE, *PRATYLENCHUS VULNUS*

21.2.1. Symptoms

P. vulnus is a true pathogen of the five tested plum rootstocks, capable of causing significant root destruction and reduction in plant growth as a result of a high and rapid population build-up in the first years. Root lesion nematodes damage roots by moving through cortical tissues and feeding in these areas. It feeds on the cortex and damages the root system chemically and mechanically. Although the plum rootstock Bruce (*P. salicina* x *P. angustifolia*) did not perform as well as Torinel and Redglow, it also appear to be poor host for *P. vulnus* (Alcañiz *et al.*, 1996).

21.2.2. Biology

Root lesion nematodes reproduce sexually, and eggs are deposited throughout the root cortex. Under optimal temperature (25 ^{0}C) the life cycle lasts about 30 days. All life stages overwinter in Michigan.

21.2.3. Ecology

Root lesion nematodes cause more damage and build up to higher population densities in sandy soils than in heavier soils.

21.2.4. Management

The plum rootstocks Torinel (*Prunus domestica*) and Redglow (*P. salicina* x *P. munsoniana* cv. Jewel) showed a moderately resistant response to *Pratylenchus vulnus*.

The Myrobalan 29 C plum rootstock is susceptible the root lesion nematode, *P. vulnus* in soils with low P availability. Both mycorrhizal treatments and P fertilization conferred a certain degree of protection against *P. vulnus* after 6 months, but after 2 years of plant growth, only *Glomus intrararadices* was still highly effective (Pinochet *et al.*, 1998).

Mycorrhizal plants colonized by *G. intrararadices* tolerated the parasitism of this nematode. Root damage caused by *P. vulnus* probably did not affect growth and development of Myrobalan 29 C plants under these circumstances due to the stimulation of plant nutrition and vegetative growth. Therefore, early mycorrhizal inoculation with *G. intrararadices* confers protection against *P. vulnus* on Myrobalan 29 C rootstock. The mycorrhizal fungus *G. mosseae*, very effective in nematode-free soil, was not a good endophyte for Myrobalan 29 C plants infected with *P. vulnus*. The selection of the proper fungus for mycorrhizal inoculation is important, especially when considering replant sites, where fruit tree plantations have been established for a long time, and where the P level in these soils is very low. The utilization of mixed inocula containing a mixture of AM fungi (*G. mosseae* and *G. intrararadices*) is probably the best option in future for practical implementation in nursey production systems using micro-propagated rootstocks (Pinochet *et al.*, 1998).

The plum hybrid 'Bruce' has shown resistance to both *M. incognita* and *P. vulnus* nematode species (Dirlewanger *et al.*, 2002; Pinochet *et al.*, 1996).

21.3. RING NEMATODE, *MESOCRICONEMA XENOPLAX*

21.3.1. Symptoms

The ring nematode, an ectoparasite of roots, is frequently recovered from orchards. Ring nematodes spend their lives in soil feeding on roots. High population densities of this nematode can inhibit normal growth and development of roots. Feeding by ring nematodes, particularly on small feeder roots, predisposes trees to bacterial canker *(Pseudomonas syringae)*. Ring nematodes can increase the susceptibility of *Prunus* trees to bacterial canker .

Orchards infested with high population levels of ring nematodes frequently exhibit symptoms associated with bacterial canker including blighted buds, blossoms, and leaves, and cankers that occur on and can result in the girdling and death of limbs and/or trees.

21.3.2. Management

21.3.2.1. Cultural methods: Cover crops such as several Bahia grasses (*Paspalum* spp.) have demonstrated some promise in suppressing *M. xenoplax*. Bahia grass must be planted between rows of trees, because it is a strong competitor for water and nutrients. If Bahia grass lowers the population of *M. xenoplax* between rows during the life of the current planting, a later planting could be established in the Bahia grass centers following herbicide treatment.

Crop rotation with 'Stacy' wheat (*Triticum aestivum*) which is a non-host to *M. xenoplax* and appeared to be antagonistic via some unknown mechanism (Nyczepir and Bertrand, 1990). The benefits to the grower in rotating with wheat are the monetary return while waiting to replant the land with the potential to avoid or reduce the need for preplant fumigation.

21.3.2.2. Host resistance: Partial tolerance to the ring nematode has been identified in 'Lovell' rootstock, and it is preferred over Nemaguard because it survives longer on peach-tree-short-life (PTSL) sites in the southeastern United States (Sharpe *et al.*, 1989).

21.4. DAGGER NEMATODE, *XIPHINEMA AMERICANUM*

21.4.1. Symptoms

The dagger nematode feeds as an ectoparasite and causes a slight swelling of root tissue, which prevents the root system from functioning in a normal manner. The dagger nematode is also a vector of the tomato ring spot virus.

Dagger nematodes reduce tree vigor with their feeding but mostly are important because they vector tomato ringspot virus. A strain of TmRSV causes girdling and stem pitting at the graft union of infected plum trees, resulting in weakened trees that may die prematurely. The disease is PBL and the vectors are *X. americanum* and *X. rivesi*.

21.4.2. Management

Management strategies include use of certified virus-free nursery stock, and planting trees propagated on the virus-immune Marianna 2624 rootstock.

21.4.2.1. Cultural methods: Whenever possible, plant new orchards in land that has previously been planted to non-woody crops for several years. California-bred alfalfa cultivars can be a good choice except where ring nematode develops well (i.e. highly porous soils). Root lesion nematodes survive well within dead root tissue or soil, with 5% lasting 5 years after tree removal.

The following measures will help to prevent spread of nematodes to uninfested fields:

- » Use certified planting stock.
- » Clean soil from equipment before moving between orchards.
- » Do not reuse irrigation tail water.
- » For new orchards, the use of certified nematode free and nematode resistant rootstocks will be increasingly important.
- » Leave sandy soils fallow during the year, but finer-textured soils may need a summer crop of Sudan grass to enable deep-drying.
- » Crop rotation with plants that are not susceptible and are not hosts to the nematodes.
- » Consider planting young trees on resistant rootstocks.

21.4.2.2. Host Resistance: Consider the use of resistant rootstocks. Because rootstocks for plums differ in response to various plant parasitic nematodes, it is important to select rootstocks that are resistant to the species of nematode present in soil.

21.5. ROOT-KNOT NEMATODES, *MELOIDOGYNE* SPP.

Plum trees and plum rootstock are susceptible to damage from root-knot nematodes. This type of nematode burrows into the cells of roots and stays there, feeding for its entire life.

21.5.1. Symptoms

Signs of root-knot nematodes in plums include a root system that is poorly developed. Above the soil, trees will show a general lack of vigorous growth, smaller leaves, and twigs or branches that die-back. By harvest time, the yield is reduced. Cankers and blighted buds, leaves, and flowers can also be seen on affected trees. It is not uncommon to see root-knot nematode signs in some trees but not others.

Young trees planted in soil that is infested with nematodes are most vulnerable to the worst effects of these pests. They can show stunted growth and either die soon after transplanting or continue to show poor growth and low fruit yield.

Root-knot nematodes take up a single feeding site within a root where they remain for their lifetime.

21.5.2. Management

In California, 85% of the plum orchards are replanted on 'Nemaguard' rootstock (*P. persica* × *P. davidiana*) since it provides field resistance to all races of root-knot nematodes (McKenry, 1987). In plums, there is root-knot nematode resistance in 'Marianna' (*P. cerasifera* × *P. munsoniana*) and 'Myrobalan' (*P. cerasifera*). The so called *Ma* genes in some 'Myrobalan' clones completely suppress root-knot nematode multiplication and confer a complete spectrum, high level and stable resistance to *M. incognita*, *M. arenaria* and *M. javanica* (Esmenjaud *et al.*, 1996; Lecouls *et al.*, 1999).

Prunus rootstocks from around the world have been screened to determine resistance to both major endoparasites, *M. incognita* and *P. vulnus*, but so far only the plum hybrid 'Bruce' has shown resistance to both nematode species (Dirlewanger *et al.*, 2002; Pinochet *et al.*, 1996).

21.6. PIN NEMATODE, *PARATYENCHUS NEOAMBLYCEPHALUS*

21.6.1. Symptoms

Addition of *P. neoamblycephalus* to Myrobalan plum seedlings at time of transplanting resulted in smaller (P = < 0.01) plants after 4 months of growth. The inverse relation of final to initial nematode population levels suggests an early limitation of root growth at higher nematode levels. *P. neoamblycephalus* is involved (Mai, 1968) in the stunting of Myrobalan plum seedlings. Nematode-infected roots were darker, and had fewer feeder roots than non-inoculated controls. nematodes were found feeding ectoparasitically, either isolated or in groups of 5-15, with their heads in the cortex. No preference for a particular region of the root was noted, but group feeding occurred at ruptures or where lateral roots emerged. Death of lateral roots and small lesions, probably caused by oxidation of phenolic compounds in the root, were associated with feeding. The nematodes observed were predominantly second and third-stage larvae. Eggs were found on the root surface. Males and fourth-stage larvae do not possess a stylet (Braun and Lownsbery, 1975).

21.6.2. Host range

Rose, apricot, peach, apple, cherry, grapes, and all *Prunus cerasifera* selections and hybrids were good hosts for *P. neoamblvcephalus*. The two rootstocks used most commonly for prunes in California, cultivars Myrobalan 29C and Marianna 2624 are included among these good hosts.

21.6.3. Ecology

The best temperature for increase of *P. neoamblycephalus* was 15 °C. Limitation of Myrobalan growth by *P. neoamblycephalus* was greatest at 20 and 27, and least at 30 °C. Lowering pH by acid amendment from 6.5 to 4.5 inhibited ($P = < 0.05$) the reproduction of *P. neoamblycephalus*. The lower number of nematodes with low pH is probably a result of acid inhibition of nitrification resulting in some accumulation of ammonia, toxic to nematodes (Eno *et al.*, 1955).

21.6.4. Management

Nursery should be raised in nematode free soil. Only healthy, nematode free planting stock should be transplanted in the field. Weed removal, nutritional supply in form of organic amendments and use of resistant rootstocks help in keeping the nematode population under control. Root dip treatment in oxamyl @ 0.024 g/L kills the nematode population.

Fallowing or planting green manure cover crops for 1 to 2 years is beneficial. Resistant cover crops include barley, oat, carrot, sunflower, French bean, *etc.* Use certified nematode-free rootstocks or seedlings to establish new orchards. When the orchard is developed, use procedures that improve soil tilth and drainage to help reduce nematode damage.

21.7. CONCLUSION

In the near future, growers will have to adopt IPM systems due to restrictions in the use of broad-spectrum soil fumigants. Resistant rootstocks reduce nematode reproduction considerably, but combination of two or more strategies will be necessary to alleviate nematode problems, since there is seldom a single effective method.

Additional research is needed to find means for interrupting nematode's life cycles, enhancing microbial activity in the rhizosphere to promote plant growth or increase its tolerance to the nematode. Any chemical, microbial, cultural or management approach that is developed must be within the capability of the grower and should meet the necessary environmental and economic requirements. The grower will benefit if these treatments are reliable, practicable and economically justified.

Chapter - 22

Olive, *Olea europaea*

22.1. INTRODUCTION

22.1.1. Origin

The geographic origin of cultivated olive (*Olea europaea*) can be traced to areas along the eastern Mediterranean coast where Turkey, Syria, Lebanon, Palestine, and Israel are currently located.

2.1.2. Area and production

Worldwide 19,270,115 tons of olive is produced per year. Spain is the largest olive producer in the world with 6,559,884 tons production volume per year. Greece comes second with 2,343,383 tons yearly production (Table 22.1).

Table 22.1. Major olive producing countries

Country	Area ('000 ha)	Production ('000 tons)	Yield (tons/ha)
Spain	2573.473	6559.884	2.549
Greece	887.177	2343.383	2.641
Italy	1165.562	2092.175	1.795
Turkey	845.542	1730.000	2.046
Morocco	1008.365	1416.107	1.404
Syria	765.603	899.435	1.174
Tunisia	1646.000	700.000	0.425
Algeria	424.028	696.962	1.643
Egypt	67.293	694.309	10.317
Portugal	355.075	617.610	1,739
Libya	357.797	188.975	0.528

Today, approximately 98% of olives are cultivated in Mediterranean Basin countries. Spain, Italy, and Greece together produce about 77% of the world's olive oil. Olive is mainly produced in Spain,

Italy, Greece, Turkey, Morocco, Syria, Tunisia, etc.

Major olive growing states in India include Rajasthan, Uttar Pradesh, Jammu and Kashmir, Himachal Pradesh. Other states like Punjab, Bihar, Rajasthan, Uttarakhand and Maharashtra also started participating rapidly in commercial olive cultivation in India.

Rajasthan cultivates the most amount of olives for India. As olive is a very hardy crop and can sustain the temperature variation from -5 to 48 ^{0}C. Thus it has the potential to be grown in an arid region like Rajasthan. There is about 800 ha in the region in different agro-climatic regions namely Sriganganagar, Nagpur, Bikaner, Jalore, Jhunjhunu, Alwar, Jaipur, *etc.*

22.1.3. Varieties

Spanish and Italian cultivars including "Arbequina," "Manzanilla," "Picual," and "Frantoio" have been extensively planted with about 70% of production being devoted to olive oil. "Arauco" is the only cultivar recognized from Argentina in the World Catalog of Olive Varieties (IOOC, 2000). Varieties suitable for temperate region and their salient features are as follows:

22.1.3.1. Coratina: A high oil yielding variety with cold tolerance characteristics. Highly self-sterile but starts bearing early and ripens late. Tree is medium vigorous, dense, spreading growth habit with very good adaptability to different soils and climates. Fruits are large, green in color, elongated ovals and asymmetrical in shape. Yields high quantity of good quality oil which is exceptionally high in polyphenols, resulting in excellent stability.

22.1.3.2. Pendolino: Moderate in vigor with dense drooping growth habit. Flowers early and profuse with long bloom period which makes it an ideal pollinizer. Self-incompatible, fairly tolerant to cold temperatures. Fruit small, delicious green and black table olive. Maturity is mid-season. Susceptible to olive knot and peacock spot. Highly susceptible to verticillium and sooty molds, moderate susceptible to the olive fly.

22.1.3.3. Cipressino: This has vigorous vegetation, rapid upward growth with a typical rising habit and gathered crown. The fruits are rounded oval shape. Maturation is spread out but is complete between the middle of November and the middle of December. It is a self-sterile variety of excellent and constant fruiting with a notable percentage of aborted ovaries (50-60%). Frantoio, Leccino, and Pendolino can be used as pollinizers. Excellent resistance to high winds, good resistance to climatic extremes and parasites.

22.1.3.4. Frontoio: A moderately vigorous dual purpose variety with spreading-drooping growth habit and medium-dense canopy. Yields fairly high and consistently, high adaptability and cold sensitive. Frantoio is self-fertile, but using pollinizers increases yields. Pendolino and Leccino can be used as pollinizers. Susceptible to peacock spot, scale insects and olive knot and also moderately susceptible to olive fly and to verticillium wilt. It is world's premier oil varieties with high oil yield. Frantoio is a late ripening variety.

22.1.3.5. Picholine: Picholine are green olives of French origin. It is partially self-fertile, Manzanillo and Leccino are good pollinizers. Resistant to leaf spot, moderately resistant to olive knot and verticillium wilt, with medium level of drought and cold tolerance and attractiveness to olive fly. Freshly brined right after being picked in citric acid has a subtle and mild salty taste.

22.1.3.6. Leccino: A rustic variety and widely planted throughout the world. The tree is quick to produce and resists well adverse climatic conditions and parasites. More recently it has been used for the production of table olives, semi-ripe or black. This variety is self-sterile and so needs a pollinator, principally Pendolino or Picholine.

22.1.3.7. Messenese: A green olives variety with partially self-fertile, Coratina and Pendolino are good pollinizers. Large fruit size with heavy bearing and has medium level of drought and cold tolerance and attractiveness to olive fly.

22.1.4. Rootstocks

Wild olive seedlings can be used for propagating commercial varieties as rootstocks.

22.1.5. Chilling requirements for flowering

Flowering occurs once buds induced the previous growing season receive sufficient chilling during the winter dormancy period to end dormancy, differentiate anatomically, and accumulate warmer temperatures adequate for budburst (Rallo and Cuevas, 2008). Malik and Bradford (2009) observed maximal flowering in "Arbequina" at minimum temperatures ranging from 4.4 to 7.8 ^{0}C. Adequate chilling (7.2 ^{0}C) during winter is essential to break rest period and to promote fruitfulness.

22.1.6. Olive oil health benefit

- » The greatest exponent of mono unsaturated fat is olive oil.
- » Olive oil is natural juice which preserves the taste, aroma, vitamins and properties of olive fruits.
- » The beneficial health effects of olive oil are due to its high content of mono unsaturated fatty acids mainly oleic acid and antioxidative substances.
- » The olive oil offers protections against heart disease by controlling LDL ("Bad") cholesterol level and raising HDL ("Good") cholesterol.
- » The olive oil provide protection against ulcers and gastritis.
- » Olive oil activates the secretion of bile and pancreatic hormones much more naturally than prescribed drugs. Consequently it lowers the incidence of gallstone formation.
- » The presence of Omega – 9 prove long lives.

22.1.7. Nematode problems

Olive trees serve as hosts to a large number of plant-parasitic nematodes, of which root-knot nematodes (*Meloidogyne* spp.), root-lesion nematodes (*Pratylenchus* spp.), and reniform nematodes (*Rotylenchulus* spp.) are widely distributed (Nico *et al.*, 2002; Kepenekci, 2001). Conversely, limited distribution on olive has been reported for the citrus *(Tylenchulus semipenetrans)* and the cyst-forming (*Heterodera mediterranea*) nematodes (Castillo *et al.*, 1999; McKenry, 1994).

22.2. ROOT-KNOT NEMATODES, *MELOIDOGYNE* SPP.

Meloidogyne species known to damage olive trees are well adapted to temperate and subtropical areas include *Meloidogyne arenaria, M. hapla, M. javanica* and *M. incognita.*

22.2.1. Symptoms

Root-knot nematodes are endoparasites, invading the root tips and establishing permanent feeding sites inside the root. *M. incognita* and *M. javanica* have been shown to reduce seedling growth drastically in inoculation trials and have been identified as a factor to reckon with in olive nurseries (Sasanelli *et al.*, 1997). Their feeding stimulates the root tissue to enlarge, forming giant cells called galls on which they continue to feed. Eventually individual galls will coalesce giving the root a bumpy swollen appearance, without any fine feeder roots. Trees become slightly weaker, especially in sandy areas, and fruit size on infected trees is reduced.

Infections by root-knot nematodes induce formation of permanent feeding cells, the giant cells, and stelar destruction in olive roots. Hyperplasia of the surrounding cells leads to the formation of typical root galls, the primary visible symptom of infection. Planting stocks of cv. Picual infected by *Meloidogyne* spp. showed a distinct yellowing affecting the uppermost leaves, followed by partial defoliation. Symptoms were more severe on *M. arenaria* and *M. javanica* infected plants than on *M. incognita* infected plants.

22.2.2. Interaction with other pathogens

M. javanica and *Verticillium dahliae* caused considerable damage to olive orchards in the northern part of Iran, and the growth of olive cv. Zard seedlings inoculated with these two pathogens was significantly reduced compared with that of seedlings inoculated with each pathogen alone (Saeedizadeh *et al.*, 2006). Similarly, the proportion of chlorotic, wilted, and necrotic leaves of seedlings was highest in plants simultaneously inoculated with both the pathogens.

22.2.3. Management

22.2.3.1. Physical methods: Solarization of piled soil provided a significant reduction of *M. incognita* eggs and egg masses in soil at a maximum depth of 40 cm after 10 days of treatment (Nico *et al.*, 2003). Furthermore, the thermal threshold reached by solarization inactivate free eggs and egg masses of *M. incognita*.

22.2.3.2. Cultural methods: Nico *et al.* (2004) tested amendment of olive potting substrate with dry cork compost at different rates for control of *M. incognita*. Nematode numbers in the amended substrate decreased exponentially as the rate of compost increased; whereas the lowest rate of amendment provided a sufficient level of control, the highest suppression occurred using pure compost as substrate.

The effect of biofumigation with leaf tissues of *Sorghum sudanense* on suppression of *M. incognita* was tested in a laboratory with different rates of artificially infested potting mixture as well as times of exposure and use of tarp. The final nematode population was significantly reduced by treatment in all cases, but the degree of suppression was adequate only when biofumigation lasted for 30 to 60 days.

22.2.3.3. Chemical methods: In Egypt, Shawky *et al.* (2004) compared the efficacy of recommended rates of five nematicides (fenamiphos, ethoprophos, carbofuran, oxamyl, and aldicarb) for the control of *M. incognita* on olive trees. All nematicides were effective in reducing the number of nematodes in both soil and roots, but fenamiphos gave the greatest reduction in nematodes in soil and the number of galls, females, and egg masses, whereas carbofuran gave the least reduction.

Rugby at 8 ppm concentration (79/24 percent reduction in nematode population) and Enzone at 8 ppm concentration (50/38 percent reduction in nematode population) had the maximum and minimum effect in the control of nematode, respectively (Soltani *et al.*, 2013).

22.2.3.4. Biological methods: Establishment of the arbuscular mycorrhizal fungi *Glomus intraradices*, *G. mosseae* or *G. viscosum* significantly increased growth of olive plants by 88·9% within a range of 11·9–214·0%, irrespective of olive cultivar (Arbequina and Picual), plant age and infection by *M. incognita* or *M. javanica*. In plants free from AMF, infection by *Meloidogyne* spp. significantly reduced the plant main stem diameter by 22·8–38·6%, irrespective of cultivar and plant age. Establishment of AMF in olive plants significantly reduced severity of root galling by 6·3–36·8% as well as reproduction of both *Meloidogyne* spp. by 11·8–35·7%, indicating a protective effect against parasitism by root-knot nematodes (Castilloa *et al.*, 2006). It is concluded that prior inoculation of olive plants with AMF may contribute to improving the health status and vigor of cvs Arbequina and Picual planting stocks during nursery propagation.

22.2.3.5. Host resistance: In Egypt, olive cultivars Meashon and Tofany were found to be resistant to *M. incognita*; while cultivars Manzanillo and Egazi were moderately susceptible to *M. incognita*, and cultivars Hamed and Bekwal were susceptible (Al-Sayed and Abdel-Hameed, 1991). Similarly, screening of olive cultivars most commonly used in Italy for resistance against root-knot nematodes showed that cultivar Coratina is resistant to *M. incognita* and *M. javanica*, whereas cultivars Leccino and Yusti are resistant to *M. javanica* and moderately susceptible to *M. incognita* (Sasanelli *et al.*, 1997). Other cultivars (Cima di Bitonto, Cellina di Nardo, Frantoio, and FS 17) were shown to be moderately susceptible to *M. incognita* (Sasanelli *et al.*, 1997), whereas the rootstock DA 12 1 was highly susceptible to *M. javanica* (Sasanelli *et al.*, 1997; 2002).

22.3. ROOT-LESION NEMATODES, *PRATYLENCHUS* SPP.

The two major species most frequently diagnosed as damaging to olive worldwide are *Pratylenchus penetrans* and *P. vulnus*, which reduced seedling growth in bioassays (Nico *et al.*, 2003).

22.3.1. Symptoms

Root lesion nematodes are migratory endoparasites. They invade and move within roots. When there is too much feeding damage to the root, lesion nematodes leave and continue feeding on noninvaded roots. Their feeding destroys root cells, causing lesions to form along the root. Lesions are usually narrow and dark and are often invaded by rot-causing pathogens.

Root-lesion nematodes are obligate migratory endoparasites that damage olive roots by feeding in the cortical parenchyma, causing extensive necrosis of cortical and endodermal cells (Castillo and Vovlas, 2007). Infection of olive roots by *Pratylenchus* spp. causes cortical lesions and reduces root number and size. 'Manzanilla' olive planting stocks infected with *P. vulnus* were partially defoliated, and weighed 42% less than noninfected plants (Lamberti and Baines, 1969). The infection by *P. vulnus* and *P. penetrans* significantly reduced the main stem diameter.

22.3.2. Management

Eradication of *P. vulnus* from infected olive planting stocks was achieved by hot water treatment of infected roots at 50 °C for 2 min, but 40 or 45 °C during 4 to 20 min.

22.4. RENIFORM NEMATODES, *ROTYLENCHULUS* SPP.

Currently, three *Rotylenchulus* species (*R. macrodoratus; R. macrosoma* and *R. reniformis*) are reported as parasites of olive (Castillo *et al.*, 2003; Vovlas and Inserra, 1976).

22.4.1. Symptoms

Reniform nematodes (*Rotylenchulus* spp.) are semiendoparasites of roots of olive. Artificial inoculation of both wild olives and olive cultivars Arbequina and Picual with *R. macrosoma* revealed reproduction factors of 2.5, 2.2, and 1.9 times the initial population density (Pi), respectively (Castillo *et al.*, 2003).

22.4.2. Management

The thermal threshold reached by solarization inactivate free eggs and control *P. vulnus* and *P. penetrans* nematodes potentially hazardous to olive plants (Nico *et al.*, 2003).

In Egypt, olive cultivars Meashon and Tofany were found to be tolerant of *R. reniformis*, and cultivars Hamed and Bekwal were susceptible (Al-Sayed and Abdel-Hameed, 1991).

22.5. CYST NEMATODE, *HETERODERA MEDITERRANEA*

22.5.1. Symptoms

The sedentary juvenile stages and adult females of *H. mediterranea* have a semi-endoparasitic feeding habit and do not completely penetrate olive roots. *H. mediterranea* reproduced on Arbequina (Castillo and Vovlas, 2002). The plant growth is severely affected by infection with *H. mediterranea* (10,000 eggs + second-stage juveniles per plant) with significant reduction of shoot and root dry and fresh weights, as well as shoot height, trunk diameter, and number of nodes in Arbequina but not in Picual (Castillo and Vovlas, 2002).

22.5.2. Histopathology

Immature females penetrate the root and become established in the endodermis (Castillo *et al.*, 2003). The feeding site induced by *R. macrosoma* in olive roots consists of a stelar syncytium, which originates from an endodermal cell enlarging near a curved sheet of pericycle cells formed by hypertrophy of pericycle cells. *R. macrodoratus* causes the formation of an enlarged uninucleate stelar or cortical "nurse cell" with dense cytoplasm and hypertrophic nucleolus (uninucleate giant cell) (Vovlas and Inserra, 1976), whereas *R. macrosoma* forms a syncytium. The kidney-shaped females produce a gelatinous matrix that covers the female body in which about 50 eggs are laid.

22.5.3. Survival

R. macrosoma survives in soil in the juvenile and adult male stages.

22.5.4. Management

Proper irrigation of infested, weakened trees with greater frequency avoids plant stress and different crucifer species used as cover vegetation can help to avoid damage from nematodes in established plantings.

22.6. CITRUS NEMATODE, *TYLENCHULUS SEMIPENETRANS*

The citrus biotype of the citrus nematode, *Tylenchulus semipenetrans* infects olive trees in California, Chile, and Italy (McKenry, 1994) and observed feeding on olive roots. Convincing evidence of pathogenicity to olive has only been demonstrated for high infection levels of *T. semipenetrans* (McKenry, 1994).

22.6.1. Symptoms

Citrus nematodes are semi-endoparasites. They bury into the root cortex with their anterior end while the posterior end remains outside the root. Roots have a dirty and enlarged look, because soil adheres to the gelatinous matrix that protects the eggs. The surface of infected roots appears rough and irregular. Feeder roots shorten and slowly disappear.

22.6.2. Management

A field trial in Chile (Jiménez *et al.*, 2002) compared the effects of fenamiphos chemigation on root growth and development of olive trees naturally infected by *T. semipenetrans*, compared with treatments with a biocontrol formulation and an amendment with organic manure. The biocontrol formulation and organic amendment treatments were as efficient as a chemigation with fenamiphos.

22.7. NEMATODE MANAGEMENT

22.7.1. Regulatory methods

The appropriate application of exclusion principles should lead to nematode-free nursery stock. Adoption of pathogen-free certification programs for olive planting stocks and rootstocks (EPPO/OEPP, 1993) should be strictly followed. Growers should also choose new plantation sites free of plant-parasitic nematodes in which to establish planting material (Lorrain, 1998).

22.7.2. Physical methods

One of the most successful is soil solarization, a hydrothermal process whereby moist soil is heated under transparent plastic tarps to temperatures detrimental to soilborne pathogens (Katan, 1981). Solarization of small piles of nursery substrates used as potting mixtures for glasshouse-grown plants is a novel approach for the management of soil-borne plant pathogens (Nico *et al.*, 2003).

22.7.3. Cultural methods

Suppression of plant pathogens in soil can be also achieved using special amendments in a process known as biofumigation (Kierkegaard *et al.*, 1993), which is based on the release of biocidal volatiles during breakdown of tissues of some plants, mainly of Brassica and Sorghum species. Biofumigation is often practiced with plastic tarping, in order to confine the volatile compounds produced and enhance their biological effect.

22.7.4. Chemical methods

Nematicides like ethoprophos, fosthiazate, fenamiphos, oxamyl, dazomet, metham sodium, metham potassium, and sodium tetrathiocarbonate (MARM. 2009) can be used to manage olive nematodes.

22.7.5. Biological methods

The use of AMF-infected planting material during olive plant propagation in nurseries may be a useful practice for minimizing the effects of root-knot nematodes, as well as for protecting these plants against future nematode infections in commercial olive orchards.

22.7.6. Host resistance

Development of resistant rootstocks may be a valuable tool for nematode management in olive, as suggested by the successful results obtained with the clonal rootstock 'Allegra' in California (McKenry, 1994).

22.8. FUTURE PROSPECTS AND CONCLUSION

The removal of many nematicides from the market due to environmental and human health concerns has focused attention on the development of alternative control strategies for managing plant-parasitic nematodes in olive. Because few resistant rootstocks have been identified in olive, exclusion should be the first and primary control strategy to consider for management of plant-parasitic nematode in olive orchards. The use of pathogen-free planting material and non-infested soils is essential for minimizing the effects of single or concomitant species of plant-parasitic nematodes and other pathogens during the early years of olive cultivation and for preventing impairment of growth and yield loss. For producing certified olive trees, the rootstocks or scion material (for grafted olive trees) should be grafted or budded only onto rootstocks of equivalent (i.e., propagation-stock) or higher certification status. Plants to be certified should be kept in nursery soil that has been proved free from the nematode species that can cause significant damage to olive growth (e.g., *X. diversicaudatum*, *M. arenaria*, and *P. vulnus*), and also from all quarantine nematodes (Anon, 2006). Finally, the lack of sufficient levels of nematode resistance or tolerance can be compensated for, to some extent, by the use of other control measures, preferably nonchemical ones, including soil solarization, biofumigation, and biological treatments. The fulfilment of this strategy would help in maintaining the critical role that olive production plays in Mediterranean farming systems as a major source of oil and fruits, as well as a net contributor to agricultural sustainability in countries with Mediterranean-type climate.

Chapter - 23

Pomegranate, *Punica granatum*

23.1. INTRODUCTION

23.1.1. Origin

Pomegranate originated from Iran.

23.1.2. Area and production

The estimated global cultivated area under pomegranate is around 3 lakh ha and production of 3.0 million tons. Among the pomegranate-growing countries, India holds the first position in terms of both areas under cultivation and total pomegranate production (Marathe *et al.*, 2022).

Iran, India, China, Turkey and the USA are major pomegranate producing countries in the world with 76% of the total production. It is also grown in Afghanistan, Bangladesh, Myanmar, Vietnam, Thailand, Kazakhstan, Turkmenistan, Armenia, Georgia, Morocco, Tunisia, Egypt, Israel, Syria, Lebanon, Greece, Cyprus, Italy, France, Spain, Portugal, Mexico, Argentina and Chile. Global area, production, and productivity of pomegranate is presented in Table 23.1.

Table 23.1. Global area, production, and productivity of pomegranate (2011-12)

Country	Area ('000 ha)	Production ('000 MT)	Productivity (MT/ha)
India	107.30	743.1	6.9
Iran	60.00	650.0	10.8
Turkey	8.00	218.0	27.25
Afghanistan	8.00	60.0	7.50
USA	6.00	100.0	16.70
Israel	1.60	20.0	12.50
South Africa	1.30	4.0	3.10
Spain	3.00	60.0	20.00

Tunisia	2.60	25.0	9.61
Australia	0.25	1.0	4.00
Argentina	0.80	2.0	2.50
Other countries	101.15	616.9	6.09
WORLD	300.00	2500.0	9.36

As per recent advance estimates for the year 2019-2020 by National Horticultural Board (NHB), total pomegranate cultivation in India spreads over an area of 2.83 lakh hectares with a record production of 31.46 lakh MT of fruits. Pomegranate is grown commercially in the states of Maharashtra, Gujarat, Karnataka, Andhra Pradesh, Tamil Nadu, Madhya Pradesh and Rajasthan. India is one of the major producer of pomegranate with largest acreage in the world. During 2016-17, pomegranate was cultivated over 2.16 lakh ha with an annual production of 24.42 lakh tons and productivity of 11.70 tons/ha in India (Table 23.2). Maharashtra is the leading state with 65.51 per cent of the area and 64.61 per cent of total production under pomegranate (Anonymous, 2017). India is the only country in the world where pomegranate is available throughout the year (January – December) (Chandra and Jadhav, 2009).

Table 23.2. State-wise area production and productivity of pomegranate in India (2017-18)

State	**Area ('000 ha)**	**Production ('000 tons)**	**Productivity (tons/ha)**
Andhra Pradesh	9.47	139.02	14.69
Arunachal Pradesh	0.76	4.34	5.69
Chhattisgarh	30.51	461.75	15.13
Gujarat	2.77	3.15	1.14
Himachal Pradesh	0.02	0.04	2.75
Jammu & Kashmir	25.97	268.23	10.33
Kerala	0.01	3.33	475.00
Madhya Pradesh	9.68	114.27	11.81
Maharashtra	147.91	1789.46	12.10
Nagaland	0.09	0.50	5.86
Odisha	0.26	0.92	3.54
Rajasthan	4.44	26.58	5.99
Tamil Nadu	0.56	13.06	23.39
Telangana	1.49	19.87	13.56
Others	0.01	0.01	1.33
TOTAL	233.93	2844.52	12.16

The important pomegranate growing areas in different states in India are listed in Table 23.3.

Table 23.3. Major pomegranate growing areas in different states of India

State	Major pomegranate growing areas
Maharashtra	Mainly in Solapur, Nashik, Sangli, Satara, Ahmednagar and Pune and to a limited extent in Osmanabad, Jalna, Beed, Aurangabad, Jalgaon and Dhule districts.
Karnataka	Bijapur, Bagalkot, Koppal, Belgaum, Gadag, Bellary, Raichur, Tumkur, Chitradurga and Davangere districts.
Andhra Pradesh	Anantapur.
Telangana	Mahbubnagar.
Gujarat	Kutch, Banaskantha, Ahmedabad, Sabarkantha and Bhavnagar.
Rajasthan	Hanumangarh and Ganganagar.
Tamil Nadu	Salem, Coimbatore and Periyakulam.
Himachal Pradesh	Solan, Kullu, Sirmour.

23.1.3. Varieties

23.1.3.1. Indigenous varieties: There are many commercial indigenous varieties which are cultivated in different parts of India (Table 23.4). Jalore Seedless is popular in Rajasthan, while Bhagwa is the predominant ruling variety of India owing to its dark red arils and attractive red color rind. Important varieties of pomegranate are Ganesh, Dholka, Bhagwa, Phule Arakta and seedless varieties such as Paper Shell, Spanish Ruby, Mascut Red and Velladu, which showed great potential. Earlier, Ganesh variety with big sized fruits was the preferred, but presently soft seeded, red peel and aril colored varieties with high per cent of juice are preferred such as Bhagwa and Mridula. In hot arid climatic condition, the varieties like Jalore Seedless, Bhagwa, G-137, Mridula, Phule Arakta and Ganesh are found promising.

Table 23.4. The major characteristics of improved indigenous varieties of pomegranate

Varieties	Major characteristics
CO-1	It is a high yielding selection. Fruits are medium-sized with attractive rind, soft seeds, higher pulp content and sweet taste. Developed by TNAU, Coimbatore
G-137	Fruits are large sized, surface smooth, yellow with red tinge, deep pink and bold aril, and soft seeds, sweet and prolific bearer. Developed by MPKV, Rahuri.
Yercaud-1	Its fruits are medium sized, with easily peelable rind. The seeds are soft with attractive, deep purple aril. Developed by TNAU, Coimbatore.
Jyoti	The fruits are medium to large sized, having attractive, yellowish red, more fleshy and pink aril. Fruits are very sweet, soft seeded and taste good. It yields moderately. Developed by UAS, Dharwad.
Jalore Seedless	Popular variety of Rajasthan, fruits round, yellow with red tinge in color, aril color light pink to pink, juicy, taste sweet, TSS 15-16° Brix and plants semi vigorous. Developed by CAZRI, Jodhpur.
Mridula	Fruits medium sized, rind smooth, dark red in color, blood red sweet arils with very soft seeds, plants are dwarf. Developed by MPKV, Rahuri.

Ruby	It has soft and red arils with good flavor. The plants are dwarf, prolific bearer, providing uniformly red fruits. Developed by IIHR, Bengaluru.
Bhagwa	Ruling variety of pomegranate in India, fruits big size, glossy red rind, thick rind, bold red arils, sweet and soft seeded, matures in 170 to 180 days after full bloom. Developed by MPKV, Rahuri.
Phule Arakta	Fruit medium sized dark red rind color, arils are also dark red, soft seeded and sweet. Developed by MPKV, Rahuri.
Phule Bhagwa Super	Very attractive, dark saffron rind color with glossiness on rind surface, good quality fruits, matures earlier than Bhagwa. Developed by MPKV, Rahuri.

23.1.3.2. Exotic varieties: The major exotic varieties of pomegranate grown in India are presented in Table 23.5.

Table 23.5. The major exotic varieties of pomegranate grown in India

Varieties	Major characteristics
Wonderful	This variety developed in California, USA. Fruits are large in size with deep uniform red color. Rind is medium thick and hard. Arils are deep crimson in color, juicy, with medium hard seeds. This variety is highly suitable for juicing with acidity 1.5-1.75% and TSS 170 Brix. The fruit matures in late September and October in India.
Granada	Early maturing variety originated in California as a bud mutation of cv. Wonderful, but displays a red crown while in the green stage, darker red in color. Seeds are medium hard and less acidity than Wonderful. The fruit ripens one month earlier than Wonderful but smaller in size. Owing to early maturity, it fetches premium price in the market.
Early Wonderful	It is high yielding variety. Fruits are large in size, thin-skinned, deep-red color. Arils are delicious with medium hard seed. It ripens about 2 weeks earlier than cv. Wonderful.

23.1.4. Export

As per recent advance estimates for the year 2019-2020 by National Horticultural Board (NHB), India exported 67.89 thousand metric tons fetching 6885 million rupees in foreign exchange. UAE is the major importing country of Indian pomegranates, accounting to nearly 32% of Indian global exports.

New export market for pomegranate include Russia, Thailand, Malaysia, Singapore, China and USA.

23.1.5. Projected opportunities

The Indian pomegranate production is growing by 20 to 25% every year. Demand is on the rise, both at domestic level and in the export market. The best-known variety in India, Bhagwa, is a huge sweet variety of really good quality. India is a growing player to destinations in Europe, the Middle East and Asia. The country is also expecting to enter the U.S. market. India competes with various players in the world, but it manages to stand out with unique varieties. The trend in India is convenience; pomegranates are widely appreciated and the seeds are really popular amongst consumers.

23.1.6. Nematode problems

The pomegranate is attacked by different nematodes like *Meloidogyne* spp., *Rotylenchulus* sp., *Pratylenchus* spp., *Hoplolaimus* sp., *Helicotylenchus* sp., *Xiphinema* spp. *etc.* Among them root-knot nematode is a key pest of pomegranate production in all over the world (Bhati, 2020).

23.2. ROOT-KNOT NEMATODE, *MELOIDOGYNE INCOGNITA*

The root-knot nematode, *Meloidogyne incognita* Race 2 is a serious pest of pomegranate (in Maharashtra, Karnataka, Gujarat and Tamil Nadu) which primarily cause galls on root system that eventually are seriously hampered in their main functions of uptake and transport of water and nutrients, reduced plant growth and productivity of the pomegranate (Khan *et al.*, 2005; Nasira *et al.*, 2011).

The area under pomegranate was increasing steadily in Coimbatore, Erode, Dindigul and Tirunelveli districts due to its higher monitory returns and export value. Growers of late in Sivagiri and Modakuruchi areas, Erode district, Tamil Nadu; are encountering problem of yellowing of leaves, stunting and less productivity of fruits. Such trees were found to be severely infested with root-knot nematodes on assaying of soil and root samples. As the presence and association of root knot nematode in pomegranate has not been documented so far in Tamil Nadu, it is presumed that the nematode could have spread from other States through saplings.

23.2.1. Distribution

Mainly distributed in the states of Maharashtra, Gujarat, Tamil Nadu, and Karnataka.

23.2.2. Economic importance

The root-knot nematodes were responsible for 17.3% loss in fruit yield of pomegranate (Singh *et al.*, 2019).

Meloidogyne incognita is responsible for annual yield loss of 23% in pomegranate amounting to Rs. 3023.44 million in India (Walia and Chakraborty, 2018).

23.2.3. Symptoms

Severely infested plants by root-knot nematode exhibited yellowing of foliage resulting in stunted plant growth. These plants produced less number of fruits or no fruits which might be due to the nematode induced nutritional deficiency. In severe cases, the galls were predominantly found on entire root system. The galls were white in color turned to light brown and hardy when they became old. The intensity of root-knot nematode damage increased with increase in age of the plant. In general, more than five year old plants were severely affected by root-knot nematode. Egg masses were observed inside as well as outside of the galls. More numbers of females were found in a single compound gall.

23.2.4. Histopathology

Histopathological studies revealed that *Meloidogyne javanica* larvae entered roots by puncturing action of the stylet. The cells adjacent to nematode were severely damaged causing cell congestion. A number of cells had egg masses and sections of larvae. The medullary cells were unequal in size. Giant cells were present with broken cell walls. The pathological symptoms observed in the present findings, especially disruption of cortical tissue ultimately affected plant growth by reducing absorption of water and nutrients (Khan *et al.*, 2008).

23.2.5. Spread

The nematodes spread through infected planting material.

23.2.6. Management

23.2.6.1. Cultural methods: Raising healthy plants (ground layers, air layers) in nematode-free soil for rooting.

i) Intercropping: Intercropping with sun hemp, marigold, mustard, fenugreek, onion, and garlic during crop rest period was found effective against root-knot nematode on pomegranate.

ii) Botanicals: Application of neem cake at 3 kg/tree was effective in reducing the nematode population (57.83%) and in increasing the fruit yield (33.64%) over control. Soil application with castor cake @ 2 t/ha is effective for the management of *M. incognita* infecting pomegranate (Darekar *et al.*, 1989).

Application of marigold flowers gave the highest reduction percentage in root galling with value of 98.8 %, followed by colocynth fruits extract which averaged 96.4 % without significant differences between them. The lowest egg-mass index (EI) was obtained from plants receiving marigold extract (0.0) with reduction percentage value of 100 % (Table 23.6) (Nour El-Deen *et al.*, 2016).

Table 23.6. Impact of certain plant extracts on development of *M. incognita* infecting Taify pomegranate under greenhouse conditions.

Treatment	No. of galls*	Root gall index (RGI)	No. of egg-masses	Egg-mass index (EI)
Citrullus colocynthis	3.0d	1.3d	2.3d	1.7b
Tagetes erecta	1.0d	0.7e	0.0e	0.0c
Rosmarinus officinalis	9.3c	2.0c	5.3c	2.0b
Thymus vulgaris	17.3b	3.0b	9.0b	2.0b
Withania somnifera	23.0b	3.0b	11.0b	2.7a
Control (Nematode lone)	82.3a	4.0a	24.0a	3.0a

Means in each column followed by the same letter (s) did not differ at $P< 0.05$ according to Duncan multiple-range test.

23.2.6.2. Chemical methods: Soil application of phenamiphos gave good control of root-knot nematodes, provided protection to roots for 60 days against nematode invasion and improved fruit yields (Siddiqui and Khan, 1986).

Treatment of pomegranate with Carbofuran reduced populations of *M. incognita*, *Xiphinema insigne* and *Helicotylencus* spp., and increased yields by one-third (Darekar *et al.*, 1989). Application of carbofuran at 4 kg a.i./ha in nursery, and 2 kg a.i./ha in the main field at transplanting and at flowering gave effective management of *M. incognita* on pomegranate.

Carbofuran 3G at 6 kg a.i. /ha was the most effective treatment in reducing nematode (*Helicotylenchus* spp., *Xiphinema insigne* and *Meloidogyne incognita*) population (55.95%) and increasing the yield (33.27%) of pomegranate (cv. G.B.G. No.1) compared with untreated control (Darekar *et al.*, 1989).

Carbofuran at 2 kg a.i. /ha was found to be significantly superior in reducing root- knot index and increasing the pomegranate yield up to 52 % (Siddiqi and Khan, 1986).

Treatment of Carbofuran 3 G and Phorate 10 G at 3 kg a.i. /ha and neem cake 3 kg/tree were equally effective in reducing root-knot nematode population (56.91 to 59.50 %) and number of galls/5 g root (42.6 to 35.24) and increasing fruit yield (32.17 to 37.8 5%) of pomegranate (Anonymous, 1993).

The treatment of Furadan 3 G at 4 kg a.i. /ha was found to be the most effective in pomegranate in reducing root-knot nematode population 56.02%, and number of root galls / 5 g roots (42.38%) and increasing the yield (37.18%) (Anonymous, 1999).

23.2.6.3. Biological methods: Soil application of 100 kg FYM enriched with *Pseudomonas. fluorescens* or *Trichoderma viride* @ 20 kg/ha at "Bahar" (blossom).

Pseudomonas fluorescens at 20 kg/ha was found to be most effective in reducing *M. incognita* population (31.28%) and number of root galls/5 g roots (29.28%) and increasing the fruit yield of pomegranate (18.99%) with 2.37 B: C ratio; followed by *Trichoderma viride* at 20 kg/ha (reduction in *M. incognita* population - 28.3%, number of root galls/5 g roots -23.9%, increasing the fruit yield -16.78%, and B: C ratio - 2.33) (Table 23.7) (Pawar *et al.*, 2013).

Table 23.7. Effect of bioinoculants on the management of root-knot nematode on pomegranate

Treatment	Decline in nematode population (%)	Reduction in galling (%)	Yield (tons/ha)	B: C ratio
Pseudomonas fluorescens at 20 kg/ha	31.2	29.2	18.4	2.37
Trichoderma viride at 20 kg/ha	28.3	23.9	18.0	2.33
Carbofuran at 3 kg a.i. /ha	29.8	25.1	17.8	2.27
Control (Untreated)	---	---	15.4	2.27
CD (P = 0.05)	2.99	2.30	1.17	---

Soil application of *Pseudomonas fluorescens* (10 kg at bahar and 10 kg/ha 90 days after bahar) recorded reduction in root-knot nematode population (39.1% and 62.60% number of root galls (37.22 and 58.90 %, respectively) and egg masses (39.20 and 46.93 %) and increased the yield of pomegranate (57.77 %) with ICBR (1:3.48 and 1:24.20) at 60 and 150 days after application, respectively (Walnuj, 2013).

The symbiotic bacterium of entomopathogenic nematode *Xenorhabdus* treatment significantly inhibited formation of galls on roots with reduction percentage of 97.9 %, followed by those treated with *Photorhabdus* with value of 96 %. Regarding egg-masses numbers, *M. incognita* was barely reproduced on plants treated with *Xenorhabdus* filtrate with reduction percentage value of 95.8 %, followed by *Photorhabdus* with value of 91.7 % (Table 23.8) (Nour El-Deen *et al.*, 2016).

Table 23.8. Effect of symbiotic bacteria of entomopathogenic nematodes on development of *Meloidogyne incognita* infecting Taify pomegranate under greenhouse conditions

Treatment	No. of galls*	Root gall index (RGI)	No. of egg-masses	Egg-mass index (EI)
Steinernema spp.	6.0bc	2.0b	3.0c	1.7b
Heterorhabditis spp.	10.0b	2.3b	6.0b	2.0b
Xenorhabdus sp.	1,7cd	1.0c	1.0c	1.0c
Photorhabdus sp.	3.3c	2.0b	2.0c	1.0c
Control (Nematode alone)	82.3a	4.0a	24.0a	3.0a

Means in each column followed by the same letter (s) did not differ at P< 0.05 according to Duncan multiple- rage test.

Apply *Purpureocillium lilacinum* or *Pseudomonas fluorescens* 10-20 g/m^2 for reduction of root-knot nematode population (for Maharashtra).

23.2.6.4. Host resistance: The five tolerant genotypes (CAZRI Sel., Co-white, Dholka, G-137, and IC-318723) had a lower number of root galls per gram of root, ranging from 31.0 to 35.2. The resistant genotype (IC-318790) showed the lowest number of galls per gram of root (18.5) (Table 23.9) (Singh *et al.*, 2022). The genotypes showing resistance/tolerance reaction against root- knot nematodes should be additionally studied to explore their suitability as a resistance source for the development of new resistant cultivar or as rootstocks for the effective long-term sustainable management of nematodes.

Table 23.9. Reaction of pomegranate genotypes against root-knot nematode, *Meloidogyne incognita*

Genotypes	Root-knot galls g^{-1} root*	Reaction
IC-318790	18 ± 4 d	Resistant
CAZRI Sel.	32 ± 2 bcd	Tolerant
Co-white	34 ± 2 bcd	Tolerant
Dholka	35 ± 2 bc	Tolerant
G-137	31 ± 3 cd	Tolerant
IC-318723	32 ± 3 bcd	Tolerant
Bhagwa	51 ± ab	Susceptible
Ganesh	52 ± 2 ab	Susceptible
Ruby	56 ± 3 a	Susceptible
Jyoti	48 ± 2 abc	Susceptible
Bassein seedless	46 ± 2 abc	Susceptible

* Within each column, different letters indicate significant differences between genotypes.

Seven genotypes namely Achik Dana, Alandi, Beseka Linsk, Jabesto, Kabul (IIHR), Kazaki Anar and Siah Shirin were moderately resistant against root-knot nematode, *Meloidogyne incognita* race 2 infecting pomegranate (Shelke and Darekar, 2000).

23.2.6.5. Integrated methods: Application of farmyard manure at 20 tons/ha, neem cake at 250 g/tree followed by bioagent, *Purpureocillium lilacinum* at 25 g/plant was found effective to manage root-knot nematode problem in pomegranate and to increase crop productivity.

Application of *Purpureocillium lilacinum/Trichoderma viride* (2 x 10^6 cfu/g) @ 20 kg/ha + Castor cake @ 2 t/ha in root zone, 30-45 cm away from tree trunk approximately 22.5 cm deep in soil at 6-monthly intervals (for middle Gujarat) was found effective.

Soil application of *P. lilacinum* (50 g/ plant) with castor cake (1 kg/plant) combination was found to be effective in reducing root-knot nematode in pomegranate and also prevented the entry of wilt pathogen, *Ceratocystis fimbriata*.

Two applications of FYM enriched with *P. lilacinum* at 15-25 kg around each plant at 3 months interval along with application of liquid formulation of *P. lilacinum* through drip; showed recovery from symptoms, and putting forth new flushes. Saprophytic microbial population in soil has multiplied many fold, reduced egg masses and females in roots and juveniles of root-knot nematodes were not seen in soil.

The farmers of North Gujarat region cultivating pomegranate are advised to apply talc based formulation of *Pochonia chlamydosporia* (10^8 spores/g) or *Purpureocillium lilacinum* (10^8 spores/ g) at 100g/ plant along with mustard cake at 1.5, 2.0 and 2.25 kg/ plant in first, second and third year, respectively; two times i.e., first during November and second during April for the effective management of root- knot nematodes.

Application of bioagents (*T. viride, P. lilacinum, Ps. fluorescens*) at 10-20 g/m^2 in nursery, and 5 kg/plant of FYM/compost enriched with bioagents in the main field at transplanting and at flowering was found effective against *M. incognita*.

Three years pooled results of experiment on biological control of root-knot nematodes on pomegranate revealed that maximum reduction in nematode population was recorded in the treatment of *P. lilacinum* @ + *P. chlamydosporia* each @ 100 gm/ plant (69.80%) followed by *P. chlamydosporia* @ 100 gm/ plant + mustard cake @ 2.5 t/ha (66.05%) over control. These treatments also had maximum reduction in root-knot index (RKI) 66.45% and 62.24%, respectively and thereby gave maximum fruit yield of 2966 kg/ha in treatment of *P. chlamydosporia* + mustard cake over other treatments (Anonymous, 2012). From economics point of view, maximum net realization of Rs. 1,24,530/ ha was recorded in the treatment of *P. chlamydosporia* + mustard cake followed by Rs. 1, 13, 525/ha in *P. lilacinum* + mustard cake.

The following INM measures have been recommended for effective management of pomogrenate root-knot nematodes:

» Intercropping with sun hemp, marigold, mustard, fenugreek, onion, and garlic during crop rest period.

» Use of biocontrol agents such as *Pseudomonas fluorescens, Purpureocillium lilacinum, Pochonia chlamydosporia, Trichoderma harzianum,* and *T. viride* at 20 kg/ha.

» Use of entomopathogenic nematodes (*Steinernema* spp., *Heterorhabditis* spp.) and their symbiotic bacteria (*Xenorhabdus, Photorhabdus*).

» INM using enriched neem cake/ mustard cake/FYM/compost/vermicompost with biocontrol agents (*Pseudomonas fluorescens, Purpureocillium lilacinum, Pochonia chlamydosporia, Trichoderma harzianum,* and *T. viride*).

» Enforcement of domestic quarantine.

23.3. WILT DISEASE COMPLEX (FUNGI, NEMATODES AND INSECT PESTS)

In recent years, pomegranate wilt complex is becoming an important disease which results in complete wilting of plant. The wilting severity is increasing at faster rate. It is caused by the association of fungi (*Ceratocystis fimbriata, Fusarium oxysporum, Rhizoctonia solani*), root-knot nematodes (*Meloidogyne incognita*) and shot hole borers (*Xyleborus fornicates* and *X. perforans*). It was first noticed in two areas of Bijapur district of Karnataka, India, in 1990 (Somasekhara and Wali, 1999). Around 1993, rapid spread of this disease was observed in the entire Bijapur district. The severity of wilt complex observed was 49.2 per cent in Maharashtra, 61.11 per cent in Karnataka and 8.69 per cent in Andhra Pradesh. In Maharashtra, among 92,185 ha area under pomegranate, 23,858 ha (25.88%) are reported to be wilt affected (Jadhav and Sharma, 2009). Despite many factors conducive for the high severity of disease, seedlings selection for planting, soil-borne nature and also association with shot hole borer and plant parasitic nematodes might be the reason for the current rampant spread of the disease in south Indian states.

23.3.1. Distribution

The disease has been reported from Karnataka, Andhra Pradesh, Tamil Nadu, North Gujarat, and Maharashtra states (Somasekhara and Wali, 1999).

23.3.2. Symptoms

When the roots and lower portion of the stems were split open, the following can be observed:

» Shows dark bluish-black/grey/brown discoloration of the wood if the pathogen is *Ceratocytis fimbriata.*

» Pin holes are observed in the bark and wood when shot hole borers are associated with wilt.

» Root galls are observed on fine roots in early stage and other roots in advance stage if it is the root-knot nematode infestation.

23.3.2.1. Wilt inducing fungi: Wilt symptoms initiate as yellowing of leaves of one or more branches, followed by drooping of foliage and complete wilting of plant within 2-3 months period. Plants may reveal drooping of leaves of the entire plant all of a sudden (within 1-2 days) resulting in complete wilting. *C. fimbriata* infection causes discoloration of vascular bundles and adjoining tissues.

23.3.2.2. Shot hole borer: Shot hole borer infest the collar region of the plant by making innumerable pin or shot holes causing discontinuity in the conducting vessels affecting conduction of water to the upper portion of the plant. Consequently, the drying of the twigs became prominent. The drying progressed from the tip towards base and in severe cases entire plant dried.

23.3.2.3. Root-knot nematode: In general, *M. incognita* infected plants were exhibited yellowing of foliage resulting in stunted plant growth. These plants produced less number of fruits or no fruits which might be due to the nematode induced nutritional deficiency. In severe cases, the galls were predominantly found in entire root system. The intensity of root-knot nematode damage increased with increase in age of the plant. In general, more than five year old plants were severely affected by root-knot nematode. Egg masses were observed inside as well as outside of the galls. More numbers of females were found in a single compound gall.

23.3.3. Predisposing factors

The wilt diseases are generally aggravated due to biotic stresses particularly drought as well as excessive rain, boron deficiency in soil resulting in increased severity of *C. fimbriata.* Wounds, natural or due to insect/ nematode or human activity like pruning and intercultural operations predispose the plants to severe infections, as the pathogen is more devastating in overindulged orchards rather than in orchards with little human activity. Stress due to flowering and fruit bearing trees also results in sudden death of the entire plant. All commercial cultivars are susceptible to wilt and it can attack plants of all ages.

23.3.4. Survival and spread

Wilt fungus *C. fimbriata* can survive in the soil and in infested plant debris for several years (5-7 yr.). Wilting caused by root-knot nematode is reported from all type of soils but more severe in sandy loam soil with more aeration.

Majority of wilt pathogen including fungus and nematodes spreads through infected planting material to new areas while intercultural operations like weeding, manure application, farm machinery and pruning tools, root grafts, flooding/runoff irrigation water and insects like shot hole borer can spread the pathogens within and nearby orchards.

23.3.5. Management

- The planting material i.e. sapling and the potting mixture in which sapling is planted should be free from wilt pathogens like fungi, nematodes and insects. Use of solarized/ sterilized soil (potting mixture) for saplings eliminates almost all soil-borne pathogens. Planting samplings on raised beds increases aeration and reduce wilt incidence in the pomegranate orchards.
- Soil solarization for 6 weeks in the hottest months (April-May) using 50-75 LLDPE (Linear Low Density Polyethylene) sheet kills the soil-borne pathogens.
- Apply promising bioformulations such as *Aspergillus niger* AN27 (IRG 07) and Mycorrhiza (*Rhizophagus irregularis* Syn. *Glomus irregularis*), @ 2.5 kg/ha and *Trichoderma harzianum*, *T. viride*, *Pseudomonas* spp. *Purpureocillium* spp. *etc.*, right from planting, every 6 months. These serve as best preventive measure for all types of wilt pathogens.
- Grow green manure crops like dhaincha (*Sesbania aculeata*) and sun hemp (*Crotalaria juncea*) during rainy season and incorporate in soil before flowering.
- On observing first symptoms of wilt, first ascertain the cause/s. If it is due to fungal pathogens *Ceratocystis*, *Fusarium*, *etc.* in the orchard, treat soil with only one of the following most promising protocols:
  - 1st drenching with Propiconazole 25 EC @ 2ml/ L + Chlorpyriphos 20 EC @ 2ml/L. After 30 days of first application, 2nd drenching with *Aspergillus niger* AN 27 (IRG 07) @ 5g/plant with 2 kg FYM/plant followed by 3rd drenching of Mycorrhiza (*Rhizophagus irregularis*) @ 25g/plant with 2 kg FYM/plant 30 days after 2nd drenching.

    OR

  - Three drenchings at 20 days interval with Propiconazole 25 EC @ 2 ml/L + Chlorpyriphos 20 EC @ 2 ml/L. Use 5-10 L solution/plant.

    OR

  - 1st and 3rd drenching with Fosetyl Al 80% WP @ 6 g/plant, 2nd and 3rd drenching with Tebuconazole 25.9% w/w EC @ 3 ml/plant. Make solution volume 10 L with water for each drenching.
  - Drenching with Metalaxyl 8% + Mancozeb 64% @ 2.0-2.5g/L will be beneficial if Phytophthora is causing any loss.
- To avoid residues in fruits, drenching should be done immediately after harvest during rest period. Irrigate the plants one day before treatment. After chemical drenching irrigation should be stopped for 2 days.
- For controlling shot hole borer (*Xyleborus* spp.) which is associated with wilt disease, 10 liters preparation (paste) containing red soil (4 kg) + Chlorpyriphos 20 EC (20 ml) + Copper oxychloride 50 WP (25 g) needs to be applied on stem surface from plant base up to 65 cm from second year onwards. Pasting should be done twice a year; once soon after the harvest and once at defoliation, before crop regulation. Shot hole borer attacks weak plants; hence, keep plant vigor through recommended nutrition and proper irrigation during as well as after crop period. Prune out affected branches and do not dump in or near orchard.
- Wilting due to root-knot nematodes can be managed with soil application of promising bioformulations along with FYM like *Purpureocillium lilacinum*, @ 2.5-5.0 kg/ha, *Aspergillus niger* (IRG07), Mycorrhiza (*Rhizophagus irregularis*) @ 2.51 kg/ha. Application should be done from planting of saplings and repeated every 6 months. Drenching with Azadirachtin

1% @ 3 ml/L can be done twice a year. If the incidence of root-knot nematode is high, first go for chemical nematicides for reducing the nematode population below the damage threshold. Drenching of nematicide like Fluopyrum 34.48% SC should be done @ 2 ml/plant. Plants should be sufficiently watered a day before drenching. Mix 2 ml of the nematicide in one liter of water and pour 250 ml per dripper (4 drippers/plant) or 500 ml per dripper (2 drippers/plant). Farmer can also use the granular nematicide Fluensulfone 2% G. In order to use the granular nematicide, make a small pit under the dripper and apply the granular chemical @ 10 gm per dripper (Maximum dose should not exceed 40 gm/plant); cover it with the soil and start watering. The planting of African marigold (*Tagetes erecta*) varieties like 'Pusa Narangi Gainda' and 'Pusa Basanti Gainda' in the space between pomegranate plants for 3-5 months gave effective results . Application of well decomposed organic manure, vermicompost, neem cake along with green manuring crops like sesbania or sun hemp and above mentioned bioformulations can manage the menace of root-knot nematode in pomegranate.

» Once the wilt disease is detected in the orchard, dig 3-4 feet long trench between healthy and diseased/wilted plant. The partial wilt affected plants showing initial symptoms should be immediately treated with suitable agrochemicals. The plants showing more than 25% dry branches or completely dead plant should be removed and brunt and should not be kept dumped as woodpile. While removing wilted plant from the orchard, due care should be taken not to spread the infected root samples and soil to nearby plants.

» Soil solarization or soil drenching with formalin should be used to disinfect such pits after removing the diseased (wilt affected) plants. For this, 30 cm deep and 60 x 60 cm pit should be made at the place where trunk has been removed. Half a liter of the formaldehyde solution (37-41% formalin) is required for 10 liters of water per pit. The solution should be made at the site of pit and poured immediately and cover it with soil and plastic cover so that vapor generated from the solution should not escape in the atmosphere. Keep the pit covered for 1 week. After 1 week, remove the plastic cover and rake the soil for next 10-15 days, till the smell for formalin is completely gone. The pit can be used for planting new sapling.

» As soon as the initial symptoms of the wilt are visible in the orchard, drenching should be taken in the root areas around the main trunk and 2-3 plants in all four sides using recommended agrochemicals.

» Pruning tools should be disinfected after each use; plants cut ends to be painted with 10% Bordeaux paste. *Azadirachta indica* (neem) oil may be added @ 50 ml/L in the paint during rainy season. Pruning should be avoided during monsoons/rainy days. Affected plants within the buffer zone should be treated with a systemic fungicide; neighboring asymptomatic apparently healthy plants should also be treated with appropriate systemic fungicides.

23.4. INTERACTION OF ROOT-KNOT NEMATODE WITH *CERATOCYSTIS* WILT

The root-knot nematode, *Meloidogyne incognita* interacts with soil-borne fungal pathogen, *Ceratocystis fimbriata* and cause wilt disease complex. The two pathogens in combination (*M. incognita, C. fimbriata*) adversely affected plant growth parameters like shoot and root length, and fresh and dry shoot and root weight. Among the two pathogens inoculated individually, maximum reduction in plant growth parameters was noticed in fungus alone (*C. fimbriata*) treatment. However, the treatment receiving the inoculum of both pathogens (*M. incognita* + *C. fimbriata*), showed a highest reduction in plant growth parameters in comparison with other treatments.

23.4.1. Distribution

The disease has been reported from Karnataka (Somasekhara and Wali, 1999), Maharashtra (Jadhav and Sharma, 2009), Andhra Pradesh, Tamil Nadu, and North Gujarat states.

23.4.2. Symptoms

23.4.2.1. Symptoms of root knot nematode infestation: General nematode infested plant shows symptoms of nutrient deficiency, yellowing of leaves and stunted growth. It is also observed that sometimes luxuriantly growing plants show reduced or no flowering for long periods (more than 1 year) due to heavy infestation of root-knot nematode, *Meloidogyne incognita*. With increase in the population of nematode due to rapid reproduction rate, feeder roots are damaged resulting in less nutrient uptake from the soil. This root damage causes losses in terms of quality (small sized fruits) and quantity (less number of fruits per plant). The knots/galls on the pomegranate roots confirm the infestation of root-knot nematode.

23.4.2.2. Symptoms of wilt complex: Root knot nematode, *Meloidogyne* spp. along with the fungus *Ceratocystis fimbriata* forms the wilt disease complex. In the orchards having wilt disease, the symptoms start with yellowing of leaves followed by gradual wilting of one or more branches culminates into death infected plants. Surveys conducted in recent years have shown heavy losses ranging from 20 to 80 percent due to wilting and death of plants. About 85% of the total wilt infected orchards surveyed, have both root knot nematode and fungi *C. fimbriata* together. It is also observed that nematode infestation increases the severity of wilt disease by providing easy sites for fungal infection as well as reducing the plants innate immunity.

When the roots and lower portion of the stems were split open, the dark bluish-black/grey/brown discoloration of the wood is observed if the pathogen is *Ceratocystis fimbriata.* Within 2-3 months after infection, plants show symptoms like leaf yellowing in one or more branches, followed by drooping of foliage, and complete wilting of plant. Plants may reveal drooping of leaves of the entire plant all of a sudden (within 1-2 days) resulting in complete wilting. *C. fimbriata* infection causes discoloration of vascular bundles and adjoining tissues

In combination with fungi, the affected roots showed brownish discoloration in fungus alone and fungus + nematode treatments. In uninoculated treatment, there was no brownish discoloration.

The two pathogens in combination (*M. incognita, C. fimbriata*) adversely affected plant growth parameters like shoot and root length, and fresh and dry shoot and root weight. Among the two pathogens inoculated individually, maximum reduction in plant growth parameters was noticed in fungus alone (*C. fimbriata*) treatment. However, the treatment receiving the inoculum of both pathogens (*M. incognita* + *C. fimbriata*), showed a highest reduction in plant growth parameters in comparison with other treatments.

23.4.3. Spread and survival

- The fungus survived in thick walled Aleurioconidia of *C. fim*briata, in soil, host and plant debris and served as a source of primary inoculum. The fungus can survive in infected wood for several years.
- Infected planting material or the soil used for raising the nursery planting material generally are the primary source of infection for wilt organisms.
- Pruning wounds are common entry points for *C. fimbriata*, and the fungus can also be carried on pruning tools and through air.

- Spread to adjacent trees can take place via root grafts generally in plantation at close spacing.
- Flood irrigation/rain water also spread the pathogens from the infected plant/s to healthy plants.
- Wounds in the roots made by insects, nematodes and rodents also predispose the plant to wilt pathogens.
- Infected seedlings, irrigation water, root contact, implements, rain water, budding knife, secateurs and wind are the major sources for the spread of disease (Somasekhara *et al.*, 2009).

23.4.4. Management

23.4.4.1. Chemical methods: Basin soil application of Carbofuran 3G at 40g/plant or Phorate 10G at 25g/plant were found effective for management of wilting due to root-knot nematodes. Drenching with Azadirachtin (1%) at 2 ml/L is also recommended. Application of Neem cake at 1-3 kg/plant depending on age is advisable twice a year. Plant *Tagetes erecta* (African marigold varieties best followed by French marigold) in between row space or in a ring, on the border of plant basin. For effective results, these should be grown for more than 4 to 5 months and incorporated in the soil. Crops like onion, tomato, chilli, potato, bell pepper, gram, legumes, cucurbits, gerbera, gladiolus *etc.* aggravate nematode infestation and hence should be avoided as intercrops. Green manuring with sesbania is beneficial.

23.4.4.2. Biological methods: Biological formulations if used should be reliable, fresh and used during rest period when no other fungicides/bactericides are used. The soil application of *Bacillus subtilis, Purpureocillium lilacinum, Trichoderma harzianum, Aspergillus niger* at 10-15g/plant along with well-decomposed farm yard manure around the trunk of pomegranate trees helps to prevent wilt infections. Neem cake at 2-3 kg/ plant effectively checks incidence of wilt complex.

Biofertilizer – Kalisena SA having *Aspergillus niger* at 2.51 kg/ha + Mychorrhizal preparation Josh at 12.5 kg/ha or Josh ultra 2.5 kg/ha gives effective control of wilt if used from beginning or before the incidence of disease. These two bioagents should be applied twice a year along with sufficient organic matter for effective wilt management. This treatment controls several soil pathogens, improves nutrient uptake, gives disease resistance and also improves yields.

23.4.4.3. Integrated methods: The following practices are recommended for the integrated management of wilt disease complex in pomegranate:

- It is advisable to take cuttings/air layers from disease-free orchards and raise saplings in sterilized soil.
- The planting material (sapling as well as soil in which it is planted) should be free from all wilt organisms.
- Soil solarization or sterilization with formalin (50 ml/L) should be done before planting
- Avoid water stagnation and create proper drainage.
- Follow recommended spacing of 4.5 m × 3.0 m in the orchard.
- The nursery beds or soil used for potting mixtures should be sterilized by soil solarization for 6 weeks using 50-100μ thick linear low density polyethylene (LLDPE) sheet during hot summer months or by using chemical sterilants at 2.5-5.0% formalin. If formalin is used, ensure that the soil is free from any formalin fumes before transplanting in the bags. Soil solarization is beneficial as it kills harmful pests and pathogens and also increases population of beneficial microorganisms which are present in the soil and are thermo-tolerant. However, formalin treatment kills both harmful and beneficial organisms.

- » The above formalin treatment can be also be used for sterilizing soil after removing infected dead plant.
- » On observing first symptoms of wilt due to fungal pathogens in the orchard, drench soil immediately with Propiconazole 25EC (2 ml/L) + Chlorpyriphos 20EC (2 ml/L) or Carbendazim 50WP (2.0 g/L) + Chlorpyriphos 20EC (2 ml/L). Use 5-10 litres solution/plant depending on plant growth to drench soil up to 30 cm depth. Drench at least 3-4 healthy plants all around the infected plant/s, and repeat drenching 3-4 times at 20-25 days interval. Drenching with Ridomil, Metalaxyl or Dithane M-45 (2g/L) will be beneficial if *Phytophthora* is involved. Propiconazole was significantly superior than remaining treatments. Propiconazole treated plants were completely recovered from wilting after 105th day followed by Difenconazole in which the plants recovered after 120th day. The control showed maximum percentage of wilting (97.61%).
- » For controlling shot hole borer (*Xyleborus* spp.) which is associated with wilt disease, apply 10 litre solution containing red soil (4kg) + Chlorpyriphos 20EC (20ml) + Copper oxychloride (25 g) at plant base up to 60 cm. diameter from second year onwards. To control stem borer, inject in the holes on the trunk with DDVP 2-3 ml and plug the holes with mud.
- » Wilt due to root-knot nematodes can be managed with soil application of Phorate 10G at 25g/plant or Carbofuran 3G at 40g/plant or other suitable nematicide in the plant basin, in a ring near root zone and cover it with soil. Drenching with Azadirachtin (1%) at 2ml/L is also recommended. Application of neem cake at 1-3 kg/plant depending on age is advisable twice a year. Plant *Tagetes erecta* (African marigold varieties best followed by French marigold) in between row space or in a ring, on the border of plant basin. For effective results, these should be grown for more than 4 to 5 months and incorporated in the soil. Crops like onion, tomato, chilli, potato, bell pepper, gram, legumes, cucurbits, gerbera, gladiolus *etc.* aggravate nematode infestation and hence should be avoided as intercrop. Green manuring with sesbania is beneficial.

23.5. FUTURE LINES OF WORK

- » The survey and surveillance of the disease may be extended to other states to know the regional severity of the wilt and mapping of the disease.
- » Broad based molecular biological studies are essential to establish pathogenic variation to device better disease management strategies.
- » Detailed studies on the influence of various ecological factors on the distribution and incidence of disease complex need to be undertaken.
- » To understand the genetic basis of mechanism of interaction, physiological and histological changes in the disease complex may be investigated.
- » Studies should be carried out regarding the compatibility of chemicals, antagonists and plant extracts for integrated disease management approach.
- » Integrated disease management approaches involving resistant and moderately resistant cultivars, bio-control agents, fungitoxicant components should be developed.

23.6. CONCLUSION

Owing to its polyphagous nature and endoparasitic feeding habit, it is very difficult to control root-knot nematodes especially on established crops. It attacks numerous economically important crops and can also be found on many weeds. The recent identification of emerging highly damaging and resistant *Meloidogyne* spp. poses a considerable challenge to formulation of effective management strategies. Lack of accurate and current data on various *Meloidogyne* spp. present in different parts of the continent and the polyphagous nature of these pathogens also poses a greater risk to the future of food production in Africa and Asian continents. To adequately address emerging and other *Meloidogyne* spp., it is imperative that resources are harnessed to drive more research aimed at assessing and understanding the species identity, genetic diversity, population structure, parasitism mechanisms, host resistance and the overall threat posed by them (Fargette *et al.*, 2010). Therefore, there is a need to embrace modern technology in conjunction with classical methods while carrying out *Meloidogyne* spp. identification.

To effectively manage the highly damaging nematode pathogens, application of biological, cultural and chemical methods should be used in line with integrated pest management (IPM) practices. This should be preceded by a thorough survey of orchards in context and an accurate diagnosis of nematode species present. This will lead to gradual management of nematodes and finally reduction in the high levels of damage that they cause on pomegranate. This strategy will eventually benefit growers and avoid high production costs. With the phasing out of various effective nematicides such as methyl bromide, the search for effective and environmentally friendly alternative products should be pursued. At the same time, more robust diagnostic techniques should be adopted to correctly identify and avoid further spread of the highly damaging, resistance breaking and emerging *Meloidogyne* spp. Growers should also be educated on proper phytosanitary procedures to avert the introduction of *Meloidogyne* spp. into their orchards.

It is concluded that though several plant parasitic nematodes are associated with pomegranate crop, the root-knot nematode, *Meloidogyne incognita* was found to be most commonly occurring as per the community analysis and the probable cause for predisposing plants to wilt fungus causing yield loss and ultimate death of plants. This knowledge gives us an insight on charting out management techniques to help the growers escape from severe yield losses caused by nematodes and the wilt fungus which very often associate themselves with nematodes infected pomegranate.

Chapter - 24

Fig, *Ficus carica*

24.1. INTRODUCTION

24.1.1. Origin

Common fig is also known as Anjeer. Fig is cultivated all over the world, originated from West Asia and spread to the Mediterranean region.

24.1.2. Area and production

Worldwide, the area under cultivation of fig trees exceeds 289,818 ha, with an estimated production of 1,315,588 t (FAOSTAT. 2019). Turkey is the biggest world producer, with 310,000 t in 2019, followed by Egypt, Morocco, Iran, Algeria and Spain. Therefore, of utmost importance for fig production is still the Mediterranean basin and the Near East (FAOSTAT. 2019). Spain is the main source of figs in Europe (51,600 t), followed by Greece (19,730 t) and Italy (11,830 t) (FAOSTAT. 2019).

The area under cultivation of fig in India is nearly 3,570 hectares with production of 14,643 tons, where in Maharashtra is the leading state followed by Karnataka (Anonymous, 2015). In India, fig is grown in Maharashtra (Pune and Aurangabad), Gujarat, Uttar Pradesh (Lucknow and Saharanpur), Karnataka (Bellary, Chitradurga and Srirangapatna), Punjab, Tamil Nadu (Coimbatore) and Andhra Pradesh (Anantapur).

The majority of producers of this fruit in Karnataka are from Kalaburgi, Raichur, Koppal districts and as per recent reports many farmers from Ballari are interested in growing the crop for export and local consumption.

24.1.3. Cultivars

There are about 20 popular varieties of fig that are being grown in different parts of the world. Some famous varieties of common fig grown in different countries are 'White Adriatic' 'Black Mission', 'Kodota' and 'Conadira' in California; 'Kalamon' in Greece; 'Sultani' in Egypt. The 'Smyrna types' includes popular Turkish cultivar known as 'Saricop' in Turkey and 'Calimyrna' in United States.

India 'Poona' is the most popular cultivar grown for consumption as fresh fruit. Most of the fig grown in Mangalore, Ballari, Coimbatore, Daulatabad, Ganjam, Lucknow and Saharanpur resemble

Poona fig in plant and fruit morphology. Recently, a variety 'Dinkar', an improvement over 'Daulatabad' variety for yield and fruit quality is gaining commercial importance.

The characteristics of three types of figs are presented in Table 24.1.

Table 24.1. The characteristics of three types of figs

Fig type	Varieties and characteristics
Common Fig	Kadota, Mission, Adriatic, Brown Turkey, Celeste and Conadria are some cultivars of this type. Poona is one of the most important commercially grown fig. Introduction and evaluation of exotic figs from California at IIHR Bangalore reveals that varieties like 'Deann', 'Conadria' and 'Excel' have superior fruit and plant characters. These new varieties when grown on 'Brown Turkey' root stock (through chip budding) hold great promise for exploiting marginal lands in arid and semiarid regions.
Smyrna fig	Calimyrna is the common cultivar of this type.
Sanpedro fig	The common cultivars of this type are Sanpedro, King and Gentile. In India, common fig is mostly grown. Some of the cultivars grown are Black Itchier, Brown Turkey, Turkish White, Kabul and Marseilles. Yercaud Timla fig is a drought tolerant cultivar. Fruits are large and reddish purple in color.

24.1.4. Exports

Major fig growing countries export fig amounting to US$ 767078 during 2021 (Tables 24.2 to 24.4)

Table 24.2. Value of global exports of fig

Country	Value of export in US$
Turkey	330,514
Afghanistan	165,928
Iran	40,436
Austria	34,951
India	738

Table 24.3. Export analysis of fig and its products (fresh and dried) (2021-22)

State	Value ('000 USD)	Destination
All India	1300	Singapore, Nepal, UAE (+9)
Tamil Nadu	600	Singapore, Malaysia, Maldives
Telangana	500	Singapore, UAE, Switzerland, UK
Bihar	10	Nepal

Table 24.4. Export analysis of fig (other than fresh) (2021-22)

State	Value ('000 USD)	Destination
All India	610,000	Nepal, USA, Maldives (+ 37)
Delhi	200,000	Nepal, UAE, Hong Kong (+ 2)

Maharashtra	90,000	Hong Kong, USA, Canada (+ 17)
Bihar	10,000	Nepal
Karnataka		UK, Oman

24.1.5. Action plan

- Upon analyzing on strengths and potential of this crop, both cultivation and processing will be independently profitable businesses. Creating suitable commercial atmosphere for prospective growers and investors in this sector is the need of the hour.
- Initially stressing on quality parameters, involving concerned departments for educating more farmers and growers to emphasize on large production of the crop is recommended. Focus on quality development and yield improvement in existing fig varieties of India.
- Later these departments and industrialists can be encouraged for establishment of processing industries in the state.
- Further when the production of fig fruits reaches a threshold mark, emphasis can be laid on exploring international market to export the dry fruit.
- State and Central Governments must sponsor financial schemes/ subsidies to aid capacity building in setting up of production of the fresh fruit, processing industries and encourage skill development programs for the youth to develop figs in the state.

24.1.6. Opportunities in Fig and related products

- As a dry fruit, fig has huge export and local market potential. It is diet rich super food according to dieticians and nutritionists.
- At present, keeping quality of fresh fruits grown in Karnataka are not satisfactory and unavailability of processing industry in the state leads to financial losses to the growers. Low shelf life of the best raw material makes it difficult to store and transport it to the processing industries located in other neighboring states. Hence, there is a grave need of having a processing industry in the state to empower this sector.
- As per export analysis of Fig, both fresh and processed, in-spite of the challenges in the sector, minute quantity of the exports from the state has fetched great rates and been appreciated for the quality.
- Presence of one or more processing industries encourages local growers and farmers to focus on the crop and establish an international market and export this fruit.

24.1.7. Nematode problems

The major nematodes associated with fig include root-knot (*Meloidogyne incognita, M. javanica*), cyst (*Heterodera fici*) and dagger (*Xiphinema index*) nematodes.

24.2. ROOT-KNOT NEMATODES, *MELOIDOGYNE INCOGNITA, M. JAVANICA*

The root-knot nematode is probably the most severe nematode problems in fig cultivation. The problem is recognized as a major limiting factor in commercial fig production in the India, USA, France and Brazil. The problem is recognized as a major limiting factor in commercial fig production in the USA (Knight, 1980), France (Scotto La Massèse *et al.*, 1984) and Brazil (Campos, 1997).

4.2.1. Distribution

Root-knot nematodes are distributed in Mediterranean, North and South American countries, India, as well as from southern Africa.

24.2.2. Crop losses

Average avoidable loss in the fruit yield of fig was recorded to be 24.92 per cent, when the crop was treated with Carbofuran 3 G @ 2 kg a.i. /ha (Jagdev *et al.*, 2020a).

24.2.3. Symptoms

The root-knot nematode is the major disease affecting figs grown in sandy soil. Damage from this root-feeding nematode is progressive and results in poor growth, low vigor, yellowing and bronzing of foliage, low yield and poor fruit quality. Root-knot nematodes on fig trees often show themselves with bumps or "knots" on the roots of the tree. Above ground, the tree has a generally stunted and unhealthy look. The above-ground symptoms are somewhat similar to nutrition deficiency. Under severe nematode infestation, severe root galling occurs which blocks the conducting tissues, deprive the plants from taking up nutrients leading to death of trees.

24.2.4. Management

24.2.4.1. Cultural methods: Choose a planting site in which a fig plant has not been previously grown, or where root-knot susceptible plants such as tomatoes, okra, or tobacco have not been recently grown.

Complete fallowing of the planting site during the summer and fall prior to planting can reduce nematode populations.

High organic matter in the soil can reduce root-knot nematodes.

Plant only nematode and disease-free plants.

24.2.4.2. Chemical methods: Soil application of carbofuran 3G at 2 kg a.i. /ha was found to be most effective in reducing root-knot nematode population (53.66%), number of root galls (50.19%) and number of egg masses (45.36%) at intermediate stage and increasing the fruit yield of fig (37.85%) with 17.79 IBCR at termination. This treatment was followed by the treatment of phorate 10 G at 2 kg a.i./ ha. The reduction in root- knot nematode population, number of root galls and egg masses and increase in fruit yield recorded in this treatment were 49.46, 50.11, 42.31 and 28.19%, respectively with 33.64 IBCR.

Nematode populations were considerably reduced in young trees by stem treatments with an experimental paste formulation of Fenamiphos (Inserra and O'Bannon, 1974).

24.2.4.3. Biological methods: At termination stage, the treatment of Phule *Trichoderma* plus at 20 kg/ha was found to be most effective in reducing nematode population (40.36%), number of root galls (37.08%) and number of egg masses (41.02%) and increasing the fruit yield of fig (13.54%) with 11.20 IBCR (Jagdev *et al.*, 2020b).

24.2.4.4. Host resistance: Four *Ficus* species (*F. racemosa*, *F. cocculifolia*, *F. gnaphalocarpa*, and *F. palmata*) showed a high degree of resistance to species of root-knot nematodes, as well as good graft compatibility with *F. carica* (Krezdorn and Glasgow, 1970). O'Rourke (1992) recommended that the commercial nematode resistant fig cv. 'LSU Purple' (from 'Hunt' an unknown California Capri fig) for planting. The rootstock 'Zidi' performed well in a number of field trials (Hosomi *et al.*, 2002). In Israel, the fig varieties 'Celeste' and 'Poulette' were considered resistant to the nematode (Gur, 1955).

'Ficus Echo' (*Ficus sycomorus*) seems to be effective as a nematode-resistant rootstock for figs for South Florida. 'Ficus Echo' propagates readily by cuttings. Fig cvs. 'Celeste', 'LSU Tiger', and 'LSU Purple' fig has grown rapidly.

24.2.4.5. Integrated methods: Carbofuran 10G was the most effective treatment in suppressing the nematode densities (mainly *Meloidogyne javanica* and *Tylenchorhynchus mediterraneus*). The most effective management combinations, next to carbofuran 10G, in suppressing the nematode densities in the rhizosphere of fig trees were *Purpureocillium lilacinum* + *Pasteuria penetrans* + urea 46%, followed by *Trichoderma harzianum* + *P. penetrans* + cow manure, *P. lilacinum* + *P. penetrans* + cow manure, and *Pochonia chlamydosporia* + *P. penetrans* + urea 46% (54.68–57.17% nematode reductions). The most effective treatments in increasing the fruit yield of fig trees were *T. harzianum* + *P. penetrans* + compost, followed by Carbofuran10G, *T. harzianum* + *P. penetrans* + urea 46%, *P. lilacinum* + *P. penetrans* + chicken manure and *T. harzianum* + *P. penetrans* + urea 46% (54.68–57.17% increase in number of fruits/tree) (Tables 24.5. and 24.6.) (Dawabah *et al.*, 2019).

Table 24.5. Effect of different combinations of integrated management on the number of plant-parasitic nematodes (PPN) associated with fig, *Ficus carica*, rhizosphere

Treatment	% Reduction/increase in nematode population/ 200 cm^3 soil
Purpureocillium lilacinum + *Pastueria penetrans* + urea 46%	- 57.17
Trichoderma harzianum + *P. penetrans* + cow manure	- 56.66
P. lilacinum + *P. penetrans* + cow manure	- 54.97
Pochonia chlamydosporia + *P. penetrans* + urea 46%	- 54.68
Carbofuran 10G	- 83.13
Control	+ 35.46

Table 24.6. Effect of different combinations of integrated management on the number of fruits per fig tree.

Treatment	Number of fruits/tree		
	1st yield Mar–May 2016	2nd yield Nov. 2016–Jan. 2017	Total yield
T. harzianum + *P. penetrans* + compost	28.75 cde	84.75 ef	156.75
Carbofuran10G	40.50 ab	103.25 bc	143.75
T. harzianum + *P. penetrans* + urea 46%	36.50 abcde	98.75 bcd	140.00
P. lilacinum + *P. penetrans* + chicken manure	38.25 abc	100.50 bcd	138.75
T. harzianum + *P. penetrans* + urea 46%	36.50 abcde	98.75 bcd	135.25
Control	16.50 f	35.75 i	52.25
CD (P ≥ 0.05)	10.05		

24.3. CYST NEMATODE, *HETERODEA FICI*

Heterodera fici is another nematode pest of fig, which is fairly widely distributed throughout the world. In particular, *H. fici* is a worldwide parasite of cultivated *Ficus* species (*F. carica*, *F. elastica*, *F. rubiginosa*,

F. benghalensis, *F. lyrata*, *F. australis*, and *F. benjamina* (Baldwin and Mundo-Ocampo, 1991, Evans and Rowe. 1998; Wohlfarter *et al.*, 2011).

24.3.1. Geographical distribution

Subbotin *et al.* (2010) summarized the known occurrence of the fig cyst nematode from Germany, Greece, Hungary, Italy, the Netherlands, Norway, Poland, Portugal, Russia, Spain, Yugoslavia, China, Georgia, Iran, Turkey, Uzbekistan, Australia, New Zealand, United States (California, Florida, Louisiana, Maryland, and Virginia), Brazil, Algeria, and South Africa. More recently, *H. fici* was also found for the first time on *F. carica* maintained in a nursery in Canada (Sun *et al.*, 2017).

Heterodera fici is fairly widely distributed throughout the world, having been reported infesting trees in several Mediterranean countries including France (Scotto La Massèse *et al.*, 1984), Spain (Bello-Perez and Jimenez-Milian, 1963), Italy (Di Vito, 1976) and Turkey (Yuksel, 1981), as well as in California (Sher and Raski, 1956), Brazil (Brancalion *et al.*, 1981), Yugoslovia (Krnjaic *et al.*, 1997), and Soviet Asia (Narbaev and Sidikov, 1985).

24.3.2. Symptoms

Poor plant growth and leaf chlorosis have been reported for infected figs (Vovlas *et al.*, 1989). The pest could lower crop yield and crop value. These symptoms are non-diagnostic and easily confused with other diseases, caused by other nematode species, pathogens, or abiotic problems.

24.3.3. Nematode population

The fig cyst nematode *H. fici* was recovered in surveys in established commercial fig orchards (more than 30 years old) as well as in private and public gardens. High infection rates were observed ranging from 44 to 180 cysts/100 ml of soil, 12–36 cysts per g of roots; 1.2–1.6 eggs – J2 /ml of soil, thus suggesting that the nematode might be causing damage.

24.3.4. Pathogenicity

The potential pathogenicity of *H. fici* on fig seedlings was demonstrated in pot trials by Di Vito and Inserra (1982) who reported 30% death of plants with an initial nematode population of 8 eggs and juveniles/cm^3 of soil, and 100% plant mortality with an initial nematode density of 64 eggs and juveniles/cm^3 of soil and higher. *H. fici* is limited to *Ficus* species only. *Heterodera fici* is reported as parasite of woody plants. The nematode has caused considerable damage to potted plants of the related *F. elastica* (Narbaev and Sidikov, 1985).

23.3.5. Host range

The host-suitability test confirmed that *H. fici* has a narrow host range limited to *Ficus* species and for this reason it is not considered a major pest, as are other species of *Heterodera*.

24.3.6. Post-invasion development

Post root-invasion development of *H. fici* on *Ficus carica* roots was completed in about 64–68 days at 26–28 °C. Invasion and nematode development stages were recorded by stereoscope observations on acid fuchsin-stained roots at seven-day intervals. The maximum developmental phase observed was that of young cysts containing more than 25 eggs. The first period (24–26 days after inoculation was utilized by the parasitic specimens for exploration, penetration, and selection of the feeding site. The second stage juveniles stop penetration and start to feed on the initial feeding cells by using a sedentary endoparasitic feeding position. Observations at 50–56 days after inoculation revealed that the 3rd molt was completed and that sexual differentiation occurred. The swollen posterior portion of sexually mature nematodes (pyriform females and 4th stage vermiform males still inside the cuticle) protrudes from the root. Fourth-stage females and males were observed one week later. Lemon-shaped mature females were observed eight weeks after invasion. The post-infection period was concluded about 8 weeks later. No new cysts were observed.

23.3.7. Histopathology

Heterodera fici establishes permanent, fully-developed feeding sites on *F. carica* roots and reaches maturity in 64–68 days. Histological examination of sectioned healthy and nematode-infected fig roots showed that infection by *H. fici* can cause cellular alterations in the cortex, endodermis, pericycle, and vascular parenchyma tissues of fig roots. Observations of cross-sections of *H. fici* infected roots indicated that the nematode can induce the formation of both cortical and endodermal syncytia. In many cases, the nematode female only penetrated one to three layers of the cortical root cells without reaching the stele. In these roots, the nematode established a permanent feeding site in a cortical cell that was fused with adjacent cells forming the syncytium.

Structurally, vascular syncytia varied in shape and expansion when induced by females or males. Those induced by females were larger and with more cytoplasmic contents compared to those induced by males. All syncytial cells were hypertrophic with dense cytoplasm and large nuclei. In some cases, nematode females penetrated into the cortex with the anterior elongated body portion where they induced a permanent feeding site and continued to remain in a semi-endoparasitic position, probably because of the nature of root tissues (woody host).

24.3.8. Management

For reducing nematode damages, it is best to combine several measures (Dawabah *et al.*, 2019). Common measures used are:

- » Selection of nematode-free planting material.
- » Weed control (Lima-Medina *et al.*, 2013).
- » Biological control agents and chemical nematicides.
- » Development of nematode-resistant rootstocks (Hosomi *et al.*, 2002; Costa *et al.*, 2015).

24.4. DAGGER NEMATODE, *XIPHINEMA INDEX*

F. carica is the type host of *Xiphinema index*, and this nematode attains extremely large populations around fig trees in the Mediterranean region.

24.4.1. Symptoms

Dagger nematode causes gall formation on root tips. The anatomical changes caused by the nematode on fig roots – in the form of terminal galls and modified cells – as well as the associated biochemical changes have been studied and described in great detail (Poehling *et al.*, 1980; Wyss *et al.*, 1980), as has the feeding behavior of the nematode on fig roots (Wyss, 1987). Fig has been shown to be a more favourable host of *X. index* than grapevine (Malan and Meyer, 1999). The nematode is considered of economic importance in California, where trees respond favorably to pre-plant fumigation (Kodira and Westedahl, 2003).

24.4.2. Virus transmission

There is no known virus transmission in fig by this nematode, which is the vector of fanleaf virus disease in grapevine. For this reason, movement of fig trees, especially to grape producing regions, should require regulatory attention (Hirata *et al.*, 2002).

24.4.3. Management

Carbofuran 10G was the most effective treatment in suppressing the nematode densities on fig trees. The most effective management combinations, next to carbofuran 10G, in suppressing the nematode densities in the rhizosphere of fig trees were *Purpureocillium lilacinum* + *Pasteuria penetrans* + cow manure, *T. harzianum* + *P. penetrans* + cow manure, *P. lilacinum* + *P. penetrans* + urea 46%, and *Pochonia. chlamydosporia* + *P. penetrans* + urea 46% (54.68–57.17% nematode reductions) (Dawabah *et al.*, 2019).

Chapter - 25

The Roadmap Ahead

25.1. INTRODUCTION

Nematodes constitute one of the major limiting factors in fruit crop production. Parasitic nematodes can cause damage to tree fruit roots. Many kinds of nematodes have been reported in and around the roots of various fruit crops, but only few are responsible for causing serious damage. They cause extensive root necrosis resulting in serious economic consequences viz., fertilizers are not effectively utilized, the period from planting to harvesting is extended, drastic reduction in fruit or bunch weights, the quality of fruits is impoverished and orchards have to be replanted every 2 to 3 years because of drastic reduction in plant numbers. Furthermore, roots damaged by nematodes are easy prey to bacteria and fungi which invade the roots and accelerate root decay. The root-knot nematode, *Meloidogyne incognita*; the burrowing nematode, *Radopholus similis*; the citrus nematode, *Tylenchulus semipentrans*; lesion nematodes, *Pratylenchus* spp.; the reniform nematode, *Rotylenchulus reniforms*; ring nematodes (*Mesocriconema* spp.), and dagger nematodes (*Xiphinema* spp.) are the major nematode pests that infect the fruit crops. Nematodes reduce yield without the production of any noticeable above-ground symptoms. Typical above-ground symptoms of nematode infections include stunting, yellowing and wilting.

The changing cropping patterns, introduction of new crops, crop diversification, and agronomic practices *etc.* also change the spectrum of plant parasitic nematodes. For example, drip irrigation in fruit crops and widespread and unchecked movement of planting materials from fruit crop nurseries, have led to emergence of new nematode problems in newer areas and intensification of existing nematode problems. Some of the emerging nematode problems due to introduction of new nematode pests under globalization of agricultural produce; impact of climate change; threat to lack of management strategy in standing crops, *etc.* include root-knot nematode problems on acid lime (*Meloidogyne indica*), pomegranate (*M. incognita* Race 2) and guava (*M. enterolobii*).

These nematodes are responsible for 30 to 40 percent yield losses in various fruit crops. The incidence of fungal pathogens would be doubled in the presence of the nematodes. These nematodes cause breakdown of resistance to fungal diseases in certain varieties of fruit crops. The nematodes spread from one area to another mainly through infested planting materials. In case of banana, paring or trimming of suckers, often done before planting is usually not sufficient to remove the infections that extend deeply into the set. This nematode is disseminated when water that drains from infested areas gets recycled into irrigation system. Soil that adheres to implements, tyres of motor vehicles and shoes of plantation workers may also spread nematodes from one area to other area.

Among the fruits, the root-knot nematode hosts include banana (*Musa* spp.), grapevine (*Vitis vinifera*), and papaya (*Carica papaya*). Root lesion nematodes (*Pratylenchus penetrans* and *P. vulnes*, primarily) are sometimes associated with apple replant disease. They are major problems causing damage in apple, peach, cherry, and grape. *Radopholus citri* and *Radopholus similis* infect citrus and banana, respectively. The citrus nematode (*Tylenchulus semipenetrans*) is a semi-endoparasitic nematode causes slow decline in citrus crops. It is also a problem in grapes, loquats, persimmon, and olive. *R. reniformis* parasitizes a large number of fruit trees. Dagger nematodes (*Xiphinema spp.*) are commonly vectors of viruses, for example the tomato ringspot virus, which causes apple union necrosis and decline (especially for Red Delicious on M106 rootstocks), and the Cherry rasp leaf virus (CRLV).

25.2. CURRANT NEMATODE MANAGEMENT STRATEGIES

Management practices should be effective, environmentally safe, and economical and must focus on reducing nematode populations to levels below the damage threshold. The common methods of nematodes management include use of resistant varieties, rotating of crops, soil amendments, soil solarization and applying pesticides. Soil solarization is very effective for control of many nematodes and soil-borne pathogens. Integration of resistant crop cultivars with crop rotation is the best management practice.

Crop rotation and cover cropping are often practices used in integrated pest management to reduce plant-parasitic nematode incidence. Leguminous cover crops *Mucuna pruriens* and *Crotalaria spectabilis* showed multiple resistance to root-knot nematodes (*Meloidogyne arenaria, M. incognita, M. javanica*). Sudan grass and corn are excellent green manure crops that provide good nematode control. Flooding and bare fallowing was shown to decrease nematode soil populations and increase yields. Marigold, Sudan grass and *Brassica* spp. can be used as green manure crops to control plant parasitic nematodes and boost free-living nematode populations in the soil. Resistant cultivar is considered as one of the best methods for nematode control and has been found to be successful in several countries.

Bacterial biological control agents such as *Pasteuria* spp. (*P. penetrans*) and *Bacillus* spp. (*B. cereus* strain S2), *Pseudomonas fluorescens* have shown great promise in nematode management. Opportunistic fungi like *Purpureocillium lilacinum*, *Pochonia chlamydosporia*, *Trichoderma harzianum* have potential as biological control agents. Organic amendments contribute to biological activity in the soil and enhance the natural activity of organisms antagonistic to nematodes.

Most commonly practiced integrated nematode management components include crop rotation, resistant and tolerant cultivars, using biocontrol agents, chemicals and cultural practices.

25.3. NEW PARADIGMS FOR NEMATODE MANAGEMENT

Biological control amplifies biotic imbalances, because the economic pressures compel drastic methods based on the physical elimination of antagonists. High grade results achieved by crop protection researches carried out on crop cultural practices, plant resistance to nematodes or biocontrol, are very diverse. But, all these control practices, included or not in integrated pest management strategies, seem not to be sufficient: they all target some nematode species (population approach), and then involve changes in nematode communities but do not necessarily modify their overall pathogenicity. So, binary researches focused on plant-nematode or nematode-parasite relationships should be extended to ecological investigations on nematode communities.

These reasons support eco-epidemiological approaches, based on the ecology of communities (interspecific competitions, biological, and edaphic constraints) and focused on the management of the natural enemy biodiversity, appearing as a new research alternative for crop protection.

Specific and functional evolution and rapid multiplication of plant-parasitic nematodes are enhanced by agriculture intensification and by environment adaptation. In fact, plant-parasitic nematodes are only considered as predators, and not as one of the soil components. Up-to-date, all control strategies developed in agriculture are focused on the eradication of target species. They induce biotic gaps, community rearrangements, insurgence of virulent races, increased aggressiveness of minor species, *etc*. Then, the "soil cleaning" strategy appears to be not sustainable.

The development of sustainable management strategies must move from the "therapeutic approach", mostly adopted in research programs carried out in the world, to the "ecological approach". This approach will seek for information and knowledge about biotic trade-offs in ecosystems, in order to introduce them in agro-systems (resilience). That questioning can be declined as follows:

» Why seek with plant-parasitic nematode eradication?
» Can agronomic problems be solved by agronomic strategies only?
» Can nematode diversity in communities be considered as an auxiliary for nematode management?

Effector biogenesis has not been studied for 30 years because the system was intractable. With recent development in functional genetics, the E-biogenesis is now well placed to dissect this strategically important knowledge gap. *E-biogenesis* aims to investigate methods of blocking crop parasitism by attacking the biogenesis machinery that produces the molecules that cause disease. The vision of the *E-biogenesis* is centred on the ability of nematodes to produce molecules that cause disease; so called "effectors". Blocking effectors blocks parasitism. However, blocking individual effectors is insufficient. Rather, there is a need to attack the features that unite most effectors; their biogenesis machinery.

This research will highlight a multiplicity of attractive targets for the biotechnological control of globally important plant nematodes. To achieve this, there is a need to open a whole new area of effector study, contribute new fundamental knowledge of strategic relevance, and provide a platform for a paradigm shift from individual effectors to a holistic view of effector biogenesis. Surely, effector biogenesis (*E-biogenesis*) is the Achilles heel of the nematode. Recently (2020), a project on *E-biogenesis has been granted by European Research Council to Dr.* Sebastian Eves-van den Akker, Crop Science Centre, University of Cambridge.

25.4. TROPICAL FRUIT CROPS

Although many species of plant parasitic nematodes are associated with tropical fruit crops, root-knot nematode *Meloidogyne incognita*, burrowing nematode *Radopholus similis* and citrus nematode *Tylenchulus semipenetrans* are the major nematode pests that infect these fruit crops. Damage due to nematodes is documented for relatively few host-parasite combinations. These should be recognized by growers of tropical fruit crops, and such combinations be avoided in selecting planting sites. Good management practices and sanitation are essential in all phases of nursery and orchard operations, because once a serious nematode problem is established on a fruit tree, it is very difficult to restore tree health and production.

Knowledge of the host-parasite associations of nematodes on tropical fruit species is essential in deciding what fruit species to plant in sites which are infested by one or more nematode species. Future studies are needed to determine the nematode population levels which are damaging to the various tropical fruit species, and control measures should be developed for the appropriate cases. Further studies needs to be directed in order to identify resistant genes and nematode effectors to raise up new management approaches. The strategy for nematode management could be shifted from the concept of control to the concept of management by combining several methods in an integrated manner at keeping the nematode population below economic injury level.

25.5. SUB-TROPICAL FRUIT CROPS

Currently nematodes have become a major problem to the commercial sub-tropical fruit growers. Diverse nematode pathogens which cause diseases of sub-tropical fruit crops can be limiting factors in the production and marketing of these commodities, and are often difficult to manage. There is adequate information available on nematode infestation in sub-tropical fruit crops. Resistance, the most effective tool with which many of these nematodes are managed, is usually available in coevolved pathosystems but may be uncommon in new encounter situations. Inadequate host resistance can be a significant barrier in the management of newly encountered nematode diseases, as well as those that are caused by generalist pathogens. Development of integrated nematode and disease management in sub-tropical fruit crops is the need of the hour. Further, there is an urgent need to do further studies on host parasite relationship and epidemiology of the nematode diseases in order to have a sound integrated management approach.

25.6. TEMPERATE FRUIT CROPS

The major nematode problems on temperate fruit crops worldwide include species of *Pratylenchus*, *Meloidogyne*, *Mesocriconema* and *Xiphinema*. Deeper and broader understanding of nematode diseases is essential to develop effective, economical and eco-friendly nematode management strategies. Work on disease complexes where nematodes and soil microorganisms are involved needs to be intensified with adequate collabration between nematologists and plant pathologists. A prior knowledge of host-parasite relationship of major nematode pests and techniques for precise determination of damage threshold that limit nematode population is a must for a successful nematode management program. Also while adopting a management strategy economy and ecology must be taken into consideration.

For food and nutrition security, management of nematodes is very essential. There is a need to develop bio-intensive integrated nematode management strategies using nematode-free planting stock by seed certification, heat treatment for denmatizing planting stock, use of crop rotation and intercropping with non-hosts or antagonistic crops, soil application of organic amendments enriched with biological control agents, treatment of nurseries and main field with safer chemicals, and development of resistant rootstocks.

In the recent years, some cultivars/rootstocks resistant to plant parasitic nematodes have been developed. However, for an encouraging result, more emphasis is required in this field by identifying new sources of plant resistance and their incorporation into crops by conventional breeding or genetic engineering biotechnology.

25.7. SEMI-ARID FRUIT CROPS

The most commonly encountered plant-parasitic nematodes in semi-arid fruit crops like pomegranate and fig include the root-knot nematode, *Meloidogyne incognita*. Complete death of the plant occurs due to wilt disease complex in pomegranate caused by the association of the root-knot nematode, *Meloidogyne incognita* and the fungus, *Ceratocystis fimbriata*. Cover crops can be important tools for sustainable nematode management in arid and semi-arid fruit crops. Using appropriate cover crops can prevent or reduce nematode population and enhance soil health. The successful system of cover cropping is limited by availability of water especially during frequent droughts. Choosing the right cover crops and managing them appropriately can provide long-term benefits for many nematode problems.

There is a need for development of suitable root-knot and wilt management schedules involving cultural practices, sanitation measures and chemical methods. Emphasis need to be given to develop root-knot-wilt resistant rootstock to evolve suitable wilt resistant variety to ensure more economical, feasible and effective wilt management schedule.

25.8. TRANSFER OF TECHNOLOGY

Generating awareness among farmers is crucial for preventing crop loss by plant parasitic nematodes (PPNs). Many important PPNs have a wide host range and can attack a number of crops, like cereals, pulses, oilseeds, fruits, fibers, fodder, plantations, spices, medicinal and aromatics, *etc.* The nematodes are hidden enemies, since they are present in soil. Hence, the farmers, administrators, and policymakers are unaware of the damage they cause to crop plants. They should be made aware of the nematode damage symptoms, biosecurity threats, and their management methods through transfer of technology by extension agencies.

Application of Information and Communication Technology (IC&T) is essential for effective nematode awareness programs. In many counties abroad, IC&T has been very successfully applied by the agriculture extension agencies. Field days and crop clinics should be organized and newsletters, newspaper articles, website news to be published for effective dissemination of information about nematodes among farmers and stakeholders. Nematode problems could be easily disseminated through IC&T approach like TV and radio programs (e.g. Krishi Darshan), newspaper articles, extension publications such as leaf-lets, book-lets, extension bulletins in regional languages and organizing Krishi Mela, seminars, exhibitions, field demonstrations, e-mail, SMS, social media (Facebook, Twitter, Blogs), *etc.*

PPNs have local or regional distribution on certain crops and their nature of damage, survival, disseminations and management practices are known. These should be made aware to the farmers through ICAR Institutes, SAUs, State Agriculture and Horticulture Departments, KVKs, and NGO's.

25.9. FUTURE APPROACHES

PPNs constitute one of the major limiting factors for cultivation of fruit crops. The changes in agricultural situations have tremendous effects on the emergence of new nematode problems in India. The recent outbreaks of *Meloidogyne indica* in kagzi lime; *M. incognita* in pomegranate in Maharashtra, Karnataka, Gujarat and Andhra Pradesh; and *M. enterolobii* on guava in 10 states including Tamil Nadu; are few examples of current threats and serious concerns. However, PPNs are neglected pests of fruit crops and considered as of low priority for crop production and protection in India. The economic significance of nematodes in pomology is very often underestimated, and their damage potential is not recognized by the plant protection specialists, scientists and administrators. As an important component of integrated pest management, nematode pathogens cannot be ignored and they could be tackled with the intelligent planning of nematode suppressive crop sequences, summer plowing, organic manuring, clean cultivation, adjusting sowing time, water and irrigation management and sensible use of nematicides. The increasing concerns on the ill effects of chemical pesticides on the environment have driven recent research interests on the use of several alternative strategies like botanicals, biopesticides and cropping system research for management of nematodes. Some of the successes have been obtained for managing plant parasitic nematodes with neem-based formulations, fungal formulation of *Purpureocillium lilacinum*, *Pochonia chlamydosporia*, and *Trichoderma* spp. and bacterial formulations of *Pasteuria penetrans*, *Bacillus subtilis*, and *Pseudomonas fluorescens* strains. Wide adoption of bioagents in field scale is still a fantasy rather than reality because of their inconsistent efficacy in different agro-ecological situations. However, the enrichment of bioagents with organic amendments like FYM, compost, oil cakes, vermicompost, *etc.* has enabled the mass multiplication of bioagents and their successful field application. Integration of more than one option can be explored on the basis of their compatibility, economic viability and availability to the farmers. Developing holistic approaches for managing orchard problems including nematodes inducing disease complexes in concert with other pathogens like fungi, bacteria and viruses is the major challenge ahead for the greater benefit of the growers.

In looking to the future, some of the things that needs to be done from a realistic and practical stand point of view include:

- » Web-based state-wise distribution maps of important fruit crops nematode fauna in the country.
- » Impact of economically important plant parasitic nematode populations on fruit crop health.
- » Population dynamics of economically important nematode pests under different cropping systems and role of cropping systems in nematode management.
- » Identification and development of nematode resistant fruit cultivars.
- » Front line demonstrations at farmers' fields to demonstrate and validate the most effective, low cost, and eco-friendly nematode management technologies generated in the project.
- » Designing nematode management technologies for organic farming systems.
- » Pest risk analysis to address phytosanitary needs against plant parasitic nematodes under WTO regime of international trade in agricultural commodities.
- » Carry out researches on use of indigenous microorganisms for their antagonistic role against plant parasitic nematodes.
- » Nematode problems and their management in protected fruit crops like strawberry.
- » Development of a national nematode diagnostic facility based on morphological and molecular characterization of nematodes.
- » New nematode control measures which are economical, eco-friendly and sustainable are required due to the likely removal of effective nematicides from the market because of increasing concerns of their possible effects on human health and the environment.

25.10. CONCLUSION

Considering the magnitude of damage caused by the important plant parasitic nematodes on fruit crops, it can be concluded that they can be no longer be ignored and the greater emphasis should be on the research that constitutes mainly on the development of the resistant varieties and ecofriendly technologies as there is increasing concern on the nematicide residue problems. Good agricultural practices with the emphasis on the cultural methods, tolerant hosts and biological agents is now being practiced and popularized in wider perspective. The aspects that needs to be emphasized in future to attain the sustainable nematode management are extensive dissemination of the diagnostic and advisory services centers trough electronic media for the farmers, data base for region or the fruit crop specific nematode management practices, search for the novel and potent green molecules of biological origin for nematode control as a curative control, development of location specific crop rotation and intercropping for nematode management, to standardize the protocols for mass production, application methods at required doses and finally organizing the workshops to frontline agricultural workers, to sensitize about the nematodes and its damage.

Crop practices favor monocultures, which reduce soil biodiversity that supports proliferation of nematodes. Imbalances induced by these situations strengthened by large inputs, would solve economic questions only. Such strategies delay the effective control of nematodes. On the other hand, taking into account the whole nematode diversity, according to an ecological approach, makes it possible to consider more suitable strategies for sustainable management of these parasites in fruit crops, based on an auto-regulation of the global pathogenic effect of the nematode communities. It is consequently critical to focus control practices on agronomic methods because they involve a great impoverishment of the inputs generated by the agro-industrial research (pesticides, biopesticides, and resistant seeds), conferring an acute brittleness in terms of sustainability at the same time.

Integrated nematode management should hence flank the development of nematodes more than killing them. We must be convinced that both biological balances in soil and environmental factors are fundamental and non-avoiding, helps for plant-parasitic nematode management. Every strategy processed in order to eradicate plant parasites would hence invariably prove to be unsustainable.

There are several genera and species of nematodes that are of economic importance. Correct nematode diagnosis can help in developing management programs. The nematodes must be eliminated to minimize the damage by selecting the appropriate method. Commonly practiced methods include crop rotation, resistant and tolerant cultivars, cultural practices, bioagents, and chemicals. The ability to reduce yield losses caused by nematodes needs understanding pathogen biology and the application of appropriate control measures. Use of chemicals is impractical commercially and cultural methods fail to give complete control. Breeding for resistance and tolerance is the major strategy for long-term and environmentally sound nematode management. It is necessary to undertake research particularly on nematode races and pathotypes, and there is a great need for global collaborative research to control these important pathogens.

In conclusion, future studies should focus on environmentally benign approaches which are based on multidisciplinary strategies that can fill the gaps of single sided management methods. Such approaches will also reduce the chance of resistance as the complexity of different nematode management components would make resistance highly improbable. Whatever strategy is devised, future attempts should focus on important factors such as synergism between nematode antagonists, environmental conditions, sustainability, studying the effect of new treatments on non-target organisms, and association of individual plants with nematode antagonists of interest. In summing up, a sustainable management of plant-parasitic nematodes is feasible when two or more compatible tactics are applied concurrently while appraising environmental protection.

References

Abad, P., Gouzy, J., Aury, J.M. *et al.* 2008. Genome sequence of the metazoan plant-parasitic nematode *Meloidogyne incognita. Nature Biotechnology* 26(8): 909-915.

Abd-Elgawad, M.M.M. and Askary, T.H. 2015. Impact of phytonematodes on agriculture economy. In: T.H. Askary and P.R.P. Martinelli (eds.) *Biocontrol Agents of Phytonematodes*, pp 3–49. CAB International, Wallingford.

Abdul Karim, S., Yuen, P.M. and Norlia, Y. 2002. Challenges in the integrated management of the root-knot nematode on guava. *Proceeding Mesyuarat Teknikal Pusat HR Port Dickson*: 27.

Abo-Korah, M.S. 2017. Biological control of root-knot nematode, *Meloidogyne javanica* infecting ground cherry, using two nematophagous and mychorrhizal fungi. *Egyptian Journal of Biological Pest Control* 27(1): 111-115.

Agu, C.M. 2008. Root-gall nematode disease of pineapple as affected by seed material, amount and type of organic soil amendment. *Plant Sci. Res.* 1(2): 36-39.

Akinbade, S.A., Mojtahedi, H., Guerra, L. *et al.* 2014. First report of *Xiphinema rivesi* (Nematoda, Longidoridae) in Washington State. *Plant Disease* 98 (7): 1018. doi: 10.1094/ PDIS-09-13-0961-PDN.

Alam, M.M. 1993. Bioactivity against phytonematodes. In N.S. Randhawa and B.S. Parmar (eds.) *Neem Research and Development*, pp. 123-143, Society of Pesticide Science, India.

Alcañiz, E., Jorge Pinochet, J. and Fernández, C. 1996. Evaluation of Prunus rootstocks for root-lesion nematode resistance. *HortScience* 31(6): 1013–1016.

Almeida, A.M., Gomes, V.M. and Souza, R.M. 2011. Greenhouse and field assessment of rhizobacteria to control guava decline. *Bragantia* 70: 837–842.

Almeida, E.J., Santos, J.M. and Martins, A.B.G. 2009. Resistance of guava and araca to *Meloidogyne mayaguensis. Pesq. Agropec. Bras.* 44: 421-423.

Almeida, E.J., Santos, J.M. and Martins, A.B.G. 2010. Population fluctuation of *Meloidogyne enterolobii* in guava (*Psidium guajava*) orchard. *Nematol. Bras.* 34:164–168.

Al-Sayed, A. and Abdel-Hameed, S.H. 1991. Resistance and susceptibility of olives to *Meloidogyne incognita* and *Rotylenchulus reniformis. Ann. Agric. Sci. Moshtohor* 29: 1221-1226.

Andret-Link, P., Schmitt-Keichinger, C., Demangeat, G. *et al.* 2003. RNA2-encoded proteins for the exclusive transmission of Grapevine fanleaf virus by its nematode vector *Xiphinema index.* In: *Proceedings of 8th International Congress of Plant Pathology*, Christchurch, 237 pp.

Anita. and Chaubey, A.K. 2003. Influence of soil temperature and moisture on population dynamics of ectoparasitic nematodes infesting *Mangifera indica. Ann. Pl. Prot. Sci.* 11: 181–183.

Anitha, D., Vijaya, T., Reddy, N.V. *et al.* 2013. Microbial endophytes and their potential for improved bioremediation and biotransformation: a review. *Indo Am J Pharmaceutical Res* 3: 6408-6417.

Annapurna, K., Bojappa, K.M. and Bhargava, B.S. 1988. Leaf sampling guide for sapota (*Manilkara achras* M. Foseberg) cv. Cricket Ball. *Crop Res.* 1: 69-75.

Anon. 1989. *Consolidated Biennial Report (1987-89) of AICRP on Plant Parasitic Nematodes with Integrated Approach for Their Control.* Dept. of Nematology, Haryana Agri. Univ., Hisar, 35 pp.

Anon. 1990. *Final Report of Research Scheme on Nematode Problems in Some Selected Temperate Fruit Crops and Their Management.* Y.S. Parmar Univ. of Hort. & Forestry, Solan, 54 pp.

Anon. 1993. *Biennial Report (1991-93) of AICRP on Plant Parasitic Nematodes with Integrated Approach for Their Control.* Dept. of Nematology, Haryana Agri. Univ., Hisar, 80 pp.

Anon. 1993. *Management of Root-Knot Nematode, M. incognita Infecting Pomegranate.* Consolidated Report (1989-

93) for Third QRT of AICRP on Project on Plant Parasitic Nematodes with Integrated Approach for their Control, M.P.K.V. Rahuri, pp. 29-30.

Anon. 1999. *Ann. Rept. National Research Centre for Pomegrenate.* Solapur (M.S.) India, pp. 21-23.

Anon. 2009. *Nematological Research in Odisha - The Tiny Feet Towards Excellence.* All India Coordinated Research Project on Plant Parasitic Nematodes with Integrated Approach for their Control, Department of Nematology, College of Agriculture, Orissa University of Agriculture and Technology, Bhubaneswar.

Anon. 2011. *Meloidogyne enterolobii. EPPO Bull.* 41: 329-339.

Anon. 2006. Schemes for the production of healthy plants for planting. Pathogen tested olive trees and rootstocks. *EPPO Bull.* 36: 77-83.

Anon. 2012. *Three decades of Nematological Research in Odisha* (1977-2007). All India Coordinated Research Project on Plant Parasitic Nematodes with Integrated Approach for their Control, Department of Nematology, College of Agriculture, Orissa University of Agriculture and Technology, Bhubaneswar.

Anon. 2012. Biological control of root-knot nematodes on Pomegranate. In *Bulletin on Experiments on Nematicides: Neem as a Biopesticide.* Nico Orgo Manures, Dakor, Gujarat.

Anon. 2014-15. *Area and Production of Different Fruits in India.* National Horticulture Board, India, www.nhb.gov.in.

Anon. 2015. *Fruit Production in India.* Agricultural Department, Govt. of Maharashtra, Pune 2015 @www. Mahaagris. Com>accessed on 24 August, 2015.

Anon. 2017. *Horticultural Statistics at a Glance, 2017.* Horticulture Statistics Division, Department of Agriculture, Cooperation & Farmers Welfare, Ministry of Agriculture & Farmers Welfare, Government of India, New Delhi. pp. 114-119.

Anon. 2022. *Directorate of Horticulture Kashmir. Horticulture Production Department, Government of Jammu and Kashmir.* http://hortikashmir.gov.in/ Area%20 Production %20 data.html.

Ansari, R.A. and Khan, T.A. 2012. Parasitic association of root-knot nematode, *Meloidogyne incognita* on guava. *e-Journal of Science & Technology (e-JST)* 5: 65-67.

Anwar, S.A., Gorsi, S., Anwar ul Haq, M. *et al.* 1991. Plant parasitic nematodes of some field, vegetable, fruit and ornamental crops. *J. Agri Res. of Lahore* 29: 233.

APEDA Database 2011-12. APEDA Annual Report 2011-12. APEDA, New Delhi.

APEDA. 2021. *Agriexchange, The Changing Face of Agribusiness.* APEDA, New Delhi.

Ardakani, A.S. 2013. Toxicity of silver, titanium and silicon nanoparticles on the root-knot nematode, *Meloidogyne incognita*, and growth parameters of tomato. *Nematology* 15(6): 671-677.

Argelis, A. 1987. Present situation of grapevine virus diseases with reference to the problems which they cause in Greek vineyards. In R. Cavalloro (ed.) *Integrated Pest Control in Viticulture*, pp. 309-312. A.A. Balkema, Rotterdam, Netherlands, 395 pp.

Ashokkumar, N. and Poornima, K. 2019. Occurrence and distribution of root knot nematode, *Meloidogyne enterolobii* on guava (*Psidium guajava* L.) in Tamil Nadu. *J. Pharmacogn. Phytochem.* 8: 1922-1924.

Ashokkumar N., Poornima K., Kalaiarasan, P. and Kavino, M. 2019. Screening and histological characterization of guava (*Psidium guajava* L.) cultivars against root-knot nematode, *Meloidogyne enterolobii. Pest Mangmt. Hort. Ecosystems* 25: 84-92.

Ashok Kumar, P. 1988. *Studies on* Hemicriconemoides mangiferae *Siddiqi, 1961.* Master of Science Dissertation. Tamil Nadu Agricultural University, Coimbatore, Tamil Nadu, India. 129 pp.

Ashok Kumar, P. and Vadivelu, S. 1990. Description of developmental stages of *Hemicriconemoides mangiferae* Siddiqi, 1961 (Nemata: Criconematidae). *Revue de Nematologie* 13: 317–322.

Ashok Kumar, P., Vadivelu, S., Mehta, U.K. *et al.* 1991. Occurrence and seasonal variation of *Hemicriconemoides mangiferae* from tropical and sub–tropical orchards in peninsular India. *Nematropica* 21: 167–176.

Askary, T.H. 2015a. Nematophagous fungi as biocontrol agents of phytonematodes. In: T.H. Askary and P.R.P. Martinelli (eds.) *Biocontrol Agents of Phytonematodes*, pp. 81–125. CAB International, Wallingford.

Askary, T.H. 2015b. Limitations, research needs and future prospects in the biological control of phytonematodes. In: T.H. Askary and P.R.P. Martinelli (eds.) *Biocontrol Agents of Phytonematodes*, pp. 446–454. CAB International, Wallingford.

Askary, T.H. and Martinelli, P.R.P 2015. *Biocontrol Agents of Phytonematodes*. CAB International, Wallingford, 470 pp.

Avelar-Mejia, J.J., Teliz-Ortiz, D. and Zavaleta-Mejia, E. 2001. Patógenos asociados con el "declinamiento del guayabo". *Revista Mexicana de Fitopatologia* 19: 223-229.

Ayala, A., Acosta, N. and Adsuar, J.A. 1971. A preliminary report on the response of *Carica papaya* to foliar application of two systemic nematicides. *Nematropica* 1: 10.

Ayala, A., Roman, J. and Gonzalez, J. 1967. Pangola grass as a rotation crop for pineapple nematode control. *Journal of Agriculture Science, University of Puerto Rico* 51: 94-96.

Ayala-Zavala, J.F., Vega-Vega, V., Rosas-Domínguez, C. *et al.* 2011. Agro-industrial potential of exotic fruit by-products as a source of food additives. *Food Research International* 44(7): 1866-1874.

Aydinli, G. and Mennan, S. 2022. Prevalence of root-knot nematodes and their effects on fruit yield in kiwifruit orchards in Samsun Province (Türkiye). *Türk. Entomol. Derg.* 46(2): 187-197.

Babu, A.M., Kumar, V., Philip, T. *et al.* 1996. Scanning electron microscope studies on mulberry root parasitised by *Meloidogyne incognita*. *Redia* 34: 195- 201.

Badra, T. and Khattab, M.M. 1982. Chemically induced resistance to *Rotylenchulus reniformis* by ethephon growth regulant and relevant pathometabolites in mango seedlings. *Nemat. Medit.* 10: 49–56.

Bafokuzara, N.D. 1982. Nematodes associated with pineapples in Uganda. *Nematropica* 12(1): 45-49.

Baghel, P.P.S. 1995. Nematode problems in citrus. In G. Swarup, D.R. Dasgupta and J.S. Gill (eds.) *Nematode Pest Management-An Appraisal of Eco-friendly Approach*, Indian Agri. Res. Inst., New Delhi, pp. 203-210.

Baghel, P.P.S. and Bhatti, D.S. 1982. Vertical and horizontal distribution of phytonematodes associated with citrus. *Indian Journal of Nematology* 12: 339-344.

Baghel, P.P.S. and Bhatti, D.S. 1983a. Evaluation of pesticides for the control of phytonematodes on citrus. *Third Nematology Symposium*, Himachal Pradesh Agricultural University, Solan, pp. 38-39.

Baghel, P.P.S. and Bhatti, D.S. 1983b. Relative efficacy of nematicides for control of phytonematodes on grapevine varieties. *Third Nematology Symposium,* Himachal Pradesh Agricultural University, Solan, p. 39.

Baghel, P.P.S. and Gupta, D.C. 1986. Effect of intercropping on root-knot nematode (*Meloidogyne javanica*) infecting grapevine (var. Pearlette). *Indian Journal of Nematology* 16: 283-285.

Baghel, P.P.S. and Bhatti, D.S. 1990. Interaction of vesicular arbuscular mycorrhizal fungus *Glomus mosseae* and *Meloidogyne javanica* on peach. *Proc. of Seminar on Bioagents in Nematode Management*, Indian Agri. Res. Inst., New Delhi, p. 41.

Baghel, P.P.S., Bhatti, D.S. and Jalali, B.L. 1990. Interaction of VA mycorrhizal fungus and *Tylenchulus semipenetrans* on citrus. *Proc. of the National Conf. on Mycorrhiza,* Haryana Agri. Univ., Hisar, p. 118.

Bajaj, H.K., Kanwar, R.S., Baghel, P.P.S. and Bhatti, D.S. 1988. The Motia (*Jasminum sambac*) - a new host record of citrus nematode, *Tylenchulus semipenetrans* Cobb, 1913. *Journal of Research, Haryana Agricultural University* 18: 146-147.

Baldwin, J.G. and Mundo-Ocampo, M. 1991. Heteroderinae, cyst and non-cyst-forming nematodes. In: W.R. Nickle (ed.) *Manual of Agricultural Nematology*, pp. 275–362. Marcel Dekker, New York.

Banihashemian, S.N., Jamali, S., Golmohammadi, M. and Ghasemnezhad, M. 2022. Isolation and identification of endophytic bacteria associated with kiwifruit and their biocontrol potential against *Meloidogyne incognita*. *Egypt J Biol Pest Control* 32: 111. https://doi.org/10.1186/ s41938-022-00601-y.

Barbercheck, M.E. and Broembsen, S.L. von. 1986. Effects of soil solarization on plant-parasitic nematodes and *Phytophthora cinnamomi* in South Africa. *Plant Dis.* 70: 945-950.

Beckman, J.G., Chaparro, J.X. and Sherman, W.B. 2012 'MP-29', a clonal interspecific hybrid rootstock for peach. *Hortscience* 47(1):128–131.

Beckman, T.G. and Nyczepir, A.P. 2011. Peach tree short life. In D. Horton and D. Johnson, (eds.) *Southeastern Peach Growers' Handbook*. University of Georgia, Cooperative Extension Service Bulletin #1384, Athens, GA.

Bello, A., Arias, M., López-Pérez, J.A. *et al.* 2004. Biofumigation, fallow, and nematode management in vineyard replant. *Nematropica* 34(1): 53-64.

Bello-Perez, A. and Jimenez Millan, F. 1963. *Heterodera fici* Kirjanova, 1954 (Nematoda), nueva especie para la fauna espanola. *Boletin de la Real Sociedad Espanola de Historia Natural. Seccion Biológica* 65: 25–28.

Bertrand, P.F. and Nyczepir, A.P. 1989. Nematodes. In: S. C. Meyers (ed.) *Peach Production Handbook*, pp. 146-151. University of Georgia Cooperative Extension Service, Athens.

Bhagyrathy, N., Siddappaji, C., Shankar, M.A. *et al.* 2000. Interaction of root knot nematode *Meloidogyne incognita* and fungus *Rhizoctonia bataticola* on mulberry plants. In K.P. Chinnaswamy, R. Govindan, N.K. Krishna Prasada and D.N.R. Reddy (eds.) *Moriculture in Tropics*, pp.152-153 University of Agricultural Sciences, GKVK. Bangalore, India.

Bhambal, S.B. 1987. *Effect of Foliar Application of Micronutrients on Growth, Yield, Fruit Quality and Leaf Nutrient Status of Pomegranate* (Punica granatum *L.*) *cv. Ganesh*. M. Sc. (Ag.) thesis, Mahtma Phule Agri. Univ., Rahuri, Maharashtra.

Bhargava, B.S. and Chadha, K.L. 1993. Leaf nutrient guide for fruit crops. In K.L. Chadha and O.P. Pareek (eds.) *Advances in Horticulture*, Vol. 2, pp. 973-1029. Malhotra Publishing House, New Delhi.

Bhargava, B.S., Raturi, G.B. and Hiwale, S.S. 1990. Leaf sampling technique in ber for nutritional diagnosis. *Singapore J. Prim. Ind.* 18: 85.

Bhati, S.S. 2020. Nematodes: The unseen enemy of pomegranate in western Rajasthan. *Agriculture and Food: E-Newsl.* 2(6): 187-188.

Bhati, S.S. and Parashar, A. 2020. First report and morphological description of *Meloidogyne enterolobii* infecting guava (*Psidium guajava* L.) in transitional plain of Luni basin of Rajasthan. *Int J Curr Microbiol App Sci.* 9(9): 2267–2274.

Bhatia, H.S. and Kumar, J. 2009. Performance of new apple cultivars on different rootstocks under high density plantation. *Agric. Sci. Digest* 29 (4) : 303-305.

Bindra, O.S., Chhabra, H.K., Chadha, K.L. and Mehrotra, N.K. 1967. A study on the correlation of citrus nematode population with decline of citrus. *Journal of Research, Punjab Agricultural University* 4: 543-546.

Blake, C.D. 1966. The histological changes in banana roots caused by *Radopholus similis* and *Helicotylenchus multicinctus. Nematologica 12:* 129–132.

Blok, V.C., Ehwaeti, M., Forgetter, M. *et al.* 1997. Evolution of resistance and virulence in relation to the management of nematodes with different biology, origins and reproductive strategies. *Nematologica* 43: 1-13.

Blok, V.C., Wishart, J., Fargette, M. *et al.* 2002. Mitochondrial DNA differences distinguishing *Meloidogyne mayaguensis* from the major species of tropical root-knot nematodes. *Nematology* 4: 773-781.

Bodine, E.W. and Newton, J.H. 1942. The rasp leaf of cherry. *Phytopathology* 32: 333-335.

Bogantes-Arias, A. and Mora-Newcomer, E. 2010. Evaluation of four rootstocks for graft in guava (*Psidium guajava* L.). *Agronomia Mesoamericana* 21: 103-111.

Brancalion, A.M., Zanon, J.I. and Zem, A.C. 1981. Ocorrencia do nematoide *Heterodera fici* no Rio Grande do Sul. *Revista de Agricultura, Piracicaba* 56: 4.

Braun, A.L. and Lownsbery, B.F. 1975. The pin nematode, *Paratylenchus neoamblycephalus*, on myrobalan plum and other hosts. *Journal of Nematology* 7(4): 336-340.

Brenner, R.J., Focks, D.A., Arbogas R.T. *et al.* 1998. Practical use of spatial analysis in precision targeting for integrated pest management. *American Entomologist* 44: 79–101.

Brito, J.A., Kaur, R., Cetintas, R. *et al.* 2008. Identification and isozyme characterization of *Meloidogyne* spp. infecting horticultural and agronomic crops, and weed plants in Florida. *Nematology 10*(5): 757-766.

Brito, J.A., Kaur, R., Cetintas, J. D. *et al.* 2010. *Meloidogyne* spp. infecting ornamental plants in Florida. *Nematropica.* 40: 87–103.

Brito, J.A., Stanley, J.D., Mendes, M.L. *et al.* 2007. Host status of selected cultivars plants to *Meloidogyne mayaguensis* from Florida. *J. Nematol.* 36: 232-240.

Brittain, J.A. and Miller, R.W. 1978. *Managing Peach Tree Short Life in The Southeast.* Bulletin 585, Clemson University Extension Service, Clemson, SC.

Brown, D.J.F., Halbrendt, J.M., Robbins, R.T. and Vrain, T.C. 1993. Transmission of nepoviruses by *Xiphinema americanum*-group nematodes. *J. Nematol.* 25: 349-354.

Burton, G.W. and Johnson, A.W. 1987. Coastal Bermuda grass rotations for control of root-knot nematode. *J. Nematol.* 18: 138-140.

Buzo, T., McKenna, Kaku, S., Anwar, S.A. and McKenry, M.V. 2009. VX211, a vigorous new walnut hybrid clone with nematode tolerance and a useful resistance mechanism. *Journal of Nematology* 41: 211–216.

CABI. 2000. *Distribution Maps of Plant Diseases No. 804.* Meloidogyne mayaguensis. CABI, Wallingford (GB).

Callan, N.W., Mathre, D.E. and Miller, J.B. 1990. Bio-priming seed treatment for biological control of *Pythium ultimum* pre-emergence damping-off in *sh2* sweet corn. *Plant Dis* 74: 368-372.

Campos, A.R., Lordellow, L.G.E., Abreu, O.C. and Olivira, D.A. 1974. Incidence of *Meloidogyne* on mulberry varieties. In L.G.B. Dordello (ed.) *Trabalhos aprsentados d seuriao de nematologia*, pp. 14-18, Piracicaba, Brazil.

Campos, V.P. 1997. Nematodes of the fig crop. *Informe Agropecuario Belo Horizonte* 18188: 33–38.

Cao, T., McKenry, M.V., Duncan, R.A. *et al.* 2006. Influence of ring nematode infestation and calcium, nitrogen, and indoleacetic acid applications on peach susceptibility to *Pseudomonas syringae* pv. *syringae*. *Phytopathology* 96(6): 608-615.

Carneiro, R.M.D.G., Almeida, M.R.A., Braga, R.S. *et al.* 2006a. Primeiro registro de *Meloidogyne mayaguensis* parasitando plantas de tomate e pimentão resistentes à meloidoginose no estado de São Paulo. *Nemat. Bras.* 30: 81-86.

Carneiro R.M.D.G., Castro J.M.C., Carneiro R.G. *et al.* 2012. Major guava nematodes and control prospects using resistance on *Psidium* spp. and non-host crops. In C.A.F. Santos *et al.* (eds.) *Proc. 3rd Int. Symp. on Guava and Other Myrtaceae. Acta Hort.* 959: 41-49.

Carneiro, R.M.D.G., Cirotto, P.A., Silva, D.B. and Carneiro, R.G. 2007. Resistance to *Meloidogyne mayaguensis* in *Psidium* spp. accessions and their grafting compatibility with *P. guajava* cv. 'Paluma'. *Fitopatol. Bras.* 32: 281-284.

Carneiro, R.M.D.G., Mônaco, A.P.A., Moritz, M.P. *et al.* 2006. Identificação de *Meloidogyne mayaguensis* em goiabeira e em plantas invasoras, em solo argiloso no Estado do Paraná. *Nemat. Bras.* 30: 293-298.

Carneiro, R.M.D.G., Moreira, W., Almeida, M.R.A. and Gomes, A.C.M.M. 2001. Primeiro registro de *Meloidogyne mayaguensis* em goiabeira no Brasil. *Nemat. Bras.* 25: 223-228.

Casassa, A.M., Crozzoli, R., Matheus, J. *et al.* 1998. Effect of the root-knot nematode, *Meloidogyne incognita*, on the growth of guava (*Psidium* spp.) in nurseries. *Nematologia Mediterranea* 26: 237-242.

Casassa, A.M., Matheus, J., Crozzoli, R. *et al.* 1997. Response of some selections of guava to *Meloidogyne incognita*, in Mara County, Zulia State, Venezuela. *Fitopatologia Venezolana* 10: 5-8.

Castagnone-Sereno, P. 2012. *Meloidogyne enterolobii* (=*M. mayaguensis*): Profile of an emerging, highly pathogenic, root-knot nematode species. *Nematology* 14: 133-138.

Castellano G., Casassa-Padrón A.M., Ramírez-Méndez R. *et al.* 2012. Nemátodos fitoparásitos asociados a frutales estratégicos em el municipio Baralt del estado Zulia, Venezuela. *Fitopatología Venezolana.* 25: 2–6.

Castillo, P. and Vovlas, N. 2002. Factors affecting egg hatch of *Heterodera mediterranea* and differential responses of olive cultivars to infestation. *J. Nematol.* 34: 146-150.

Castillo, P. and Vovlas, N. 2007. Pratylenchus *(Nematoda: Pratylenchidae): Diagnosis, Biology, Pathogenicity and Management.* Brill, Leiden, Netherlands.

Castillo, P., Nico, A.I., Azcón Aguilar, C. *et al.* 2006. Protection of olive planting stocks against parasitism of root-knot nematodes by arbuscular mycorrhizal fungi. *Plant Pathol.* 55: 705-713.

Castillo, P., Vovlas, N., Nico, A.I. and Jiménez-Díaz, R.M. 1999. Infection of olive trees by *Heterodera mediterranea* in orchards in southern Spain. *Plant Dis.* 83: 710-713.

Castillo, P., Vovlas, N. and Troccoli, A. 2003. The reniform nematode, *Rotylenchulus macrosoma*, infecting olive in southern Spain. *Nematology* 5: 23-29.

Castro, A.P.G., Cares, J.E., Carvalho, D.D.C. *et al.* 2010. Resistência de genótipos comerciais e silvestres de *Passiflora* spp. a *Meloidogyne incognita* em condições de casa de vegetação. *Revista da Faculdade de Zootecnia, Veterinária e Agronomia* 17: 186-198.

Caveness, F.E. 1967. Shade house host ranges of some Nigerian nematodes. *Plant Disease Reporter* 51: 33–37.

Cetintas, R., Kaur, R., Brito, J.A. *et al.* 2007. Pathogenicity and reproductive potential of *Meloidogyne mayaguensis* and *M. floridensis* compared with three common *Meloidogyne* spp. *Nematropica* 37: 21–31.

Chadha, K.L. and Reddy, B.M.C. 1993. Pineapple nutrition. In K.L. Chadha and O.P. Pareek (eds.) *Advances in Horticulture,* Vol. 2, pp. 899-905. Malhotra Publishing House, New Delhi.

Chandra, R. and Jadhav, V.T. 2009. Pomegranate research and development in India and future thrusts. In: S. Dhole (ed.) *Preparation for The Sources of The 2nd Green Revolution in Indian Agriculture, Vol. II*, pp. 39-44. Magnum Foundation, Nagpur, Maharashtra.

Channabasappa, B.S., Krishnappa, K. and Reddy, B.M.R. 1995. Utilization of ecofriendly biological agents and biocomponents in the integrated management of *Radopholus similis* on banana. *Natl. Symp. on Nematode problems of India – An Appraisal of the Nematode Management with Ecofriendly Approaches and Biocomponents,* Indian Agri. Res. Inst., New Delhi.

Charles, J.S.K. 1989. *Bioecology of the Cyst Nematode Infecting Banana.* Ph. D. thesis, Kerala Agricultural University, Trichur, 122 pp.

Charles, J.S.K. and Venkitesan, T.S. 1984. New hosts of *Heterodera oryzicola* Rao & Jayaprakash, 1978 in Kerala, India. *Indian Journal of Nematology* 14: 181-182.

Charles, J.S.K. and Venkitesan, T.S. 1993. *Status Report on the Nematological Investigations on Banana in Kerala Agricultural University.* Kerala Agri. Univ., Vellanikkara, Kerala, 43 pp.

Charles, J.S.K., Venkitesan, T.S., Thomas, Y. and Varkey, P.A. 1985. Correlation of plant growth components to bunch weight in banana infested with burrowing nematode, *Radopholus similis* (Cobb). *Indian J. Nematol.* 15: 186-190.

Chawla, M.L., Samathanam, G.J. and Sharma, S.B. 1980. Occurrence of citrus nematode (*Tylenchulus semipenetrans* Cobb, 1913) on roots of mangosteen (*Garcinia mangostena*). *Indian Journal of Nematology* 10: 240-242.

Chawla, M.L. and Sharma, S.B. 1984. Horizontal and vertical distribution of citrus nematode, *Tylenchulus semipenetrans. Indian Journal of Nematology* 14: 193-195.

Chen, P. and Tsay, T.T. 2006. Effect of crop rotation on *Meloidogyne* spp. and *Pratylenchus* spp. populations in strawberry fields in Taiwan. *J Nematol* 38(3): 339–344.

Chen, Z.X. and Dickson, D.W. 1998. Review of *Pasteuria penetrans*: Biology, ecology, and biological control potential. *J. Nematol.* 30: 621-625.

Chhabra, H.K. 1972. Citrus nematode and its control. *Pesticides* 6(2): 88-89.

Chhabra, H.K. 1978. Distribution pattern of *Tylenchulus semipenetrans* and determination of suitable distance and depth for soil sampling. *Indian Journal of Nematology* 7: 66-68.

Chhabra, H.K. and Bindra, O.S. 1971. Screening of nematicides against the citrus nematode, *Tylenchulus semipenetrans* Cobb. *Indian Journal of Entomology* 33: 472-473.

Chhabra, H.K. and Bindra, O.S. 1974. Screening of citrus rootstocks against the citrus nematode, *Tylenchulus semipenetrans* Cobb, 1913 in the Punjab. *Indian Journal of Horticulture* 31: 194-195.

Chhabra, H.K., Bindra, O.S. and Singh, I. 1977. Further studies on the control of the citrus nematode, *Tylenchulus semipenetrans* Cobb, 1913 infecting *Citrus maxima* Merrill (Abstr.). *Proceedings of the International Symposium on Citriculture,* Bangalore, P. 38.

Chitambar, J.J., Westerdahl, B.B. and Subbotin, S.A. 2018. Plant parasitic nematodes in California Agriculture. In S. Subbotin and J. Chitambar (eds.) *Plant Parasitic Nematodes in Sustainable Agriculture of North America. Sustainability in Plant and Crop Protection.* Springer, Cham.

Chitwood, B.G. and Toung, M.C. 1960. Host parasite interactions of the Asiatic pyroid citrus nematode. *Plant Disease Reporter* 48: 848-854.

Chona, B.L., Sethi, C.L., Swarup, G. and Gill, J.S. 1965. Citrus nematode (*Tylenchulus semipenetrans* Cobb, 1913) associated with die-back disease of citrus. *Indian Journal of Nematology* 22: 370-373.

Ciancio, A., Fazio, P. and Lamberti, F. 1996. Nematodes occurring in the rhizosphere of walnut in Italy. *Nematol. Medit.* 24: 129-133.

CITH. 2010. Annual Report of Central Institute of Temperate Research for the year 2010, Srinagar.

CITH. 2013-14. Annual Report of Central Institute of Temperate Research for the year 2013-14, Srinagar.

CITH. 2014-15. Annual Report of Central Institute of Temperate Research for the year 2014-15, Srinagar.

Cobb, N.A. 1893. *Nematodes, Mostly Australian and Fijian.* Linn. Soc. New South Wales, Macleay Memorial Vol., 59 pp.

Cohen, M. F., Yamasaki, H. and Mazzola, M. 2005. *Brassica napus* seed meal soil amendment modifies microbial community structure, nitric oxide production and incidence of Rhizoctonia root rot. *Soil Biol. Biochem.* 37: 1215-1227.

Cohn, E. 1965. The development of the citrus nematode on some of its hosts. *Nematologica* 11: 593-600.

Cohn, E. 1968. *Nematodes as Possible Vectors of Virus Diseases of Citrus and other Fruit Crops.* Second Annual Report to USDA, Project No. A10-CR-72, Bet Dagan, Israel.

Collett, R.L., Marais, M., Daneel, M. *et al.* 2021. *Meloidogyne enterolobii*, a threat to crop production with particular reference to sub-Saharan Africa: an extensive, critical and updated review. *Nematol.* 23(3): 247–285.

Corbett, D.C.M. 1974. *Pratylenchus vulnus.* C.I.H. *Descriptions of Plant-parasitic Nematodes, Set 3, No. 37.* CAB International, Wallingford, UK.

Costa, D.C. and Matos, A.P. 2000. *Nematoses. Empresa Brasileira de Pesquisa Agropecuária.* Embrapa Mandioca e Fruticultura Tropical. Circular 2333. Cruz das Almas, BA, Brazil.

Costa, M.G.S., Correia, E, Dos Reis, L.L. and Wilcken, S.R.S. 2015. Reaction of fig tree to three species of root-knot nematodes. *Revista Brasileiera de Fruticultura* 37: 617-622.

Costa, S.R., Santos, C.A.F. and Castro, J.M.C. 2012. *Tolerance of* Psidium guajava × P. guineense *Hybrids to* Meloidogyne enterolobii. Série Documentos 247, Petrolina, PE, Brazil, 39 pp.

Criancio, A. and Grasso, G. 1998. Endomigratory feeding behaviour of *Mesocriconema xenoplax* parasitizing walnut (*Juglans regia* L.). *Fundam. Appl. Nematol.* 21(1): 63-68.

Cropley, R. 1964. Further studies on European rasp leaf and leafroll diseases of cherry trees. *Ann. Appl. Biol.* 53: 333-341.

Crossa Raynand, P. and Audergon, J. 1987. Apricot rootstocks. In R.C. Rom and R.F. Carlson (eds.) *Rootstocks for Fruit Crops*, pp. 295–320. Wiley, New York.

Cuadra, R. and Quincosa, A. 1982. Potential of different *Psidium gujava* species as sources for resistance of guava to *Meloidogyne. Ciencias-de-la-Agricultura* 13: 19-26.

D'Souza, G.I., Kumar, A.C., Viswanathan, P.R.K. and Shamanna, H.V. 1970. Relative distribution and prevalence of plant parasitic nematodes in the coffee tracts of South-Western India. *Indian Coffee* 34: 329-330, 342.

Daneel, M., Engelbrecht, E., Fourie, H. and Ahuja, P. 2018. The host status of Brassicaceae to *Meloidogyne* and their effects as cover and biofumigant crops on root-knot nematode populations associated with potato and tomato under South African field conditions. *Crop Protection* 110: 198–206.

Daneel, M.S. 2017. Nematode pests of minor tropical and subtropical crops. In: H. Fourie, V. Spaull, R. Jones,

M.S. Daneel and D. De Waele (eds.) *Nematology in South Africa: A View from The 21st Century*, pp. 373–393. Springer, Cham, Switzerland.

Daneel, M.S. and Steyn, W.P. 2022. Litchi and guava nematode challenges in South Africa: Can we change nematode communities and minimize the problems? In R.A. Sikora *et al.* (eds.) *Integrated Nematode Management: State-of-The-Art and Visions for The Future*, pp. 224-230. CAB International.

Darekar, K.S., Mhase, N.L. and Shelke, S.S. 1989. Management of nematodes infesting pomegranate. *International Nematology Network Newsletter* 6: 15-17.

Dariva, J.M. 2011. *Fusarioses do Maracujazeiro: Etiologia e Sintomatologia*. 2011. 71 f. Dissertação (Mestrado em Produção Vegetal: Área de Concentração em Fitossanidade) - Universidade Estadual de Montes Claros, Janaúba.

Datta, S.C. 2005. Possible use of amaranth as catch crop for root-knot nematodes intercropped with mulberry. *J. Environ. Sociobiol.* 2: 61-65.

Datta, S.C. and Datta, R. 2007. Intercropping amaranth with mulberry for managing root-knot nematodes and improving sericulture. *Sericol.* 47: 297-306.

Dawabah, A.A.M., Al-Yahya, F.A. and Lafi, H.A. 2019. Integrated management of plant parasitic nematodes on guava and fig trees under tropical field conditions. *Egyptian J.Biol. Pest Control* 29: 29-37.

De Leij, F.A.A.M. and Kerry, B.R. 1991. The nematophagous fungus *Verticillium chlamydosporium* as a potential biological control agent against *Meloidogyne arenaria. Revue de Nematologie* 14: 157-164.

Debrot, E.A. 1964. Studies on a strain of raspberry ringspot virus occurring in England. *Annals of Applied Biology* 54: 183-191.

Deshmukh, M.R., Karmakar, S.P. and Patil, S.G. 2004. Screening of grape varieties for resistance to root-knot nematode (*Meloidogyne incognita*). *Indian Journal of Nematology* 34: 224-226.

Devarajan, K. and Rajendran, G. 2001. Effect of fungus *Paecilomyces lilacinus* (Thom.) Samson on the burrowing nematode *Radopholus similis* (Cobb) Thorne in banana. *Pest Management in Horticultural Ecosystems* 7: 171-173.

Di Vito, M. 1976. *Heterodera fici* su *Ficus carica* in Basilicata. *Nematologia Mediterranea* 4: 107–108.

Di Vito, M. and Inserra, R.N. 1982. Effects of *Heterodera fici* on the growth of commercial fig seedlings in pots. *Journal of Nematology* 14: 416–418.

Di Vito, M., Vovlas, N. and Simeone, A.M. 1988. Effect of root-knot nematode *Meloidogyne incognita* on the growth of kiwi (*Actinidia deliciosa*) in pots. *Advances in Horticultural Science* 2: 109–112.

Dias, W.P., Freitas, V.M., Ribeiro, N.R. *et al.* 2010. Reação de genótipos de milho a *Meloidogyne mayaguensis* e *M. ethiopica*. *Nemat. Bras.* 34(2): 99-105.

Diksha, B. 2018. *Economic Analysis of Cherry Cultivation in Shimla District of Himachal Pradesh*. M Sc (Ag) Thesis, Dr Yashwant Singh Parmar University of Horticulture and Forestry, Solan, HP, India.

Dirlewanger, E., Salesses, G., Bonnet, A. *et al.* 2002. Breeding for Prunus rootstocks cumulating resistance to root-knot nematodes and favorable traits under Meditarrenean environments: A European project. *Acta Horticulturae* 592: 61-67.

Domínguez, P., Miranda, L., Medina, J.J. *et al.* 2016. Evaluation of non-fumigant alternative soil treatments for strawberry production in Huelva (Spain). *Int. J. Fruit Sci.* 16: 28–36.

DuCharme, E.P. 1959. Morphogenesis and histopathology of lesions induced on citrus roots by *Radopholus similis*. *Phytopathology* 49: 388-395.

DuCharme, E.P. 1968. Burrowing nematode decline of citrus, a review. In G.C. Smart and V.G. Perry (eds.) *Tropical Nematology*, pp. 20-37. Univ. of Florida Press, Gainesville, USA.

DuCharme, E.P. and Suit, R.F. 1953. Nematodes associated with avocado roots in citrus spreading decline areas. *Plant Disease Reporter* 37: 427–428.

DuCharme, E.P. and Suit, R.F. 1956. Immunity of the 'Lychee' from the burrowing nematode. *Proceedings of Florida State Horticultural Society* 68: 270-272.

Duncan, L.W. and Cohn, E. 1986. Nematode parasites of citrus. In M. Luc, R.A. Sikora and J. Bridge (eds.) *Plant Parasitic Nematodes in Subtropical and Tropical Agriculture*, pp. 321-346. CAB International, Wallingford, UK.

Dybas, R.A. and Green, A.S.J. 1984. Avermectins: Their chemistry and pesticidal activity. *British Crop Protection Conference, Pest and Diseases 9B* 3: 947-954.

Edward, J.C., Misra, S.L. and Singh, G.R. 1965. Qualitative and quantitative studies of plant parasitic nematodes associated with the rhizosphere of some citrus plants. *Punjab Horticultural Journal* 5: 29-36.

Edward, J.C. and Rai, B.B. 1970. Plant parasitic nematodes associated with hill oranges (*Citrus reticulata* Blanco) in Sikkim. *Allahabad Farmer* 44: 251-254.

Edwards, M. 1991. Control of plant parasitic nematodes in sultana grapevines (*Vitis vinifera*) using systemic nematicides. *Australian J. Expt. Agri.* 31: 579-584.

Egunjobi, O.A and Larinde, M.A 1975. Nematodes and maize growth in Nigeria. II. Effects of soil amendments on populations of *Pratylenchus brachyurus* and on the growth and production of maize (*Zea mays*) in Nigeria. *Nematologia mediterranea* 3: 65-73.

Eissa, M.F.M. and Abd-Elgawad, M.M.M. 2015. Nematophagous bacteria as biocontrol agents of phytonematodes. In: T.H. Askary and P.R.P. Martinelli (eds.) *Biocontrol Agents of Phytonematodes*, pp. 217–243. CAB International, Wallingford.

El Borai, F.E. and Ducan, L.W. 2005. Nematodes parasites of subtropical and tropical fruits tree crops. In: M. Luc, R.A. Sikora and J. Bridge (eds.) *Plant Parasitic Nematodes in Subtropical and Tropical Agriculture*, pp. 467-492. CAB International, Wallingford.

El-Habashy, D.E.M.F. 2010. *Pathological Studies on Parasitic Nematodes Affecting Strawberry*. M. Sc. Thesis, Fac Agric, Damanhour, Alexandria Univ, Egypt.

Elling, A.A. 2013. Major emerging problems with minor *Meloidogyne* species. *Phytopathology* 3(11): 1092–1102.

El-Moor, R.D., Peixoto, J.R., Ramos, M.L.G. and Mattos, J.K.A. 2009. Reaction of genotypes of passion fruit plant sour to the root-knot nematodes (*Meloiodgyne incognita* and *Meloidogyne javanica*). *Bioscience Journal* 25: 53-59.

El-Tanany, M.M., El-Shahaat, M.S. and Khalil, M.S. 2018. Efficacy of three bio-pesticides and oxamyl against citrus nematode (*Tylenchulus semipenetrans*) and on productivity of Washington navel orange trees. *Egyptian J Horticulture* 45: 275-287.

Endo, B.Y. 1971. Nematode induced syncitia (giant cells). Host parasite relationships of Heteroderidae. In B.M. Zuckerman, W.F. Mai and R.A. Rhode (eds.) *Plant Parasitic Nematodes, Vol. II*. Academic Press, New York.

Eno, C.F., Blue, W.G. and Good, J.M. Jr. 1955. The effect of anhydrous ammonia on nematodes, fungi, bacteria, and nitrification in some Florida soils. *Soil Sci. Soc. Am. Proc.* 19: 55- 58.

EPPO. 2017. *EPPO Global Database* (available online). https://gd.eppo.int/taxon/ MELGMA [accessed on 05 September 2018].

EPPO/CABI. 1996. Tomato black ring nepovirus. In: I.M. Smith, D.G. McNamara, P.R. Scott and M. Holderness (eds.) *Quarantine Pests for Europe*. 2nd edition. CAB International, Wallingford, UK.

EPPO/CABI. 1996. *Xiphinema americanum sensu lato*. In: I.M. Smith, D.G. McNamara, P.R. Scott and M. Holderness (eds.) *Quarantine Pests for Europe*. 2nd edition. CAB International, Wallingford, UK.

EPPO/OEPP. 1993. Certification schemes. No. 7. Nursery requirements-recommended requirements for establishments participating in certification of fruit or ornamental crops. *OEPP/EPPO Bull.* 23: 249-252.

Ertian, H. 2003. *Protection of Mulberry Plants* (Translate from Chinese). Oxford and IBH Publishing Co. Pvt. Ltd., New Delhi, India.

Esmenjaud, D. 1986. Les nématodes de la vigne. *Phytoma* 374: 24-27.

Esmenjaud, D., Minot, J.C., Voisin, R. *et al.* 1996. Inheritance of resistance to the root-knot nematode *Meloidogyne arenaria* in Myrobalan plum. *Theoretical and Applied Genetics* 92: 873-879.

Esmenjaud, D., Minot, J.C., Voisin, R. *et al.* 1997. Differential response to root-knot nematodes in Prunus species and correlative genetic implications. *J. Nematology* 29: 370-380.

Esmenjaud, D., Voisin, R., Van Ghelder, C. *et al.* 2009. Genetic dissection of resistance to root-knot nematodes *Meloidogyne* spp. in plum, peach, almond, and apricot from various segregating interspecific *Prunus* progenies. *Tree Genet. Genomes* 5: 279–289.

Esmenjaud, D., Walter, B., Minot, J. C. *et al.* 1993. Biotin-avidin ELISA detection of grapevine fanleaf virus in the vector *Xiphinema index*. *J. Nematol.* 25: 401-405.

Esnard, J., Marben, M.N. and Zuckerman, B.M. 1998. Effects of three microbial broth cultures and an organic amendment on growth and population of free living and plant parasitic nematodes on banana. *European J. Pl. Path.* 104: 457-463.

Estupinan-Lopez, L., Campos, V.P., Silva, A.P. *et al.* 2017. Volatile organic compounds from cottonseed meal are toxic to *Meloidogyne incognita*. *Tropical Plant Pathology* 42: 443–450.

Evans, K. and Rowe, J.A. 1998. Distribution and economic importance. In: S.B. Sharma (ed.) *The Cyst Nematodes*, pp. 1–30. Kluwer Academic Publishers, Dordrecht.

FAOSTAT. 2019. Available online: http://www.fao.org/faostat/en/#data/QCL (accessed on 30 March 2021)

Fargette, M., Berthier, K., Richaud, M. 2010. Crosses prior to parthenogenesis explain the current genetic diversity of tropical plant-parasitic *Meloidogyne* species (Nematoda: Tylenchida). *Infection, Genetics and Evolution* 10: 807–814.

Faske, T.R. and Starr, J.L. 2007. Cotton root protection from plant parasitic nematodes by abamectin treated seed. *J Nematol* 39(1): 27–30.

Fernández, C., Pinochet, J., Esmenjaud, D. *et al.* 1994. Resistance among new Prunus rootstocks and selections to root-knot nematodes in Spain and France. *HortScience* 29: 1064- 1067.

Ferreira, T.D.F., Souza, R.M., Idalino, W.S.S. *et al.* 2014. Interaction of *Pratylenchus brachyurus* and *Helicotylenchus* sp. with mealybug wilt of pineapple in microplots. *Nematropica* 44: 181-189.

Ferris, H. 2017. *Xiphinema americanum*, 11/29/2016 2016 [cited February 26 2017].

Ferris, H., Venette, R.C. and Scow, K.M. 2004. Soil management to enhance bacterivore and fungivore nematode populations and their nitrogen mineralization function. *Applied Soil Ecology* 25: 19-35.

Fischer, I.H., Bueno, C.J., Garcia, M.J.M. and Almeida, A.M. 2010. Reaction of passion fruit plant yellow to the complex of fusariose-nematode of branch knot. *Acta Scientiarum Agronomy* 2: 223- 227.

Fischer, I.H. and Rezende Jorge, A.M. 2008. Diseases of passion flower (*Passiflora* sp.). In *Pest Technology*. Global Science Books.

Ford, H.W. 1967. Burrowing nematode resistant rootstocks as biological barriers in citrus groves. *Proceedings of Florida State Horticultural Society* 80: 56-59.

Ford, H.W. and Feder, W.A. 1964. *Three Citrus Rootstocks Recommended for the Trial in Spreading Decline Areas*. Univ. of Florida Agri Expt. Stn. Circular S-151, 8pp.

Forge, T.A. and Kempler, C.2009. Organic mulches influence population densities of root-lesion nematodes, soil health indicators, and root growth of red raspberry. *Canadian Journal of Plant Pathology* 31(2): 241-249.

Freitas V.M., Silva J.G.P., Gomes C.B. *et al.* 2017. Host status of selected cultivated fruit crops to *Meloidogyne enterolobii*. *Eur J Plant Pathol.* 148(2):307–319.

Galkovskyi, T., Mileyko, Y., Bucksch, A. *et al.* 2012. GiA Roots: software for the high throughput analysis of plant root system architecture. *BMC Plant Biol.* 12: 116.

Ganesh Babu, M.M. and Gunasekaran, P. 2009. Production and structural characterization of crystalline silver nanoparticles from *Bacillus cereus* isolate. *Colloids Surf B Biointerfaces* 74: 191-195.

Ganeshan, K., Poornima, K., Renukadevi, P. and Nakkeeran, S. 2019a. Characterisation of antinemic and antimicrobial compounds of *Bacillus amyloliquefaciens* (vb7), against nematode-disease complex causing guava decline in Tamil Nadu. *Int. J. Agric. Sci.* 11: 9191–9196.

Ganeshan, K., Poornima, K., Renukadevi, P. and Nakkeeran, S. 2019b. Synergistic effect of root-knot nematode,

Meloidogyne enterolobii and *Fusarium oxysporum* in causing guava decline in Tamil Nadu. *Pest Mgmt. Hort. Ecosys.* 25: 252–256.

Ganga, G. and Sulochana, C.J. 1997. *An Introduction to Sericulture.* 2nd Edition. Oxford & IBH Publishing Co. Pvt. Ltd. New Delhi. India.

Gao, J., Wang, G., Ma, S. *et al.* 2015. CRISPR/Cas9-mediated targeted mutagenesis in *Nicotiana tabacum. Plant Mol. Biol.* 87: 99-110.

Garabedian, S. and Van Gundy, S.D. 1983. Use of avermectins for the control of *Meloidogyne incognita* on tomatoes. *Journal of Nematology* 15: 503-510.

Gatoo, M.A., Naseem, S., Arfat, M.Y. *et al.* 2014. Physicochemical properties of nanomaterials: implication in associated toxic manifestations. *BioMed Res Int*: 1-8.

Ghini, R., Hamada, E. and Bettiol, W. 2008. Climate change and plant diseases. *Scientia Agricola* 65: 98–107.

Ghorab, A.I., Riad, F.W. and Shohla, G.S. 1987. Seasonal fluctuation of plant parasitic nematodes associated with mango. *Agri. Res. Rev.* 65: 315–323.

Ghosh, B., Bose, T.K. and Mitra, S.K. 1986. Chemical induction of flowering and control of fruit drop in litchi (*Litchi chinensis* Sonn.). *Proc. 22nd Int. Hort. Cong.*, California, Abstr. 1189.

Gnanaprakash, S., Madhumitha, B., Jayapradha, C. *et al.* 2016. Identification of resistance in mulberry, *Morus* spp. for root knot nematode, *Meloidogyne incognita. International Journal of Plant Sciences* 11(2): 262-264.

Gomes, M.G., Souza, R.M., Dias, V.M. *et al.* 2011. Guava decline: the complex disease involving *Meloidogyne mayaguensis* and *Fusarium solani. Journal of Phytopathology* 159: 45-50.

Gomes, V.M., Souza, R.M., Corrêa, F.M. and Dolinski, C. 2010. Management of *Meloidogyne mayaguensis* in commercial guava orchards with chemical fertilization and organic amendments. *Nematologia Brasileira* 34: 23-30.

Gomes, V.M., Souza, R.M., Midorikawa, G. *et al.* 2012. Guava decline: Evidence of nationwide incidence in Brazil. *Nematropica* 42: 153-162.

Gomes, V.M., Souza, R.M., Mussi-Dias, V. *et al.* 2011. Guava decline: A complex disease involving *Meloidogyne mayaguensis* and *Fusarium solani. J. Phytopathol.* 159: 45-50.

Gomes, V.M., Souza, R.M., Silva, M.M. and Dolinski, C. 2008. Caracterizac ̧a˜o do estado nutricional de goiabeiras em decłı´nio parasitadas por *Meloidogyne mayaguensis. Nematol. Bras.* 32: 154–160.

Gonsalves, D. 1988. Tomato ringspot virus decline. In R.C. Pearson and A.C. Goheen (eds.) *Compendium of Grape Diseases*, pp. 49-50. American Phytopathological Society Press, St Paul, Minnesota, 93 pp.

Gonzalez, H.R. 1993. Effects of an organic nematicide and the control of parasitic nematodes of kiwi crop. *Revista Fruticola* 14: 67–71.

Goswami, B.K. and Singh, S. 2002. Effect of *Aspergillus niger* and *Cladosporium oxysporum* on plant vigour and root-knot nematode, *Meloidogyne incognita* multiplication on egg plant. *Indian J. Nematol.* 32: 94-96.

Govindaiah. 1990. *Studies on The Root-Knot Nematode* Meloidogyne incognita *Infesting Mulberry.* Ph. D. Thesis, University of Mysore, Mysore, India

Govindaiah., Dandin, S.B., Giridhar, K. and Datta, R.K. 1994. Efficacy of different doses of neem oil cake on *Meloidogyne incognita* infesting mulberry. *Sericologia* 34: 717-721.

Govindaiah., Dandin, S.B. and Madhavarao, Y.R. 1989. Host range of *Meloidogyne incognita* causing root knot disease in mulberry (*Morus alba* L.). *Indian J. Serie.* 28: 121-126.

Govindaiah., Dandin, S.B., Philip, T. and Datta, R.K. 1991b. Effect of marigold *(Tagetes parula)* intercropping against *Meloidogyne ingognita* infecting mulberry. *Indian J. Nematol.* 21: 96-99.

Govindaiah., Dandin, S.B. and Sharma, D.D. 1991. Pathogenicity and avoidable leaf yield loss due to *Meloidogyne incognita* in mulberry (*Marus alba* L.). *Indian J. Nematol.* 21: 52-57.

Govindaiah., Dandin, S.B., Sharma D.D. and Datta, R.K. 1993b. Efficacy of different doses of Carbofuran on *Meloidogyne incognita* infesting mulberry. *Indian J. Serie.* 32: 99-101.

Govindaiah., Sharma, D.D., Bajpai, A.K. and Datta, R.K. 1993a. Identification of races of *Meloidogyne incognita* in mulberry. *Indian J. Serie.* 32, 91-93.

Govindaiah., Sharma, D.D., Gunasekhar, V. and Datta, R.K. 1996. Screening of mulberry genotypes against root knot nematode disease *(Meloidogyne incognita). Indian J. Nematol.* 26: 108-111.

Govindaiah., Sharma, D.D., Hemantharaj, R.K. *et al.* 1997. Nematicidal efficacy of organic manures, inter crops, mulches and nematicide against root knot nematode in mulberry. *Indian J. Nematol.* 27: 28-35.

Gowda, D.A., Swamy, B.C.N. and Setty, K.G.H. 1982. Nematophagous fungi from cultivated soils in Karnataka. *Journal of Soil Biology and Ecology* 2: 95-96.

Gowen, S.R. and Queneherve, P. 1990. Nematode parasites of bananas, plantains and abaca. In: *Plant Parasitic Nematodes of Subtropical and Tropical Agriculture*, pp. 431-460. CAB International, Walligford, UK.

Gowen, S. and Quénéhervé, P. 2005. Nematode parasites of bananas and plantains. In: M. Luc, R.A. Sikora and J. Bridge (eds.) *Plant Parasitic Nematodes in Subtropical and Tropical Agriculture*, pp. 61-643. CAB International, Wallingford.

Grandison, G.S. 1983. Root-knot nematode control in kiwi fruit, *Actinidia chinensis* by chemical bare root dip. *Plant Disease* 67(8): 899-900.

Griesbach, J.A. and Maggenti, A.R. 1989. Vector capability of *Xiphinema americanum sensu lato* in California. *Rev. de Nematol.* 13: 93-103.

Guérout, R. 1975. Nematodes of pineapple: a review. *Pest Articles & News Summaries* 21(2): 123-140.

Gupta, J.C. and Atwal, A.S. 1971. Biology and ecology of *Hoplolaimus indicus* (Hoplolaiminae: Nematoda). II. The influence of various environmental factors and host plants on the reproductive potential. *Nematologica 17*(2): 277-284.

Gupta, J.C. and Atwal, A.S. 1972. Biology and ecology of *Hoplolaimus indicus* (Hoplolaiminae: Nematoda). 3. Influence of population density on tomato and citrus plants. *Proceedings of XIth International Symposium on Nematology,* Reading, U. K., p. 26.

Gupta, N.K. and Gupta, J.C. 1966. On a nematode species of the genus *Hoplolaimus* Daday, 1905 from a citrus plant in the Punjab. *Research Bull., Punjab University of Science* 17: 211-213.

Gur, A. 1955. *The Fig.* Departments of Horticulture and Agricultural Education, Ministry of Agriculture, Israel, Bulletin No. 57 (in Hebrew).

Halbrendt, J.M. and Brown, D.J.F. 1993. Aspects of biology and development of *Xiphinema americanum* and related species. *J. Nematol.* 25: 355-360.

Halbrendt, J.M., Podleckis, E.V., Hadidi, A. *et al.* 1994. A rapid protocol for evaluating Prunus germplasm for tomato ringspot virus resistance. *HortScience* 29: 1068- 1070.

Hammam, M.M.A., El-Nagdi, W.M.A., Abd-El-Khair, H. and Abd-Elgawad, M.M.M. 2019. Biological management of the root-knot nematode on strawberry in Egypt. *Egypt J Agron* 18(1): In Press.

Hansen, A.J., Nylland, G., McElroy, F.D. and Stace-Smith, R. 1974. Origin, cause, host range and spread of cherry rasp leaf disease in North America. *Phytopathology* 64: 721-727.

Hanumashetti, S.I., Rao, M.M. and Bankapur, V.M. 1981. Preliminary studies on the effect of growth regulators and chemicals on the improvement of colouration in Gulabi grapes. *Curr. Res.* 10: 45-46.

Haque, M.S. 1976. *Studies on The Parasitic Nematodes of West Bengal.* Ph. D. Thesis, Burdwan University, West Bengal, India.

Hardin, M.R., Benrey, B., Coll, M. *et al.* 1995. Arthropod pest resurgence: an overview of potential mechanisms. *Crop Protection* 14: 3–18.

Hardman, R.A. 2006. Toxicologic review of quantum dots: Toxicity depends on physicochemical and environmental factors. *Environ. Health Perspect.* 114: 165–172.

Harrison, B.D. 1964. Specific nematode vectors for serologically distinctive forms of raspberry ringspot and tomato black ring viruses. *Virology* 22: 554-550.

Harrison, B.D. and Murant, A.F. 1996. Nepoviruses: ecology and control. In: B.D. Harrison, A.F. Murant (eds.) *The Plant Viruses. Polyhedral Virions and Bipartite Genomes*, pp. 211-228. Plenum Press, New York, USA.

Haseeb, A., Alam, M.M. and Khan, A.M. 1986. Suitability of fruit trees to certain plant parasitic nematodes. *Indian Journal of Nematology* 16(1): 124-126.

Haseeb, A. and Khan, M.R. 2012. Prospects of Trichoderma in nematode management. In: M. Nagesh, B.S.B. Rajkumar and N.K. Krishnakumar (eds.) *Status and Prospects for Enhancing the Uptake of Antagonistic Organisms for Nematode Management in India*. National Bureau of Agriculturally Important Insects (ICAR), Bangalore.

Heald, C.M. and Inserra, R.N. 1988. Effect of temperature on infection and survival of *Rotylenchulus reniformis*. *J. Nematology* 20(3): 356-361.

Hendrix, F.F., Powell, W.M., Owen, J.H. and Campbell, W.A. 1965. Pathogens associated diseased peach roots (Abstr.). *Phytopathology* 55: 1061.

Hewitt, W.B., Raski, D.J. and Goheen, A.C. 1958. Nematode vector of soil-borne fanleaf virus of grapevines. *Phytopathology* 48: 586-595.

Hirata, K., Suzuki, K., Hisai, J. *et al.* 2002. Morphology and morphometrics of juvenile stages of *Xiphinema index* Thorne and Allen (Dorylaimida: Longidoridae) intercepted in import plant quarantine in Japan. *Research Bulletin of the Plant Protection Service Japan* 38: 21–26.

Hoitink. H.A.J. and Boehm, M.J. 199. Biocontrol within the context of soil microbial communities: a substrate-dependent phenomenon. *Annu Rev Phytopathol.* 37: 427–446.

Horton, D., Brannen, P., Bellinger, B. and Ritchie, D. 2013. 2013 *Southeastern Peach, Nectarine, and Plum Pest Management and Culture Guide*. University of Georgia, Cooperative Extension Service Bulletin #1171, Athens, GA.

Hosomi, A., Dan, M. and Kato, A. 2002. Screening of fig varieties for rootstocks resistant to soil sickness. *Journal of the Japanese Society for Horticultural Science* 71: 171–176.

Howland, A.D., Schreiner, R.P. and Zasada, I.A. 2014. Spatial distribution of plant-parasitic nematodes in semi-arid *Vitis vinifera* vineyards in Washington. *Journal of Nematology* 46 (4): 321-330.

Hoy, J.W. and Mircetich, S.M. 1984. Prune brown line disease: susceptibility of prune rootstocks and tomato ringspot virus detection. *Phytopathology* 74: 272-276.

Huang, C.S. and Maggenti, A.R. 1969. Wall modifications in developing giant cells of *Vicia faba* and *Cucumis sativus* induced by root knot nematode, *Meloidogyne javanica*. *Phytopathology* 59: 931-937.

HuDie, S., WanRong, L., CuiMing, T. *et al.* 2019. Evaluation of mulberry varieties resistance to root-knot nematode disease and screening of control agents. *Journal of Zhongkai University of Agriculture and Engineering* 32(1): 9-12.

Humphreys, D.A., Williamson, V.M., Salazar, L. *et al.* 2012. Presence of *Meloidogyne enterolobii* Yang & Eisenback (=*M. mayaguensis*) in guava and acerola from Costa Rica. *Nematology* 14: 199-207

Hussey, R.S. and Williamson, V.M. 1997. Physiological and molecular aspects of nematode parasitism. In: K.R. Barker, G.A. Pederson and G.L. Windham. (eds.) *Plant and Nematode Interactions*, pp. 87-108. Amer. Soc. of Agron. Madison, W1.

Inomoto, M.M. and Monteiro, A.R. 1989. Thermal treatment of suckers of 'Giant Cavendish' banana for the eradication of plant-parasitic nematodes. *Nematol. Bras.* 13: 139-150.

Inserra, R.N. and O'Bannon, J.H. 1974. Systemic activity of phenamiphos for control of *Meloidogyne arenaria* on *Gardenia jasminoides* and *Ficus carica*. *Plant Disease Reporter* 58: 1075–1077.

Inserra, R.N., Stanley, J.D., Ochoa, A. *et al.* 2014. *Hemicriconemoides Species as Crop Damaging Parasitic Nematodes in Florida*. Nematology Circular No. 223, Florida Department of Agriculture and Consumer Services., Division of Plant Industry.

Inserra, R.N., Vovlas, N. and O'Bannon, J. 1980. A classification of *Tylenchulus semipenetrans* biotypes. *Journal of Nematology* 12: 283-287.

Inserra, R.N., Vovlas, N. and Morgan Golden, A. 1979. *Helicotylenchus oleae* n. sp. and *H. neopaxilli* n. sp. (Hoplolaimidae), two new spiral nematodes parasitic on olive trees in Italy. *J. Nematol.* 11: 56-62.

IOOC. 2000. *World Catalogue of Olive Varieties*. International Olive Oil Council, Madrid.

Isutsa, D.K. and Merwin, IA. 2000. Malus germplasm varies in resistance or tolerance to apple replant disease in a mixture of New York orchard soils. *Hortscience* 35: 262–268.

Itoh, Y., Ohshima, Y. and Ichinohe, M. 1969. Apple root-knot nematode, *Meloidogyne mali* n. sp. on apple-tree from Japan (Tylenchida: Heteroderidae). *Appl. Entomol. Zool.* 4: 194-202.

Jack Pinkerton, J.N. and Finn, C.E. 2005. Responses of strawberry species and cultivars to the root-lesion and northern root-knot nematodes. *American Soc. Hort. Hort. Sci.* 40(1): 33-38.

Jadhav, V.T. and Sharma, K.K. 2009. Integrated management of diseases in pomegranate (Lead Paper). *Int. Symp. on Pomegranate and other Minor Fruits including Mediterranean Fruits.* Univ. of Agri. Sci., Dharwad, Karnataka, India, pp. 48-52.

Jaffee, B.A., Gaspard, J.T., Ferris, H. and Muldoon, A. 1988. Quantification of parasitism of the soil-borne nematode *Criconemella xenoplax* by the nematophagous fungus *Hirsutella rhossiliensis*. *Soil Biol. Biochem.* 20: 631-636.

Jagdev, G.H., Hole, U.B., Kulkarni, S.R. and Mhase, N.L. 2020a. Assessment of avoidable yield losses due to root-knot nematode, *Meloidogyne incognita* infesting fig under field conditions. *The Pharma Innovation Journal* 9(2): 13-16.

Jagdev, G.H., Hole, U.B., Kulkarni, S.R. and Mhase, N.L. 2020b. Management of root-knot nematode, *Meloidogyne incognita* infesting fig under field conditions. *International Journal of Chemical Studies* 8(1): 1592-1597.

Jasy, T. and Koshy, P.K. 1992. Pathogenicity of the burrowing nematode, *Radopholus similis*, on avocado. *Indian Journal of Nematology* 22, 122–124.

Jatala, P. 1986. Biological control of nematodes. *Annual Review of Phytopathology* 24: 453-489.

Jiménez, R.M., Gallo, D.P. and Sotomayor, L.E. 2002. Experimental control of *Tylenchulus semipenetrans* Cobb. affecting olive-tree crop. *IDESIA* 20: 91-102.

Jindapunnapat, K., Chinnasri, B. and Kwankuae, S. 2013. Biological control of root-knot nematodes (*Meloidogyne enterolobii*) in guava by the fungus *Trichoderma harzianum*. *J. Dev. Sustain. Agr.* 8:110–118.

Jonathan, E.I. 1994. *Studies on the Root-knot Nematodes (*Meloidogyne incognita*) (Kofoid and White, 1919) Chitwood, 1949, in Banana (*Musa *sp.) cv. Poovan (AAB)*. Ph. D. Thesis, Tamil Nadu Agri. Univ., Coimbatore, Tamil Nadu, India.

Jonathan, E.I. and Cannayane, I. 2002. Field application of biocontrol agents for management of spiral nematode, *Helicotylenchus multicinctus* in banana (Abstr.). *Global Conf. on Banana and Plantain,* Bangalore, p. 193.

Jonathan, E.I. and Rajendran, G. 2000. Assessment of avoidable yield loss in banana due to root-knot nematode, *Meloidogyne incognita. Indian J. Nematol.* 30: 162-164.

Jonathan, E.I. and Umamaheswari, R. 2006. Biomanagement of nematodes infesting banana by bacterial endophytes. *Indian J. Nematol.* 36: 234-240.

Jones, A.T., Mayo, M.A. and Henderson, S.J. 1985. Biological and biochemical properties of an isolate of cherry rasp leaf virus from red raspberry. *Annals of Applied Biology* 106: 101-110.

Jones, J.T., Haegeman, A., Danchin, E.G.J. *et al.* 2013. Top 10 plant-parasitic nematodes in molecular plant pathology. *Molecular Plant Pathology* 14: 946–961.

Jones, R.K. and Milne, D.L. 1982. Nematode pests of bananas. In: D.P. Keetch and J. Heyns (eds.) *Nematology in Southern Africa.* Science Bulletin No. 400, pp. 30-37. Department of Agriculture and Fisheries, Pretoria.

Kalishwaralal, K., Deepak, V., Ramkumarpandian, S. *et al.* 2008. Extracellular biosynthesis of silver nanoparticles by the culture supernatant of *Bacillus licheniformis. Mater Lett* 62: 4411-4413.

Kalt, W. 2002. Health functional phytochemicals of fruits. *Horticulture Review* 27: 269-315.

Kang, H., Jongmin Seo, J. and Ko, H.R. 2022. First report of the apple root-knot nematode, *Meloidogyne mali*, on maple trees in the Republic of Korea. *Plant Disease* 106: 2021.

Kanyagia, S.T. 1993. *Nematodes Attacking The Purple Passion Fruit* (Passiflora edulls) *Plants In Kenya and Studies on The Biology of Two Common Species,* Meloidogyne javanica *and* M. incognita *with Special Reference to their Pathogenicity and Chemical Control.* Ph. D. thesis, Univ. of Nairobi, Kenya.

Karuna, K., Ravichandra, N.G., Krishnappa, K. and Sreenivasa, K.R. 2004. Management of root-knot nematode, *Meloidogyne incognita* on banana by biocontrol agents. *National Symp. on Paradigms in Nematol. Res. for Biodynamic Farming,* Univ. of Agri. Sci., Bangalore, p. 70.

Katan, J. 1981. Solar heating (solarization) of soil for control of soilborne pests. *Annu. Rev. Phytopathol.* 19: 211-236.

Keetch, D.P. 1977. Nematodes in pineapple. In: *Pineapple Cultivation in South Africa.* Citrus and Subtropical Fruit Research Institute, Nelspruit. sh. H-18.

Keetch, D.P. 1982. Nematode pests of pineapple. In: D.P. Keetch and J. Heyns (eds.) *Nematology in South Africa,* pp. 19-29. Department of Agriculture and Fisheries, Pretoria, Republic of South Africa.

Kepenekci, I. 2001. Preliminary list of Tylenchida (Nematoda) associated with olive in the Black Sea and the Mediterranean regions of Turkey. *Nematol. Mediterr.* 29: 145-147.

Kepenekci I., Dura O. and Dura S. 2017. Determination of nematicidal effects of some biopesticides against root-knot nematode (*Meloidogyne incognita*) on Kiwifruit. *Journal of Agricultural Science and Technology* 7: 546-551.

Khan, A., Bilqees, F.M., Khatoon, N. and Shaukat, S.S. 2008. Histopathology of pomegranate roots infected with *Meloidogyne javanica. Int. J. Biol. Biotech.* 5 (3-4): 231-234.

Khan, A., Khatoon, N., Bilqees, F.M. and Mehboob, S. 2010. Histopathology of apple (*Malus pumila* Mill.) roots infected with root-knot nematode (*Meloidogyne incognita*). *Sarhad J. Agric.* 26(1): 61-64.

Khan, A., Shaukat, S.S. and Siddiqui, I.A. 2005. A survey of nematodes of pomegranate orchards in Balochistan province, Pakistan. *Nematol Medit* 33: 25-28.

Khan, A.M., Siddiqi, M.R., Khan, E. *et al.* 1964. *List of Stylet Bearing Nematodes Reported from India-I.* Nematology Publication 1, Aligarh Muslim University, Aligarh, U. P., India, 19 pp.

Khan, M.R., Pal, S., Manohar, G.T. *et al.* 2017. Detection, diagnosis and pathogenic potential of *Meloidogyne incognita* on passion fruit from Mizoram, India. *Pakistan Journal of Zoology* 49(4): 1151-1546.

Khan, R.M., Poornima, K., Somvanshi, V.S. and Walia, R.K. 2022. *Meloidogyne enterolobii*: a threat to guava (*Psidium guajava*) cultivation. *Archives of Phytopathology and Plant Protection* 55:17: 1961-1997.

Khan, T., Khan, S.T., Fazal, M. and Siddiqi, Z.A. 1997. Biological control of *Meloidogyne incognita* and *Fusarium solani* disease complex in papaya using *Paecilomyces lilacinus* and *Trichoderma harzianum. International Journal of Nematology* 7: 127-132.

Khan, T.A. 1991. *Studies on A Disease Complex of Papaya Caused by* Meloidogyne incognita *and* Fusarium solani. Final Tech. Rept., DST Project, Aligarh Muslim Univ., Aligarh, 47 pp.

Khan, T.A. and Khan, S.T. 1998. Nematode diseases of papaya. In P.C. Trivedi (ed.) *Nematode Diseases in Plants,* pp. 344-360. CBS Publishers and Distributors, New Delhi.

Khan, T.A., Khan, S.T. and Saxena S.K. 1992. Effect of interaction of *Meloidogyne incognita* and *Fusarium solani* on the relative response of papaya cvs. *First Afro-Asian Nematology Symposium*, Aligarh Muslim Univ., Aligarh, p. 14.

Khan, T.A., Khan, S.T. and Saxena, S.K. 1995. Relative response of papaya cultivars to the interaction of *Meloidogyne incognita* and *Fusarium solani. Annals of Plant Protection Sciences* 3: 46–51.

Khanduja, S.D. and Garg, Y.K. 1984. Macro-nutrient element composition of leaves from jujube (*Zizyphus mauritiana* Lamk.) tree in north India. *Indian J. Hort.* 41: 22-24.

Khuntia, N. and Das, S. N. 1969. Plant parasitic nematodes associated with fruit trees in Orissa (Abstract). *All India Nematology Symposium,* Indian Agri. Res. Inst., New Delhi, pp. 22-23.

Kierkegaard, J.A., Gardner, P.A., Desmarchelier, J.M. and Angus, J.F. 1993. Biofumigation: Using Brassica species to control pests and diseases in horticulture and agriculture. In: N. Wratten and R.J. Mailer (eds.) *9th Australian Research Assembly on Brassicas*, pp. 77-82. Agricultural Research Institute, Wagga Wagga, Australia.

Kirby, M.F. 1978. Reniform and root-knot nematodes on passionfruit in Fiji. *Nematropica* 8: 21–25.

Kiriga, A.W. 2013. *Occurrence of Plant Parasitic Nematodes in Commercial Pineapple Fields and Effect of Biocontrol Agents on* Meloidogyne *Species in Kenya*. M. Sc. Thesis, School of Agriculture and Enterprise Development, Kenyatta University, Kenya.

Kirkegaard, J.A., Gardner, P.A., Desmarchelier, J.M. and Angus, J.F. 1993. Biofumigation using Brassica species to control pests and diseases in horticulture and agriculture. In: N. Wratten and R.J. Mailer (eds.) *Proceedings 9th Australian Research Assembly on Brassicas*, pp 77–82. Agricultural Research Institute, Wagga Wagga.

Kishore, R., Kamalwanshi, R.S. and Pandey, M. 2005. Effect of root-knot nematode with *Fusarium solani* on damping-off of papaya. *Indian J. Nematol.* 35: 96-97.

Kleynhans, K.P.N., Van den Berg, E. and Swart, A. 1996. *Plant Nematodes in South Africa*. Plant Protection Research Institute Handbook No 8. Agricultural Research Council–Plant Protection Research Institute, Pretoria.

Kluepfel, D.A., Nyczepir, A.P., Lawrence, J.E. 2002. Biological control of the phytoparasitic nematode *Mesocriconema xenoplax* on peach trees. *Journal of Nematology* 34(2): 120–123.

Knight, K.W.L. 2001. Plant parasitic nematodes associated with six subtropical crops in New Zealand. *New Zealand Journal of Crop and Horticultural Science* 29: 267–275.

Kodira, U.C. and Westerdahl, B.B. 2003. *UC IPM Pest Management Guidelines: Fig*. UCPMG Publication 3447, Division of Agriculture and Natural Resources, University of California (ANR Publications), Oakland. Available at: www.ipm.ucdavis.edu/PMG/ r261200111.htm.

Kommineni, K.V., Gillett, J.M. and Ramsdell, D.C. 1998. A study of tomato ringspot virus and prune brown line resistance in twenty-five rootstock-scion combinations. *HortTechnology* 8(3): 349-353.

Korayem, A.M. and Koura, F.H.F. 1993. Population studies of plant parasitic nematodes associated with mango in Giza, Egypt. *Afro Asian Journal of Nematology* 3: 148–151.

Koshy, P.K., Sosamma, V.K. and Faleiro, J.R. 1987. Occurrence of *Heterodera oryzicola* on banana in Goa. *Indian Journal of Nematology* 17: 334.

Krishnamurthy Rao, B.H. and Thammiraju, N.B. 1975. Nematodes associated with citrus. In G.S. Reddy and V. D. Murthy (eds.) *Citrus Decline in Andhra Pradesh*, pp. 55-58. Andhra Pradesh Agri. Univ., Hyderabad.

Krnjaic, D., Krnjaic, S. and Bacic, J. 1997. Distribution and population density of fig cyst nematode (*Heterodera fici* Kirjanova) in the region of SR Yugoslavia. *Zastita Bilja* 48: 245–251.

Krueger, R., Dover, K.E., McSorley, R. and Wang, K.H. 2010. *Marigolds (Tagetes spp.) for Nematode Management*. Electronic Data Information Source, University of Florida, Institute of Food and Agricultural Sciences, Gainesville, FL. ENY 056.

Kumar, S. 2009. Management of root-knot nematode in papaya by *Pochonia chlamydosporia*. *Indian Journal of Nematology* 39 (1): 105-107.

Kumutha, K. 2001. *Symbiotic Influence of AM Fungi and Rhizobacteria on Biochemical and Nutritional Changes in Mulberry* (*Morus alba*). Ph. D. thesis, Tamil Nadu Agri. Univ., Coimbatore.

Ladner, D.C., Tchounwou, P.B. and Lawrence, G.W. 2008. Evaluation of the effect of ecologic on root knot nematode, *Meloidogyne incognita*, and tomato plant, *Lycopersicon esculenum*. *Int J Environ Res Public Health* 5: 104-110.

Lakshmana Murthy, K.V. 1983. *Studies on Burrowing Nematode,* Radopholus similis *(Cobb, 1983) Thorne, 1949 on Banana in Karnataka*. M. Sc. (Agri.) thesis, Univ. of Agri. Sci., Bangalore.

Lamberti, F. 1981. Combating nematode vectors of viruses. *Plant Disease* 65: 113-117.

Lamberti, F. and Baines, R.C. 1969. Effect of *Pratylenchus vulnus* on the growth of "Ascolano" and "Manzanillo" olive trees in a glasshouse. *Plant Dis. Rep.* 53: 557-558.

Lamberti, F. and Bleve-Zacheo,T. 1979. Studies on *Xiphinema americanum sensu lato* with descriptions of fifteen new species (Nematoda: Longidoridae). *Nematol. Medit.* 7: 51-106.

Lamberti, F. and Ciancio, A. 1993. Diversity of *Xiphinema americanum*-group species and hierarchical cluster analysis of morphometrics. *J. Nematol.* 25: 332-343.

LaMondia, J.A. 1999. The effects of *Pratylenchus penetrans* and *Rhizoctonia fragariae* on vigor and yield of strawberry. *J Nematol* 31: 418 –423.

Laqman Khan, M. 2001. Dr. Y.S. Paramar University of Horticulture and Forestry, Solan, Himachal Pradesh. In S.C. Dhawan *et al.* (eds.) *Indian Nematology-Progress and Prospectives*, pp. 153-158. Division of Nematology, Indian Agri. Res. Inst., New Delhi.

Lear, B.S., Miyagawa, S. and Sciaroni, R. 1970, *November Rose Nematode Survey*. Flower and Nursery Report for Commercial Growers, p. 8.

Lecouls, A.C., Rubio-Cabetas, M.J., Minot, J.C. *et al.* 1999. RAPD and SCAR markers linked to the Ma1 root-knot nematode resistance gene in Myrobalan plum (*Prunus cerasifera* Her.). *Theoretical and Applied Genetics* 99: 328-335.

Lee, M.D., Chen, C.H., Tsay, T.T. and Lin, Y.Y. 1998. Survey and control of guava nematode diseases. *Plant Protection Bulletin (Taipei)* 40: 265-276.

Leinfelder, M.M. and Merwin, I.A. 2006. Rootstock selection, preplant soil treatments, and tree planting positions as factors in managing apple replant disease. *Hortscience* 41: 394-401.

Lele, V.C., Durgapal, J.C., Agarwal, D.K. and Sethi, C.L. 1978. Crown and root gall of grape (*Vitis vinefera* L.) in Andhra Pradesh. *Curr. Sci.* 47: 280.

Liberato J.R. Zambolim, L., Vale, F.X.R. *et al.* 2002. *Controle de Doenças de Plantas – Fruteiras.* Controle das doenças causadas por fungos, bactérias e nematoides em maracujaze, iro; Viçosa: Livraria e Editora UFV, pp. 699–811.

Lim, D., Roh, J.Y., Eom, H.J. *et al.* 2012. Oxidative stress-related PMK-1 P38 MAPK activation as a mechanism for toxicity of silver nanoparticles to reproduction in the nematode *Caenorhabditis elegans. Environmental Toxicology and Chemistry* 31: 585-592.

Lima-Medina, I., Somavilla, L., Carneiro, R. and Gomes, G.B. 2013. Species of *Meloidogyne* associated with fig (*Ficus carica*) and host weeds. *Nematropica* 43: 56-62.

Lima, I.M., Souza, R.M., Silva, C.P. and Carneiro, R.M.D.G. 2005. *Meloidogyne* spp. from preserved areas of Atlantic Forest in the State of Rio de Janeiro, Brazil. *Nemat. Bras.* 29: 31-38.

Linford, M.B. and Yapp, F. 1939 Root knot injury restricted by a fungus. *Phytopathology* 29: 596-609.

Loos, C.A. 1959. Symptom expression of *Fusarium* wilt disease of the Gros Michel banana in the presence of *Radopholus similis* (Cobb, 1893) Thorne, 1949 and *Meloidogyne incognita acrita* Chitwood, 1949. *Proc. Helminth. Soc. Wash. 26:* 103–111.

Loos, C.A. 1961. Eradication of the burrowing nematode, *Radopholus similis*, from bananas. *Pl. Dis. Reptr.* 45: 457-461.

Loos, C.A. and Loos, S.B. 1960. Preparing nematode free banana "seed". *Phytopathology* 50: 383-386.

Lorrain, R. 1998. Moyens de contrôle du risque nématologique. L'arboric. *Fruit* 514: 37-41.

Lownsberry, B.F., English, H., Noel, G.R. and Schick. 1977. Influence of Nemaguard and Lovell rootstocks and *Macroposthonia xenoplax* on bacterial canker of peach. *J. Nematol.* 9: 221-224.

Lownsberry, B.F., Moody, E.H., Moretto, A. *et al.* 1978. Pathogenicity of *Mesocriconema xenoplax* to walnut. J. Nematol. 10: 232-236.

Ma, K., Cho, Y., Jeong, B. and Choi, D. 2011. Control effect on root-knot nematodes by hot water dipping treatment in kiwi fruit. *Korean Journal of Organic Agriculture* 19: 255-258.

MacGowen, J.B. 1984. Hemicriconemoides mangiferae *Siddiqi 1961*. Nematology Circular 110.

Machon, J.E. and Bridge, J. 1996. *Radopholus citri* n. sp. (Tylenchida: Pratylenchidae) and its pathogenicity on citrus. *Fund. Appl. Nematol.* 19: 127–133.

Mai, W.F. (Chairman). 1968. *Control of Plant Parasitic Nematodes. Principles of Plant and Animal Pest Control. Vol. 4.* Nat. Acad. Sci. (USA) Publication 1696. 172 pp.

Malan, A.P. and Meyer, A.J. 1999. The reproduction and life cycle of a South African population of *Xiphinema index. South African Journal for Enology and Viticulture* 20: 57–60.

Malik, N.S.A. and Bradford, J.M. 2009. Inhibition of flowering in 'Arbequina' olives from chilling at lower temperatures. *J. Food Agric. Environ.* 7: 429–431.

Mangat, B.P.S. and Sharma, N.K. 1981. Influence of host nutrition on multiplication and development of citrus nematode. *Indian Phytopathology* 34: 90-91.

Mangeiro, M.Z., Nunes, R.A. and Vieira, J.O.L. Jr. *et al.* 2022. Characterizing the interaction between root-knot nematodes and soil-borne fungi which are pathogenic to passion fruit (*Passiflora edulis*). *J. Nematol.* 54(1). 20220023. doi: 10.2478/jofnem-2022-0023.

Mani, A. 1986. Occurrence of *Meloidogyne javanica* on citrus in Andhra Pradesh (India). *International Nematology Network Newsletter* 3(2): 9-10.

Mani, A. 1988a. Studies on the bacterial parasite *Pasteuria penetrans:* I. Spore viability after storage. II. Culture on citrus nematode *Tylenchulus semipenetrans. International Nematology Network Newsletter* 5(2): 24-25.

Mani, A. 1988b. Effect of interculture of marigold and mustard with acid lime on citrus nematode, *Tylenchulus semipenetrans. International Nematology Network Newsletter* 5(4): 14-15.

Mani, A. 1990. Control of citrus nematode, *Tylenchulus semipenetrans*, by bare-root dip treatment of acid lime seedlings in pesticides. *International Nematology Network Newsletter* 6 (3): 20-21.

Mani, A. 1994. Occurrence and distribution of *Tylenchulus semipenetrans* in Andhra Pradesh. *Indian Journal of Nematology* 24: 106-111.

Mani, A. and Al Hinai, M.S. 1995. Mango, a new host of *Meloidogyne incognita. Nematologia Mediterranea* 23: 267.

Mani, A. and Murty, V.D. 1986. Citrus nematode and root-knot nematodes: Distribution, crop loss, symptoms and management. In "*Plant Parasitic Nematodes - Problems and Progress*" (G. Swarup and D.R. Dasgupta, eds.), pp. 249-260. Indian Agricultural Research Institute, New Delhi.

Mani, A. and Murty, V.D. 1989. Role of biopesticides in the management of nematodes and associated diseases. *South Indian* Hort. *49: 296–300* .

Mani, A., Muzna, S.A. and Handoo, Z.A. 1998. Plant–parasitic nematodes of crops in Dhofar Governorate, Sultanate of Oman. *Nematropica* 28: 61–69.

Mani, M.P. 1996. *Effect of* Pasteuria penetrans *(Thorne) Sayre & Starr and* Pseudomonas fluorescens *(Migula) Against* Meloidogyne incognita *(Kofoid & White) Chitwood in Grapevine* (*Vitis vinefera* Linn.). M. Sc. (Agri.) thesis, Tamil Nadu Agri. Univ., Coimbatore, 46 pp.

Manici, L.M., Kelderer, M., Franke-Whittle, I.H. *et al.* 2013. Relationship between root-endophytic microbial communities and replant disease in specialized apple growing areas in Europe. *Appl Soil Ecol.* 72: 207–214.

Maqbool, M.A. 1991. *Nematode Pests of Economic Significance Affecting Major Crops of The Countries in Asia and The Pacific Region.* Technical Document Asia and Pacific Plant Protection Commission, No. 140. FAO Regional Office for Asia and the Pacific (RAPA), Bangkok, Thailand.

Maranhão, S.R.V.L., Moura, R.M. and Pedrosa, E.M.R. 2001. Reação de indivíduos segregantes de araçazeiro a *Meloidogyne incognita* raça 1, *M. javanica* e *M. mayaguensis. Nemat. Bras.* 27: 173-178.

Marathe, R.A., Mallikarjun, M.H, Manjunatha, N. and Venkata Rao B. 2022. *High-tech Pomegranate Production Practices for Export Quality Pomegranate Production and Value Addition.* National Institute of Agricultural Extension Management (MANAGE), Hyderabad & ICAR- National Research Centre on Pomegranate, Kegaon, Solapur.

MARM (Ministerio de Medio Ambiente y Medio Rural y Marino). 2009. *Registro de Productos Fitosanitarios.* http://www.mapa.es/es/agricultura/pags/fitos/registro/menu. asp.

Martelli, G.P. and Savino, V. 1988. Fanleaf degeneration. In R.C. Pearson and A.C. Goheen (eds.) *Compendium of Grape Diseases*, pp 48-49. American Phytopathological Society Press, St Paul, Minnesota, 93 pp.

Marull, J., Pinochet, J., Felipe, A. and Cenis J.L. 1994. Resistance verification in Prunus selections to a mixture of thirteen *Meloidogyne* isolates and resistance mechanisms of a peach-almond hybrid to *M. javanica. Fundamental & Applied Nematology* 17: 85-92.

Mathur, R.L., Mathur, B.N and Handa, D.K. 1969. Addition to host records of root knot nematodes. *Nematologica* 15: 160-161.

Matthiessen, J.N. and Kirkegaard, J.A. 2006. Biofumigation and enhanced biodegradation: opportunity and challenge in soil-borne pest and disease management. *Crit Rev Plant Sci* 25: 235–265.

Mazzola, M., Brown, J., Zhao, X. *et al.* 2009. Interaction of Brassicaceous seed meal and apple rootstock on recovery of *Pythium* spp. and *Pratylenchus penetrans* from roots grown in replant soils. *Plant Dis.* 93: 51–57.

Mazzola, M., Hewavitharana, S.S. and Strauss, S.L. 2015. Brassica seed meal soil amendments transform the rhizosphere microbiome and improve apple production through resistance to pathogen reinfestation. *Phytopathology* 105: 460–469.

Mazzola, M. and Manici, L.M. 2012. Apple replant disease: role of microbial ecology in cause and control. *Annu Rev Phytopathol.* 50: 45–65.

McClure, M.A., Ellis, K.C. and Nigh, E.L. 1974. Post-infection development and histopathology of *Meloidogyne incognita* in resistant cotton. *J. Nematol.* 6(1): 21-26.

McElroy, F.D. 1991. Nematode parasites, In: M.A. Ellis, R.H. Converse, R.N. Williams, and B. Williamson (eds.). *Compendium of Raspberry and Blackberry Diseases and Pests*, pp. 59-62. APS Press, St. Paul, MN.

McElroy, F.D. 1992. A plant health care program for brambles in the Pacific Northwest. *J. Nematol.* 24: 457-462.

McElroy, F.D. and Van Gundy, S.D. 1968. Observations on the feeding processes of *Hemicycliophora arenaria. Phytopathology* 58: 1558-1565.

McInnis, T., Klupfel, D. and Zehr, E. 1990. Suppression of *Criconemella xenoplax* on peach by rhizosphere bacteria. In: *Proc. of the 2nd Int. Nematol. Cong.*, Veldhoven.

McKenry, M.V. 1987. Control strategies in high value crops. In R.H. Brown and B.R. Kerry (eds.), *Principles and Practice of Nematode Control in Crops*, pp. 329-349. Academic Press, Australia.

McKenry, M.V. 1989. Damage and development of several nematode species in a plum orchard. *Applied Agricultural Research* 4: 10–13.

McKenry, M.V. 1994. Nematodes of Olive. In: L. Ferguson, G.S. Sibett and G.C. Martin, (eds.) *Olive Production Manual*, pp. 97-99. Publ. 3353, University of California, Oakland.

McKenry, M.V. 1999. *The Replant Problem and its Management.* Catalina Publishing, Catalina Cr., Fresno, CA, USA.

McKenry, M.V. 2004. Three nematode genera and the damage they cause for plum producers. *J. Nematol.* 36: 333.

McSorley, R. 1981. *Plant Parasitic Nematodes Associated with Tropical and Subtropical Fruits.* Tech. Bull. No. 83, Agri. Expt. Stn., Univ. of Florida, Gainesville, 49 pp.

McSorley, R. 1992. Nematological problems of tropical and sub–tropical fruit tree crops. *Nematropica* 22: 106–110.

McSorley, R. and Gallaher, R.N. 1991. Nematode population changes and forage yields of six corn and sorghum cultivars. *Suppl J Nematol* 23: 673 –677.

McSorley, R. and Parrado, J.L. 1983. Effect of Nemacur on phytoparasitic nematodes of mango. *1982 Fungicide and Nematicide Tests, American Phytopathological Society* 39: 92–93.

McSorley, R., Parrado, J.L. and Goldweber, S. 1981. Plant parasitic nematodes associated with mango and relationship to the tree condition. *Nematropica* 11: 1–9.

Meland, M. 1998. Yield and fruit quality of Van sweet cherry in four high density production systems over seven years. *Acta Horticulturae* 468: 425-433.

Meland, M. 2005. High density planting systems of European plums – the effect of growth and productivity of three cultivars after nine years. *Acta Agriculturae Scandinavica, Section B – Soil and Plant Science* 55: 51–57.

Mendoza, A., Sikora, R.A. and Kiewnick, S. 2004. Efficacy of *Paecilomyces lilacinus* strain 251 for the control of *Radopholus similis* in banana, *Commun. Agric. Appl. Biol. Sci.* 69: 365–372.

Meyer, S.L.F. 2003. United States Department of Agriculture – Agricultural Research Service research programs on microbes for management of plant-parasitic nematodes. *Pest Manag Sci* 59: 665–670.

Milan, A.R. 2007. Breeding of *Psidium* species for root knot nematode resistance in Malaysia. *Acta Hort.* 735: 61-69.

Milan, A.R. 2010. Collection and evaluation of guava (*Psidium guajava* L.) for nematode resistance in Malaysia. *Acta Hort.* 849: 357-362.

Milne, D.L. 1982. Nematode pests of litchi. In: D.P. Keetch and J. Heyns (eds.) *Nematology in Southern Africa Science Bulletin*. Department of Agriculture and Fisheries in Republican South Africa, Pretoria 0001, Bulletin No. 400, pp. 38–41.

Milne, D.L., Appleton, M.R. and Holtzhausen, L.C. 1975. The anatomical reaction of *Litchi chinensis* roots attacked by three nematode species and a soil mite. *Phytophylactica* 7: 15-20.

Milne, D.L., Devilliers, E.A. and Van Den Berg, E. 1975. Mango nematodes. *Citrus Subtrop. Fruit J.* 502: 17, 19, 21.

Milne, D.L. and Keetch, D.P. 1976. Some observations on the host plant relationships of *Radopholus similis* in Natal. *Nematropica* 6: 13–17.

Milne, D.L. and Kuhne, FA. 1968. Nematodes attack bananas. *Farming in South Africa* 44: 5-9.

Milne, D.L., Villiers, E.A. de and Holtzhausen, L.C. 1971. Litchi tree decline caused by nematodes. *Phytophylactica* 3(1): 37-44.

Minamizawa, K. 1997. *Moriculture - Science of Mulberry Cultivation* (Translated from Japanese), pp. 139-164. Oxford and IBH Publishing Co. Pvt. Ltd., New Delhi, India.

Misra, A.K. 2005. Important diseases of guava in India with special reference to wilt. *Acta Hortic.* 735: 507-523.

Misra, A.K. and Pandey, B.K. 2000b. Progressive natural wilting of guava plants during different months. *Indian Phytopath.* 53: 423-427

Misra, A.K., Pandey B.K., Prasad B. and Shukla S.K. 2003a. Pathogenic diversity in the cause of wilt disease of guava. *Indian Phytopath.* 56: 314-314.

Misra, A.K., Prakash O, and Sen B. 2000a. Biological control of guava wilt by *Aspergillus niger* strain AN17 (Pusa Mrida). *Proceedings of the National Seminar on Hi-tech Horticulture*, pp. 149-149. Bangalore.

Misra, A.K., Prasad D., Prasad B. and Shukla S.K. 2004. Effective management of wilt disease of guava. *Proceedings of the National Symposium on Crop Surveillance: Disease Forecasting and Management*, pp. 92-93. Div. of Plant Pathology, IARI, New Delhi.

Misra, A.K., Rajan S., Prasad B. *et al.* 2003b. Resistant source of wilt disease of guava. *Seminar on Plant Diseases of National Importance with Special Reference to Guava Wilt and Mango Malformation* (Souvenir and Abstract), pp. 51-52. Zonal Conference (East Zone), Indian Society of Mycology and Plant Pathology, CISH and IISR, Lucknow.

Misra, S.L. and Edward, J.C. 1977. Observations on histopathology of some of the more important rootstocks for commercial citrus affected by citrus nematodes. *Allahabad Farmer* 53: 7-18.

Mobin, M. and Khan, A.M. 1969. Effect of organic amendments on the population of rhizosphere fungi and nematodes around the roots of some fruit trees (Abstr.). *All India Nematol. Symp.*, Indian Agri. Res. Inst., New Delhi, p. 68.

Mojtahedi, H. and Lownsberry, B.F. 1976. The effects of ammonia-generating fertilizer on *Criconemoides xenoplax* in pot cultures. *J. Nematol.* 8: 306-309.

Molero, T., Molina, J. and Casassa-Padron, A.M. 2010. Advances in the genetic study of resistance of cultivars of *Psidium* spp. to *Meloidogyne* spp. in a tropical dry forest. *Acta Hort.* 849: 309-318.

Mortensten, J.A. 1985. Potential new rootstocks for Florida viticulture. *Proceedings of Florida State Horticulture Society* 98: 166-169.

Moseley, D, Patton, A., Bateman, R. and Kirkpatrick, T. 2010. *Controlling Nematodes on Golf Courses*. Cooperative Extension Service, University of Arkansas, MP481, 12 pp.

Moyle, R.L. and Botella, J.R. 2014. EST sequencing of *Meloidogyne javanica* infected pineapple root tissues reveals changes in gene expression during root-knot nematode induced gall formation. *Tropical Plant Biology* 7(2): 43-52.

Mukhopadhyaya, M.C. 1970. Efficacy of D-D, EDB, DBCP and VC-13 in controlling plant parasitic nematodes. *Journal of Research, Punjab Agricultural University* 7: 659-662.

Mukhopadhyaya, M.C. and Dalal, M.R. 1971. Effect of two nematicides on *Tylenchulus semipenetrans* and on sweet lime yield. *Indian Journal of Nematology* 1: 95-97.

Mukhopadhyaya, M.C. and Suryanarayana, D. 1969. Citrus decline in Haryana: Role of *Tylenchulus semipenetrans* and its control. *Indian Phytopathology* 22: 495-497.

Murant, A.F. 1970. Soil-borne viruses and diseases in Rubus. In: N.W. Frazier (ed.) *Virus Diseases of Small Fruits and Grapevines, A Handbook*, pp. 132-136. University of California, Berkeley, California, USA.

Murant, A.F. and Lister, RM, 1967. Seed-transmission in the ecology of nematode-borne viruses. Annals of Applied Biology, 59:63-76.

Murant, A.F. and Lister, R.M. 1987. European nepoviruses in strawberry [1987]. Journal Agriculture handbook - United States Department of Agriculture, Combined Forest Pest Research and Development Program (USA).

Murthy, N.C.V. 1986. Root knot nematode disease of mulberry and its control measures. *Indian Silk* 25: 33.\

Muthukrishnan, T.S. and Sivakumar, C.V. 1973. Some observations on nematodes of citrus, banana and papaya in Tamil Nadu. *Proc. Workshop on Fruit Research*, Coimbatore.

Muthulakshmi, M., Devrajan, K. and Jonathan, E.I. 2010a. Biocontrol of root knot nematode, *Meloidogyne incognita* (Kofoid and White) Chitwood in mulberry (*Morus alba* L.). *J. Biopest.* 3(2): 479 - 482.

Muthulakshmi M., Jonathan. E.I. and Devrajan. K. 2010b. Interaction of *Meloidogyne incognita* and *Macrophomina phaseolina* on Mulberry *(Morus alba* L.). *National Conf. on Innovations in Nematol. Res. for Agri. Sustainability - Challenges and A Roadmap Ahead*, p. 108. Nematol. Soc. of India & Tamil Nadu Agri. Uni., Coimbatore.

Nair, K.K.R. 1979. Studies on the chemical control of banana nematodes. *Agric. Res. J. Kerala* 17: 232-235.

Nair, M.R.G.K. 1965. On the occurrence of the citrus nematode, *Tylenchulus semipenetrans* Cobb, in Kerala. *Agriculture Research Journal Kerala* 3: 105.

Nair, M.R.G.K., Das, N.M. and Menon, M.R. 1966. On the occurrence of the burrowing nematode *Radopholus similis* (Cobb, 1893) Thorne, 1949 on banana in Kerala. *Indian J. Entomol.* 28: 553-554.

Nakamura, M., Tahara, Y., Fukata, S. *et al.* 2017. Significance of optimization of phospholipid polyethylene glycol quantity for coating carbon nano-horns to achieve low cytotoxicity. *Bulletin of the Chemical Society of Japan* 90: 662-666.

Nancy, W., Don Mathre, E., James, B. and Charles, S. 1997. Biological seed treatments: Factors involved in efficacy. *Hort Sci* 32: 179-183.

Narayanan, E.S., Kashiviswanathan, K. and Iyenger, M.N.S. 1966. A note on the occurrence of root knot nematode, *Meloidogyne incognita* (Kofoid and White) in local mulberry. *Indian J. Serie.* 5: 33-34.

Narbaev, Z.N. and Sidikov, D.T. 1985. The possible parasitization of *Heterodera fici* on other crops. *Uzbekskii Biologicheskii Zhurnal* 4: 57–58.Nasira, K., Shaheen, N. and Shahina, F. 2011. Root-knot nematode *Meloidogyne incognita wartellei* on pomegranate in Swat, KPK. *Pakistan J. Nematol.* 29(1): 117-118.

Nassar, A.M. 2016. Effectiveness of silver nano-particles of extracts of *Urtica urens* (Urticaceae) against root-knot nematode *Meloidogyne incognita*. *Asian J. Nematol.* 5(1): 14-19.

Nayak, G.M. 1986. *Studies on Papaya in Relation to Certain Plant Diseases.* M. Sc. Thesis, University of Agricultural Sciences, Bangalore, 106 pp.

Newhall, A.G. 1958. The incidence of Panama disease of banana in the presence of the root-knot and burrowing nematodes (*Meloidogyne* and *Radopholus*). *Pl. Dis. Reptr.* 42: 853-856.

NHB. 2010. *Indian Horticulture Database.* National Horticultural Board, Ministry of Agriculture, Government of India.

Nico, A.I., Jiménez-Díaz, R.M. and Castillo, P. 2003. Host suitability of the olive cultivars Arbequina and Picual for plant-parasitic nematodes. *J. Nematol.* 35: 29-34.

Nico, A.I., Jiménez-Díaz, R.M. and Castillo, P. 2003. Solarization of soil in piles for the control of *Meloidogyne incognita* in olive nurseries in southern Spain. *Plant Pathol.* 52: 770-778.

Nico, A.I., Jiménez-Díaz, R.M., and Castillo, P. 2004. Control of root-knot nematodes by composted agro-industrial wastes in potting mixtures. *Crop Prot.* 23: 581-587.

Nico, A.I., Rapoport, H.F., Jiménez-Díaz, R.M. and Castillo, P. 2002. Incidence and population density of plant-parasitic nematodes associated with olive planting stocks at nurseries in southern Spain. *Plant Dis.* 86: 1075- 1079.

Nicotra, A. and Moser, L. 1997. Two new rootstocks for peach and nectarines: Penta and Tetra. *Acta Horticulturae* 451: 269-271.

Nigh, E.L. 1966. Incidence of crown gall infection in peach as affected by the Javanese root-knot nematode. *Phytopathology* 56: 150.

Nirmal Johnson, S.B. and Devarajan, K. 2004. Effect of *Pseudomonas fluorescens* on banana nematodes. *National Symp. on Paradigms in Nematol. Res. for Biodynamic Farming*, Univ. of Agri. Sci., Bangalore, pp. 81-82.

Nishitha Naik, V., Sharma, D.D., Govindaiah and Chowdary, N.B. 2003. Occurrence of root knot disease complex in mulberry due to nematode and fungal association. In R. Govindan (ed.) *Disease and Pest Management in Sericulture*, pp. 7. Sericulture College, Chintamani, India.

Noriega, C.D.H., Marban, M.N., Rodriguez, A.J. and De Zarate, L.G.P. 1988. Effect of chemical products on phytonematodes associated with the root and on the mite *Eriophyes mangiferae* involved in witches broom of the mango cultivar Haden in Iguala Guerrero Mexico. *Revista Mexicana de Fytopatologia* 6: 61–62.

Nour El-Deen, A.H., Al-Barty Amal, F., Darwesh Hadeer, Y. and Al-Ghamdi, A.S. 2016. Eco-friendly management of root-knot nematode, *Meloidogyne incognita* infecting pomegranate at Taif Governorate, KSA. *Res. J. Pharma. Biol. Chem. Sci.* 7(1): 1070-1076.

Nyczepir, A.P. 1989. *Peach Tree Short Life: A Nematode Associated Disease.* Fla. Dept. Agric. & Consumer Serv. Nematology Circular No. 163.

Nyczepir, A.P. 1990. Influence of *Criconemella xenoplax* and pruning time on short life of peach trees. *Journal of Nematology* 22: 97-100.

Nyczepir, A.P. 2011. Host suitability of an endophyte-friendly tall fescue grass to *Mesocriconema xenoplax* and *Pratylenchus vulnus*. *Nematropica* 41: 45-51.

Nyczepir, A.P. 2011. Nematodes. In D. Horton, and D. Johnson (eds.) *South-eastern Peach Growers' Handbook.* University of Georgia, Cooperative Extension Service Bulletin #1384, Athens, GA.

Nyczepir, A.P. and Beckman, 2000. Host status of Guardian peach rootstock to *Meloidogyne* sp. and *M. javanica*. *HortScience* 35: 772.

Nyczepir, A.P. and Bertrand, P.F. 1990. Host suitability of selected small grain and field crops to *Criconemella xenoplax*. *Plant Disease* 74: 698-701.

Nyczepir, A.P. and Halbrendt, J.M. 1993. Nematode pests of deciduous fruit and nut trees. *In* K. Evans, D.L. Trudgill, and J.M. Webster (eds.) *Plant Parasitic Nematodes in Temperate Agriculture*, pp. 381-425. CAB International, Wallingford, UK.

Nyczepir, A.P. and Kluepfel, D.A. 2007. Impact of solarization, rootstock and *Pseudomonas synxantha* on

Criconemoides xenoplax populations and tree growth in a peach tree short life site. *Journal of Nematology* 39(1): 75-75.

Nyczepir, A.P. and Meyer, S.L.F. 2010. Host status of endophyte-infected and noninfected tall fescue grass to *Meloidogyne* spp. *Journal of Nematology* 42: 151-158.

Nyczepir, A.P., Nagel, A.K. and Schnabel, G. 2009. Host status of three transgenic plum lines to *Mesocriconema xenoplax*. *HortScience* 44:1932-1935.

Nyczepir, A.P. and Rodriguez-Kabana, R. 2004. Effectiveness of biofumigation for ring nematode control in a young peach orchard. *First Int. Biofumigation Symp.*, Florence, Italy.

O'Bannon, J.H. and Nemec, S. 1979. The response of *Citrus limon* seedlings to a symbiont, *Glomus etunicatus*, and a pathogen, *Radopholus similis*. *J. Nematol.* 11: 270-275.

O'Bannon, J.H. and Tomerlin, A.T. 1973. Citrus tree decline caused by *Pratylenchus coffeae*. *Journal of Nematology* 5: 311-316.

O'Rourke, E.N. 1992. LSU Purple, a new fig for the Gulf South. *Louisiana Agriculture* 35: 19–20.

Olowe, T. and Corbett, D.C.M. 1976. Aspects of the biology of *Pratylenchus brachyurus* and *P. zeae*. *Nematologica* 22 : 202-211.

Osei, K., Moss, R., Nafeo, A. *et al.* 2011. Managing nematode pests and improving yield of pineapple with *Mucuna pruriens* in Ghana. *Journal of Entomology and Nematology* 3(1): 20-24.

Osman, H.A., El-Gindi, A.Y., Taha, H.S. *et al.* 2008. Biological control of root-knot nematode *Meloidogyne incognita*. 2-Evaluation of the nematicidal effects of *Tagetes erecta* tissue-culture under laboratory and greenhouse conditions. *Egypt J Phytopathol* 36: 33–44.

Parvatha Reddy, P. 1986. Chemical control of plant parasitic nematodes infecting citrus and grapevine. *Indian Journal of Horticulture* 43: 303-305.

Parvatha Reddy, P. 1988. *Nematode Management in Horticultural Crops.* Tech. Bull. No. 4, Indian Inst. of Hortil. Res., Bangalore, 52 pp.

Parvatha Reddy, P. and Agarwal, P.K. 1987. Resistance in citrus rootstocks to the citrus nematode, *Tylenchulus semipenetrans*. *Indian J. Hort.* 44: 111-114.

Parvatha Reddy, P., Iyer, C.P.A. and Subramanyam, M.D. 1988. Evaluation of papaya cultivars and hybrids against *Meloidogyne incognita*. *Indian Journal of Nematology* 18: 381-382.

Parvatha Reddy, P., Khan, R.M. and Agarwal, P.K. 1987. Selection of citrus rootstocks and hybrids resistant to the citrus nematode, *Tylenchulus semipenetrans*. *Pakistan Journal of Nematology* 5(2): 69-72.

Parvatha Reddy, P., Khan, R.M. and Rao, M.S. 1991a. Evaluation of chemicals for the management of the burrowing nematode, *Radopholus similis* infecting banana. *Curr. Nematol.* 2: 1-2.

Parvatha Reddy, P., Khan, R.M. and Rao, M.S. 1991b. Integrated management of the citrus nematode, *Tylenchulus semipenetrans* using oil cakes and *Paecilomyces lilacinus*. *Afro-Asian Journal of Nematology* 1: 221-222.

Parvatha Reddy, P. and Nagesh, M. 2000. Integrated management of the citrus nematode using bacterial (*Pasteuria penetrans)* and fungal (*Paecilomyces lilacinus)* biocontrol agents. *Proceedings of International Symposium on Citriculture*, Nagpur, pp. 825-829.

Parvatha Reddy, P., Nagesh, M., Rao, M.S. and Rama, N. 2000. Management of *Tylenchulus semipenetrans* by integration of *Pseudomonas fluorescens* with oil cakes. *Proc. International Symposium on Citriculture*, Nagpur, pp. 830-833.

Parvatha Reddy, P., Rao, M.S., Mohandas, S. and Nagesh, M. 1995. Integrated management of the citrus nematode, *Tylenchulus semipenetrans* Cobb using VA mycorrhiza, *Glomus fasciculatum* (Thaxt.) Gerd & Trappe and oil cakes. *Pest Management in Horticultural Ecosystems* 1: 37-41.

Parvatha Reddy, P., Rao, M.S. and Nagesh, M. 1996a. Management of the citrus nematode, *Tylenchulus semipenetrans* by integration of *Trichoderma harzianum* with oil cakes. *Nematologia Mediterranea* 24: 265-267.

Parvatha Reddy, P., Rao, M.S. and Nagesh, M. 1996b. Management of the citrus nematode on acid lime by integration of parasitic fungi and oil cakes. *Pest Management in Horticultural Ecosystems* 2: 15-18.

Parvatha Reddy, P., Rao, M.S. and Nagesh, M. 1996c. Integrated management of the citrus nematode, *Tylenchulus semipenetrans* using pesticides and parasitic fungus, *Paecilomyces lilacinus. Pest Management in Horticultural Ecosystems* 2: 61-63.

Parvatha Reddy, P., Rao, M.S. and Nagesh, M. 1996d. Crop loss estimation in banana due to the burrowing nematode, *Radopholus similis. Pest Mangmt. Hort. Ecosystems* 2: 85-86.

Parvatha Reddy, P., Rao, M.S. and Nagesh, M. 2002. Integrated management of burrowing nematode (*Radopholus similis)* using endomycorrhiza (*Glomus mosseae)* and oil cakes. In H.P. Singh and K.L. Chadha (eds.) *Banana-Improvement, Production and Utilization*, pp. 344-348. AIPUB, Trichy.

Parvatha Reddy, P. and Singh, D.B. 1978a. The association of the citrus nematode with grape roots in a commercial orchard. *Current Science* 47: 640-641.

Parvatha Reddy, P. and Singh, D.B. 1978b. Evaluating the reaction of some species and varieties of *Citrus* and *Poncirus* to the citrus nematode. *Indian Journal of Nematology* 8: 82-84.

Parvatha Reddy, P. and Singh, D.B. 1984. Reaction of selected cultivars/ rootstocks of grapevine to *Meloidogyne incognita. Indian J. Nematol.* 14: 73-74.

Pasha, M.J., Siddiqui, Z.A., Khan, M.W. and Qureshi, S.I. 1987. Histopathology of eggplant roots infected with root-knot nematode, *Meloidogyne incognita. Pak. J. Nematol.* 5: 27-34.

Patel, D., Patel, B.A., Patel, S.K. *et al.* 1999. Root-knot nematode, *Meloidogyne indica* on kagzi lime in North Gujarat. *Indian J. Nematol.* 29(2): 185-205.

Patel, H.V., Patel, D.J. and Patel, B.A. 1989. Reaction of papaya varieties to *Rotylenchulus reniformis. International Nematology Network Newsletter* 6 (3): 24.

Patil, N.M., Dharne, P.K. and Deshmukh, S.S. 1999. Sucker dip treatment in monocrotophos with intercrops for the management of plant parasitic nematodes in banana. *Seminar on Technological Advancement in Banana Production, Handling and Processing Management*, Jalgaon, Maharashtra, pp. 126–128.

Pawar, S.A., Mhase, N.L., Kadam, D.B. and Chandele, A.G. 2013. Management of root-knot nematode, *Meloidogyne incognita*, Race-II infesting pomegranate by using bioinoculants. *Indian J. Nematol.* 43(1): 92-94.

Peacock, F.C. .1956. The reniform nematode in the Gold Coast. *Nematologica* 1: 307–310.

Peregrine, W.T.H. and Yunton, B. 1980. A preliminary note on nematode pests in Brunei. *Tropical Pest Management* 26: 416–419.

Pereira, F.O.M., Souza, R.M., Souza, P.M. *et al.* 2009. Estimativa do Impacto Econômico e Social Direto de *Meloidogyne mayaguensis* na Cultura da Goiaba no Brasil. *Nemat. Bras.* 33: 176-181.

Petit, R.P.A. 1990. Survey of plant parasitic nematodes associated with fruit trees of economic importance in Venezuela. *Fitopatologia Venezolana* 3: 2–5.

Philip, T., Govindaiah, Bajpai, A.K. *et al.* 1997. A preliminary survey on mulberry diseases in South India. *Indian J. Serie.* 36: 128-132.

Philippi, I. and Budge, A. 1992. Effects of *Meloidogyne hapla* on young kiwi plants. *Nematropica* 22: 47–54.

Phillips, T., and Aiken, G. 2009. Novel endophyte-infected tall fescues. *Forage and Grazinglands.* doi:10.1094/FG-2009-1102-01-RV.

Pinkerton, J.N., Vasconcelos, M.C., Sampaio, T. and Shaffer, R.G. 2005. Reaction of grape rootstocks to ring nematode *Mesocriconema xenoplax. American Journal of Enology and Viticulture* 56(4): 377-385.

Pinochet, J., Anglés, M., Dalmau, E. *et al.* 1996. Prunus rootstock evaluation to root-knot and lesion nematodes in Spain. *Journal of Nematology* 28: 616-623.

Pinochet, J., Calvet, C., Camprubi, A. and Fernandez, C. 1995. Growth and nutritional response of Nemared peach rootstock infected with *Pratylenchus vulnus* and the mycorrhizal fungus, *Glomus mosseae. Fundam. Appl. Nematol.* 18(3): 205-210.

Pinochet, J., Camprubi, A., Calvet, C. *et al.* 1998. Inducing tolerance to the root lesion nematode *Pratylenchus vulnus* by early mycorrhizal inoculation of micro-propagated Myrobalan 29 C plum rootstock. *J. Amer. Soc. Hort. Sci.* 123(3): 342-347.

Pinochet, J., Fernandez, C., Cunill, M. *et al.* 2002. Response of new interspecific hybrids for peach to root-knot and lesion nematodes, and crown gall. *Acta Hort.* 592(2): 707-716.

Pinochet, J., Verdejo, S., Soler, A. and Canals, J. 1992. Host range of a population of *Pratylenchus vulnus* in commercial fruit, nut, citrus and grape rootstocks in Spain. *Journal of Nematology* 24: 693–698.

Poehling, H.M., Wyss, U. and Neuhoff, V. 1980. Microanalysis of free amino acids in the aseptic host–parasite system: *Ficus carica–Xiphinema index* (Nematoda). *Physiological Plant Pathology* 16: 49–61.

Ponte, J.J. Franco, A., Aguiár de Holanda, Y.C. and Filho, S. 1998. Calagem, adubação orgânica e fungicida de solo no controle da podridão do pé (*Fusarium solani*) d maracujá amarelo. *Fitopatologia Venezolana* 12: 30-31.

Ponte, J.J.D.A. 1980. *Meloidogyne* - importanica and control no nordeste. *Sociedade Brasileira Nematologia* 4: 1-14.

Poornima, K, Ashokkumar, N, Kalaiarasan, P, and Mathiyazhagan, K. 2019. Screening and histological characterization of guava (*Psidium guajava*) cultivars against root-knot nematode, *Meloidogyne enterolobii*. *Pest Mgmt. Hort. Ecosys.* 25(1): 84–92.

Poornima, K., Kumar, N. and Balamohan, T.N. 2004. Biological control of lesion nematode in banana. *National Symp. on Paradigms in Nematol. Res. for Biodynamic Farming,* Univ. of Agri. Sci., Bangalore, p. 99.

Poornima, K., Suresh, P., Kalaiarsan, P. *et al.* 2016. Root knot nematode, *Meloidogyne enterolobii* in guava (*Psidium guajava* L.) - A new record from India. *Madras Agric. J.* 103: 359-365.

Poornima, K. and Walia, R.K. 2021. Horticultural nurseries and nematode dissemination. *Curr. Sci.* 119: 277–284.

Poucher, C., Ford, H.W., Suit, R.F. and DuCharme, E.P. 1967. *Burrowing Nematode in Citrus.* Florida Dept. of Agri., Divn. of Plant Industry, Bull. No. 7, 63 pp.

Prasad, M.B.N.V., Parvatha Reddy, P., Rao, M.S. and Rekha, A. 1998. Evaluation of some intergeneric citrus rootstock hybrids for resistance to the citrus nematode *Tylenchulus semipenetrans*. In "*Advances in IPM for Horticultural Crops*" (P. Parvatha Reddy, N. K. Krishna Kumar and A. Verghese, eds.), pp. 328-331. Association for Advancement of Pest Management in Hortil. Ecosystems, Bangalore.

Prasad, M.B.N.V., Sawant, S.D., Parvatha Reddy, P. *et al.* 1999. Citrus rootstock hybrids resistant to *Phytophthora* and citrus nematodes and also tolerant to salinity and draught. *Proceedings of National Symposium on Citriculture*, Nagpur, pp. 59-62.

Prasad, S.K. and Chawla, M.L. 1965. Observations on the population fluctuations of citrus nematode, *Tylenchulus semipenetrans* Cobb, 1913. *Indian Journal of Entomology* 27: 450-455.

Prasad, S.K., Dasgupta, D.R. and Mukhopadhyaya, M.C. 1964. Nematodes associated with commercial crops in Northern India. I. Uttar Pradesh. *Indian Journal of Nematology* 3: 8-23.

Pritts, M. and Nonnecke, G. 2004. *Preplant Cover Crops for Strawberries.* Cornell Univ., Dept. of Hort., Ithaca http://www.fruit.cornell.edu/berry/production/pdfs/strcovercrops.pdf. Accessed 1 April 2018.

Quénéhervé, P. and Cadet, P. 1985. Localisation des nematodes dans les rhizomes du bananier cv Poyo. *Revue de Nématologie* 8: 3-8.

Rabie, E.C. 1991. *Die Invloed van* Meloidogyne javanica *en* M. incognita *op die Voorkoms van Valspanamasiekte by Piesangs* [MSc dissertation]. University of Pretoria, Pretoria.

Radunic, M., Jazbec, A., Pecina, M. *et al.* 2011. Growth and yield of the sweet cherry (*Prunus avium* L.) as affected by training system. *African Journal of Biotechnology* 10(24): 4901-4906.

Rajagopalan, P. and Naganathan, T.G. 1977. Studies on nematode parasites of banana. *Tamil Nadu Agri. Univ. Ann. Rept.* 6: 131.

Rajendran, G. and Naganathan, T.G. 1981. Evaluation of systemic chemicals for the control of *Rotylenchulus reniformis* on papaya. *Indian Journal of Nematology* 11: 130.

Rajendran, G., Naganathan, T.G. and Sivagami, V. 1979. Studies on banana nematodes. *Indian J. Nematol.* 9: 54.

Rallo, L. and Cuevas, J. 2008. Fructificación y producción. In D. Barranco, R. Fernández-Escobar, and L. Rallo (eds.) *El cultivo del olivo*, pp. 127–162. Ediciones Mundi Prensa, Madrid.

Ramakrishnan, S. and Rajendran, G. 1998a. Influence of *Meloidogyne incognita* on yield components and physiological functions of papaya. *Nematologia Mediterranea* 26: 225–228.

Ramakrishnan, S. and Rajendran, G. 1998b. Assessment of yield loss due to *Meloidogyne incognita* in papaya under field conditions. *Nematologia Mediterranea* 26: 229–230.

Ramaswamy, N. 1971. *Studies on the Effect of Nitrogen on the Growth and Development of 'Robusta'* (Musa cavendishi *L.*). M. Sc. (Ag.) thesis, Annamalai Univ., Annamalinagar.

Ramsdell, D.C. 1988. Peach rosette mosaic virus decline. In R.C. Pearson and A.C. Goheen (eds.) *Compendium of Grape Diseases*, pp 51-52. American Phytopathological Society Press, St Paul, Minnesota, 93 pp.

Rana, M.K. 2010. Fruits and Vegetables: a potential source of non-nutrients bioactive substances (Functional foods). *Processed Food Industries* 27: 26-33

Rao, G.M.V., Prasada and Ganguly, S. 1999. Mode of reproduction and interbreeding capabilities of two host races of reniform nematode, *Rotylenchulus reniformis*. *Indian Journal of Nematology* 29(2): 174-177.

Rao, M.M. 1997. Studies on the improvement of finger size in Munavalli (Musa AAB) banana. In M.M. Rao and G.S. Sulikeri (eds.) *Research and Development in Fruit Crops in North Karnataka*, pp. 59-61. Univ. of Agri. Sci., Dharwad, Karnataka.

Rao, M.M., Narasimhan, P., Nagaraja, N. and Anandaswamy, B. 1968. Effect of naphthalene acetic acid and p-chlorophenoxy acetic acid on control of berry drop in Anab-e-Shahi grape. *J. Food Sci. Tech.* 5: 127-128.

Rao, M.M. and Shankarnarayanan, H.N. 1982. Effect of horticultural practices and growth regulators on colouration and ripening of Gulabi grapes. *Mahrashtra J. Hort. Sci.* 1: 41-42.

Rao, M.S. 2007. Papaya seedlings colonized by the bio-agents *Trichoderma harzianum* and *Pseudomonas fluorescens* to control root-knot nematodes. *Nematol. Medit.* 35: 199–203.

Rao, M.S. and Naik, D. 2003. Effect of *Trichoderma harzianum* and *Paecilomyces lilacinus* on *Meloidogyne incognita* on papaya (*Carica papaya* L.) nursery seedlings. *Pest Mangmt. Hort. Ecosystems* 9: 155-160.

Rao, M.S., Umamaheswari, R., Priti, K. *et al.* 2016. Role of biopesticides in the management of nematodes and associated diseases in horticultural crops. In: K. Hakeem, M. Akhtar, S. Abdullah (eds.) *Plant, Soil and Microbes*. Springer, Cham.

Rashid, A., Khan, F.A. and Khan, A.M. 1973. Plant parasitic nematodes associated with vegetables, fruits, cereals and other crops in North India. I. Uttar Pradesh. *Indian Journal of Nematology* 3: 8-23.

Rashid, M., Ahmad, I., Kumar, K.H. *et al.* 2019. Infestation of *Meloidogyne enterolobii* in newly established/old guava orchards and nurseries in Madhya Pradesh, Rajasthan and Uttar Pradesh. *Ann. of Plan. Prote. Scien.* 27(1): 170–171.

Raski, D.J., Hart, W.H. and Kasimatis, A.N. 1973. *Nematodes and their Control in Vineyards*. California Agricultural Experiment Station circular 533 (revised). 20 pp.

Raski, D.J., Hewitt, W.B., Goheen, A.C. *et al.* 1965. Survival of *Xiphinema index* and reservoirs of fanleaf virus in fallowed vineyard soil. *Nematologica* 11: 349-352.

Ravi, K., Nanje Gowda, D.N. and Parvatha Reddy, P. 2001. Integrated management of disease complex in banana involving borrowing nematode and Panama wilt—an ecofriendly approach. In A. Verghese and P. Parvatha Reddy (eds.) *IPM in Horticultural Crops: Emerging Trends in The New millennium*. Association for Advancement of Pest Management in Horticultural Ecosystems, Bangalore, p. 154.

Rebois, R.V. 1973. Effect of soil temperature on infectivity and development of *Rotylenchulus reniformis* on resistant and susceptible soybeans, *Glycine max*. *Journal of Nematolology* 5: 10-13.

Reuther, W., Batchelor, H.J. and Webber, H.J. 1962. *Citrus Industry, Vol. II.* Univ. of California Press, Berkeley, California.

Richard E. 2009. *The Mango*. 2nd ed. CABI.

Risbeth, J. 1960. Factors affecting the incidence of banana with panama disease. *Emp. J. Expt. Agric*. 28: 109-113.

Ritchie, D.F. and Clayton, C.N. 1981. Peach-tree short life - a complex of interacting factors. *Plant Disease* 65: 462-469.

Ritzinger, C.H.S.P. and Costa, D.C. 2004. *Nematoide das lesões* (Pratylenchus *spp.*) *em abacaxizeiro*. Abacaxi em foco. Circ. 31, Embrapa Cruz das Almas, Bahia, Brasil.

Roberts, I.M. and Brown, D.J.F. 1980. Detection of six nepoviruses in their nematode vectors by immunosorbent electron microscopy. *Ann. Appl. Biol.* 96: 187-192.

Robinson, T.L., Lakso, A.N., Carpenter, S.G. 1991. Canopy development, yield, and fruit quality of 'Empire' and 'Delicious' apple trees grown in four orchard production systems for ten years. *Journal of the American Society for Horticultural Science* 116: 179-187.

Rocha. L.S., Santana, R.F., Soares, A.C.F. and Haddad, F. 2018. Reaction of cultivars of banana tree to the complex *Meloidogyne javanica* x *Fusarium oxysporum* f. sp. *cubense*. *Revista Caatinga* 31: 572 -583.

Rocha, L.S., Xavier, A.P. and Rebeiro, R.C.F. 2021. Reaction of passion fruit genotypes to the complex *Meloidogyne incognita* and *Fusarium solani*. *Rev. Caatinga, Mossoró* 34(3): 605-613.

Rodriguez-Kabana, R., Robertson, D.G., Wells, L. *et al.* 1991. Cotton as a rotation crop for the management of *Meloidogyne arenaria* and *Sclerotium rolfsii* in peanut. *Suppl J Nematol* 23: 652 –657.

Roh, J.Y., Sim, S.J., Yi, J. *et al.* 2009. Ecotoxicity of silver nanoparticles on the soil nematode *Caenorhabditis elegans* using functional ecotoxicogenomics. *Environ Sci Technol.* 43(10): 3933-3940

Rohrbach, K.G. and Apt, W.J. 1986. Nematode and disease problems of pineapple. *Plant Disease* 70(1): 81-87.

Rom, R.C., Dozier, W.A, Knowles, J.W. *et al.* 1985. *Compact Fruit Tree* 18: 85-91.

Roman, J. and Triantaphyllou, A.C. 1969. Gametogenesis and reproduction of seven species of *Pratylenchus*. *Journal of Nematology* 1 : 357-362.

Roy, S., Ojha, P.K., Ojha, K.L. *et al.* 1998. Integrated management of wilt complex diseases of banana (*Musa* sp.). *J. Applied* Biol. 8: 46–49.

Roy, S., Ojha, P.K., Ojha, K.L. *et al.* 1999. Effects of combined application of fertilizers and organic amendments on the disease intensity of wilt complexes in banana (*Musa* sp.). *J. Appl. Biol.* 9: 84-86.

Saeed, M. 1974. *Studies on Some Stylet–Bearing Nematodes Associated with Sapodilla* (Achras zapota *L.*) *with Special Reference to* Hemicriconemoides mangiferae *Siddiqi, 1961*. Ph. D. thesis, Univ. of Karachi.

Saeed, N., Khan, M.S., Khayyam. *et al.* 2022. Histopathology of walnut (*Juglans regia* L.) root infested with root-knot nematode (*Meloidogyne incognita*). *Journal of Xi'an Shiyou University, Natural Science Edition* 18(12): 1671-1680.

Saeedizadeh, A., Kheiri, A., Okhovvat, S.M. *et al.* 2006. A study on the growth of one-year-old seedlings of olive cv. Zard in the presence of two different soil borne pathogens, *Meloidogyne javanica* and *Verticillium dahliae*. *Iran. J. Agric. Sci.* 37: 793-800.

Saltaren, G., Luis, F., de Varon Agudelo Francia, H. and de la Torre Fernando, M. 1999. Nematodes associated with crops of avocado (*Persea americana* Mill). *Fitopatologia Colombiana* 22: 68–73.

Samra, J.S., Thakur, R.S. and Chadha, K.L. 1978. Comparison of some mango cultivars in terms of their macronutrient status in fruiting and non-fruiting terminals. *Indian J. Hort.* 35: 144-187.

Sasanelli, N., D'Addabbo, T. and Moura Lemos, R. 2002. Influence of *Meloidogyne javanica* on growth of olive cuttings in pots. *Nematropica* 32: 59-63.

Sasanelli, N., Fontanazza, G., Lamberti, F. *et al.* 1997. Reaction of olive cultivars to *Meloidogyne* species. *Nematol. Mediterr.* 25: 183- 190.

Sasser, J.N. and Freckman, D.W. 1987. A world perspective on Nematology: The role of the Society. In J.A. Veech and D.W. Dickson (eds.) *Vistas on Nematology*, pp. 7-14. Society of Nematologists, Hyattsville, Maryland.

Saxena, M. 2018. *Horticultural Statistics at a Glance 2018*. Horticulture Statistics Division, Department of Agriculture, Cooperation & Farmers' Welfare, Ministry of Agriculture & Farmers' Welfare, Government of India, New Delhi.

Sayeed, M., Khan, A., Khatoon, N. *et al.* 2010. Histopathology of mango roots infected by root–knot nematode. *Pak. J. Nematol.* 28: 335–340.

Scherer, A. 2009. *Ocorrência e hospedabilidade de Meloidogyne mayaguensis em goiabeiras e em plantas de cobertura de solo no Paraná (Tese de Doutorado em Agronomia).* Universidade Estadual de Londrina, Londrina, 75pp.

Scotto La Massèse, C., Deportes, L., Mercier, S. and Roger, J. 1984. Les principaux ennemis du figuier: les nematodes et maladies. *Phytoma – Défense des Cultures* 354: 39–41.

Sehgal, M., Chawla, M.L., Singh, S. and Balakrishnan, K. 1990. Studies on interactive effect of moisture and temperature on *Tylenchulus semipenetrans. Current Nematology* 1: 35-36.

Senthilkumar, M., Jonathan, E.I. and Samiyappan, R. 2008a. Bioefficacy of *Pseudomonas fluorescens* on burrowing nematode, *Radopholus similis* in banana. *Indian J. Nematol.* 38: 46-52.

Senthilkumar, M., Jonathan, E.I. and Samiyappan, R. 2008b. Field application of biocontrol agent, *Pseudomonas fluorescens* for the management of burrowing nematode, *Radopholus similis* in banana. *Indian J. Nematol.* 38: 57-61.

Senthilkumar, T. and Rajendran, G. 2003. Studies on wilt disease expression in the presence of *Meloidogyne incognita* and *Fusarium moniliforme* in grapevine. *Nematol. Medit.* 31: 187-189.

Senthil Kumar, T. and Rajendran, G. 2004. Biocontrol agents for the management of disease complex involving root-knot nematode, *Meloidogyne incognita* and *Fusarium moniliforme* on grapevine (*Vitis vinefera*). *Indian Journal of Nematology* 34: 49-51

Seshadri, A.R. 1964. Investigations on the biology and life cycle of *Criconemoides xenoplax* Raski, 1952 (Nematoda: Criconematidae). *Nematologica* 10: 540-562.

Sethi, C.L. and Swarup, G. 1971. Plant parasitic nematodes of North-Western India. III. The genus, *Pratylenchus. Indian Phytopathology* 24: 410-412.

Shanthi, A. 2003. *Management of* Radopholus similis, Helicotylenchus multicinctus *and* Pratylenchus coffeae *in Banana.* Ph. D. thesis, Tamil Nadu Agri. Univ., Coimbatore.

Shanthi, A., Rajendran, G. and Sivakumar, M. 2004. Influence of biofertilizers on lesion nematodes in banana. *National Symp. on Paradigms in Nematol. Res. for Biodynamic Farming,* Univ. of Agri. Sci., Bangalore, p. 97.

Shanthi, A., Rajeswari, S. and Sivakumar, C.V. 1998. Soil application of *Pseudomonas fluorescens* (Migula) for the control of root-knot nematode (*Meloidogyne incognita*) on grapevine (*Vitis vinefera* Linn.). In Usha K. Mehta (ed.) *Nematology – Challenges and Opportunities in 21st Century,* pp. 203-206. Sugarcane Breeding Institute, Coimbatore.

Sharma, B.B., Sharma, H.C. and Khan, E. 1969. Nematodes associated with sweet orange and mandarin plants in relation to die-back (Abstr.). *All India Nematology Symposium,* New Delhi, p. 16.

Sharma, D.D. 1998. Distribution of nematode *(Meloidogyne incognita)* population in affected mulberry gardens. *Intl. J. Tropical Plant Diseases* 16: 73-79.

Sharma D.D. 1999. Root-knot disease of mulberry and its management. *Indian Farming* 49: 20-24.

Sharma, D.D., Chandrashekar, D.S., Gunasekhar, V. *et al.* 2001. Comparative efficacy of different control measures against root knot nematode disease of mulberry. *Indian J Serie.* 40: 151-157.

Sharma, D.D., Chandrashekar, D.S., Srikantaswamy, K. and Govindaiah. 1998. On-farm evaluation of cultural and chemical methods for the control of root-knot nematode disease of mulberry. *Indian J. Serie.* 37: 61-63.

Sharma, D.D. and Sarkar, A. 1998. Incidence and intensity of species/races of root knot nematode associated with mulberry under different farming systems and soil types in Mysore region, Karnataka State, India. *Indian J. Serie.* 37: 137-141.

Sharma, N.K. and Sharma, S.K. 1977. Spatial distribution of *Tylenchulus semipenetrans* in citrus rhizospheres (Abstr.). *Proceedings of the International Symposium on Citriculture*, Bangalore, p. 37.

Sharma, R.D. 1978. Nematodes associated with avocado (*Persea americana* Mill.) in Cerrado soils. *Societa Brasileira Nematologia Publicaciones* 3: 65–70.

Sharon, M., Choudhary, A.K. and Kumar, R. 2010. Nanotechnology in agricultural diseases and food safety. *Journal of Phytology* 2(4): 83-92.

Sharpe, R.H., Hesse, C.O., Lownsbery, B.F. *et al.* 1969. Breeding peaches for root-knot nematode resistance. *J. Amer. Soc. Hort. Sci.* 94: 209-212.

Sharpe, R.R., Reilly, C.C., Nyczepir, A.P. and Okie, W.R. 1989. Establishment of peach in a replant site as affected by soil fumigation, rootstock, and pruning date. *Plant Disease* 73: 412-415.

Shawky, S.M., El-Sawi, M.A.M. and Nabil, Y.M. 2004. Effect of some nematicides on controlling root knot nematodes and determine their residues in olive fruits and soil. *Ann. Agric. Sci. (Cairo)* 49: 389-399.

Shelke, S.S. and Darekar, K.S. 2000. Reaction of pomegranate germplasm to root-knot nematode. *Journal of Maharashtra Agricultural Universities* 25(3): 308-310.

Shelton, M.D., and Edwards, C.R. 1983. Effects of weeds on the diversity and abundance of insects in soybeans. *Environmental Entomology* 12: 296–298.

Shende, D.G. 1977. *Effect of N, P and K on Growth, Yield and Quality of Pomegranate* (Punica granatum *L.*). M. Sc. (Ag.) thesis, Mahtma Phule Agri. Univ., Rahuri, Maharashtra.

Shepard, D.P., Zehr, E.I. and Bridges, W.C. 1999. Increased susceptibility to bacterial spot of peach trees growing in soil infested with *Criconemella xenoplax*. *Plant disease* 83: 961-963.

Sher, S.A. 1955. Nematodes attacking avocado. *California Citrograph* 40: 198.

Sher, S.A., Foote, F.J. and Boswell, S.B. 1959. A root lesion nematode disease of avocados. *Plant Disease Reporter* 43: 797–800.

Sher, S.A. and Raski, D.J. 1956. *Heterodera fici* Kirjanova 1954 in California. *Plant Disease Reporter* 40: 700.

Sherman, W.B., Lyrene, P.M. and Sharpe, R.H. 1991. Flordaguard peach rootstock. *HortScience* 26: 427–428.

Shirley, A.M. 2013. *Management of Plant-Parasistic Nematodes on Peach Utilzing Postplant Nematicides and Crop Rotations*. Ph. D. thesis, The University of Georgia, Athens, Georgia.

Siddiqi, K.S. and Husen, A. 2017. Plant response to engineered metal oxide nanoparticles. *Nanoscale Res Lett* 12: 92.

Siddiqi, M.R. 1961. Occurrence of the citrus nematode, *Tylenchulus semipenetrans* Cobb. 1913 and the reniform nematode, *Rotylenchulus reniformis* in India. *Proceedings of 48th Indian Science Congress* 3: 504.

Siddiqi, M.R. 1964. Studies on nematode root-rot of citrus in Uttar Pradesh, India. *Proceedings of Zoological Society, Calcutta* 17: 67-75.

Siddiqi, M.R. 1977. *Hemicriconemoides mangiferae*. C.I.H. Descriptions of Plant Parasitic Nematodes, Set 7, No. 99, Commonw. Agri. Bureau, London, 4 pp.

Siddiqui, M.W., Ayala-Zavala, J.F. and Dhua, R.S. 2015. Genotypic variation in tomatoes affecting processing and antioxidant attributes. *Critical Review in Food Science and Nutrition* 55: 1819-1835.

Siddiqui, M.W. and Dhua, R.S. 2010. Eating artificially ripened fruits is harmful. *Current Science* 99(12): 1664-1668.

Siddiqui, M.W., Momin, C.M., Acharya, P. *et al.* 2013. Dynamics of changes in bioactive molecules and antioxidant potential of *Capsicum chinense* Jacq. cv Habanero at nine maturity stages. *Acta Physiologea Plantarum* 35 (4): 1141-1148.

Siddiqui, Z.A. and Khan, M.W. 1986. A survey of nematodes associated with pomegranate in Libya and evaluation of some systemic nematicides for their control. *Pakistan Journal of Nematology* 4: 83-90.

Siddiqui, Z.A. and Mahmood, I. 1996. Biological control of plant parasitic nematodes by fungi: A review. *Bioresour Technol* 58: 229–239.

Silva, G.S. and Pereira, A.L. 2008. Efeito da incorporação de folhas de nim ao solo sobre o complexo *Fusarium* x *Meloidogyne* em quiabeiro. *Summa phytopathologica* 34: 368-370.

Silveira, A., and Araújo, D. D. 2011. Nematoses em Passiflora spp. In M.M. Pires, A.B.S. José, and A.O. Conceição (eds.) *Maracujá – Avanços Técnológicos e Sustentabilidade*, pp. 147–173. Editora da UESC, Ilhéus.

Simeone, A.M. and Di Vito, M. 1992. Reactions to nematodes of selections of peach rootstocks. *Acta Horticulturae* 315: 197-202.

Singh, A., Kaul, R.K., Khapte, P.S. *et al.* 2022. Root-knot nematode presence and its integrated management in pomegranate orchards located in Indian arid areas. *Horticulturae* 8: 160. https://doi.org/10.3390/horticulturae8020160.

Singh, N. 2020. Emerging problem of guava decline caused by *Meloidogyne enterolobii* and *Fusarium oxysporum* f. sp. *psidii*. *Indian Phytopathol.* 73(2): 373–374.

Singh, N.P. 1981. Studies on guava nutrition with special reference to leaf analysis. *National Symp. Trop. Subtrop. Fruit Crops,* Bangalore, p. 75.

Singh, R.S. and Sitaramaiah, K. 1973. *Control of Plant Parasitic Nematodes with Organic Amendments of Soil.* Final technical report on effect of organic amendments, green manuring and inorganic fertilizers on root-knot of vegetable crops, G.B. Pant Univ. of Agri. & Technol., Bull. No. 6, 289 pp.

Singh, S., Dhawan, S.C. and Goswami, B.K. 2007. Histopathology of cowpea roots co-infected with root-knot nematode, *Meloidogyne incognita* with wilt fungus, *Fusarium oxysporum. Indian J. Nematol.* 37: 156-160.

Singh, Shyam. 1997. Status of citrus decline in India: A review. In: *Nat. Symp. Citriculture, Souvenir,* pp. *42-55.*

Singh, S.H., Ved Ratan, Gaur, G.S. and Gangwar, D. 2003. Evaluation of antagonists against guava wilt pathogen. *Farm Sci. J.* 12: 58-59.

Singh, T., Prajapati, A., Maru, A.K. *et al.* 2019. Root-knot nematodes (*Meloidogyne* spp.) infecting pomegranate: A review. *Agricultural Reviews* 40(4): 309-313.

Sipes, B.S, Caswell-Chen, E.P., Sarah, J.L. and Apt, W.J. 2005. Nematode parasites of pineapple. In M. Luc, R. A. Sikora, and J. Bridge (eds.) *Plant Parasitic Nematodes in Subtropical and Tropical Agriculture*, pp. 709-731. CABI Publishing, Wallingford, UK.

Siqueira, K.M.S., Freitas, V.M., Almeida, M.R.A. *et al.* 2009. Detecção de *Meloidogyne mayaguensis* em goiabeira e mamoeiro no estado de Goiás, usando marcadores moleculares. *Trop. Plant Pathol.* 34: 256-260.

Sivakumar, C.V. and Seshadri, A.R. 1972. Histopathology of infection by the reniform nematode, *Rotylenchulus reniformis* Linford and Oliveira, 1940 on castor, papaya and tomato. *Indian Journal of Nematology* 2: 173-181.

Sivakumar, M. and Vadivelu, S. 1999. Management of *Meloidogyne incognita* on grapevine using biocontrol agents, botanicals and biofertilizers. *Pest Mangmt. Hort. Ecosystems* 5: 127-131.

Smith, P.T. and Cobb, B.C. 1991. Physiological and enzymatic activity of pepper seeds during priming. *Physiologia Plantarum* 83: 433-439.

Soltani, T., Nejad, R.F., Ahmadi, A.R. and Fayazi F. 2013. Chemical control of root-knot nematode (*Meloidogyne javanica*) on olive in the greenhouse conditions. *J Plant Pathol Microb* 4: 183 doi:10.4172/2157-7471.1000183.

Somasekhara, Y.M. and Wali, S.Y. 1999. New record of *Ceratocystis fimbriata* causing wilt of pomegranate. *Plant Dis.* 83 (4): 400.

Somasekhara, Y.M., Wali, S.Y. and Shaik, M.K. 2009. Studies and the management of pomegranate *(Punica granatum)* wilt *(Ceratocystis fimbriata,* Latin American group*).* Paper Presented In: *2nd Intn. Symp. Pomegranate and Minor Fruits including Mediterranean Fruits.* Univ. Agric. Sci., Dharwad, pp. 132-133.

Sonyal, S. 2010. *Studies on Pomegranate Wilt Complex.* M. Sc. (Agri.) thesis, Dept. of Plant Path., Univ. of Agri. Sci., Dharwad, India.

Sosamma, V.K., Koshy, P.K. and Samuel, R. 1998. Effect of vesicular arbuscular mycorrhizae, *Glomus mosseae* on nematodes infecting banana under coconut based high density multispecies cropping system. In Usha K. Mehta (ed.) *Nematology – Challenges and Opportunities in 21st Century*, pp. 94-97. Sugarcane Breeding Institute, Coimbatore.

St. Laurent, A., Merwin, I.A., and Thies, J.E. 2008. Long-term orchard groundcover management systems affect soil microbial communities and apple replant disease severity. *Plant Soil.* 304: 209–225

Stapleton, J.J., DeVay, J.E. and Elmore, C.L. (eds.). 1998. *Soil Solarization and Integrated Management of Soil-borne Pests.* Plant Production and Protection Paper 147. FAO/UN, Rome, pp. 78-88.

Stellmach, G. and Goheen, A.C. 1988. Other virus and virus-like diseases. In R.C. Pearson and A.C. Goheen (eds.) *Compendium of Grape Diseases*, pp 53-54. American Phytopathological Society Press, St Paul, Minnesota. 93 pp.

Stirling, G.R. 1991. *Biological Control of Plant-parasitic Nematodes: Progress, Problems and Prospects*. CAB International, Wallingford, UK, 282 pp.

Stirling, G.R., McKenry, M.V. and Mankau, R. 1979. Biological control of root-knot nematode (*Meloidogyne* spp.) on peach. *Phytopathology* 69: 806-809.

Stone, O.M., Brunt, A.A. and Hollings, M. 1975. The production, propagation and distribution of virus-free clones of *Narcissus tazetta* cv. Grand Soleil d'Or. *Acta Horticulturae* 47: 77-81.

Strobel, N.E., Hussey, R.S. and Roncadori, R.W. 1982. Interactions of vesicular-arbuscular mycorrhyzal fungi, *Meloidogyne incognita*, and soil fertility on peach. *Phytopathology* 72: 690-694.

Subbotin, S.A., Mundo-Ocampo, M. and Baldwin, J.G. 2010. Description and diagnosis of *Heterodera* species. In *Systematics of Cyst Nematodes (Nematoda: Heteroderinae)*, Vol. 8B: 35–449. Koninklijke Brill NV, https://doi.org/10.1163/ej.9789004164345.i-512.14.

Suit, R.F. and DuCharme, E.P. 1953. The burrowing nematode and other parasitic nematodes in relation to spreading decline of citrus. *Plant Disease Reporter* 37: 379-383.

Sukumar, J., Padma, S.D. and Bongale, U.D. 2004. Biological control of mulberry root-knot nematode *Meloidogyne incognita* by *Trichoderma harzianum*. *Indian J. Indust. Entomol.* 8(2): 175-179.

Sukumar, J., Padma, S.D., Prasad, K.V. and Bongale, U.D. 2000. Studies on some aspects of black root rot disease of mulberry caused by *Lasiodiplodia theobromae*. In K.P. Chinnaswamy, R. Govindan, N.K. Krishna Prasada and D.N.R. Reddy (eds.) *Moriculture in Tropics*, pp.121-123. University of Agricultural Sciences, GKVK. Bangalore, India.

Sun, F., Henry, N. and Yu, Q. 2017. First report of the fig cyst nematode, *Heterodera fici* Kirjanova, on fig tree, *Ficus carica*, in Ontario, Canada. *Journal of Nematology* 49: 131–132.

Sundarababu, R., Mani, M.P. and Sivakumar, C.V. 1999. A new biocontrol agent for plant parasitic nematodes. *Indian Horticulture* 12: 25.

Sundararaju, P., Cannayane, I. and Sathiamoorthy, S. 2004. *Nematode Management in Banana and Plantains*. Tech. Bull. No. 10, National Res. Centre for Banana, Thiruchirapalli, Tamil Nadu, India, 14 pp.

Sundararaju, P., Cannayane, I. and Thangavelu, R. 2002. Biomanagement of *Pratylenchus coffeae* and *Meloidogyne incognita* on banana by using endophytic fungi. *Global Conf. on Banana and Plantain*, Bangalore, p. 191.

Sundararaju, P. and Kumar, V. 2003. Management of root-lesion nematode, *Pratylenchus coffeae* in six commercial cultivars of banana through organic and inorganic amendments. *Infomusa* 12: 21-25.

Sundararaju, P. and Thangavelu, R. 2009. Influence of *Pratylenchus coffeae* and *Meloidogyne incognita* on the Fusarium wilt complex of banana. *Indian J. Nematol.* 39: 71-74.

Swamy, B.C.N. and Govindu, H.C. 1965. A preliminary note on the plant parasitic nematodes of the Mysore state. *Indian Phytopathol.* 19: 233-240.

Swamy, B.C.N., Rajagopal, B.K. and Anahosur, K.H. 1973. Citrus nematode, a new record from Mysore State. *Mysore Journal of Agricultural Sciences* 7: 138-139.

Swarup, G. and Seshadri, A.R. 1974. Nematology in India - Problems and Progress. In *Current Trends in Plant Pathology* (S.P. Raychaudhuri and J. P. Verma, eds.), pp. 303-311. Dept. of Botany, Univ. of Lucknow, Lucknow.

Takács-György, K., Turek, R., Maria, M. and Takács, I. 2014. Sustainable new agricultural technology – economic aspects of precision crop protection. *1st International Conference 'Economic Scientific Research - Theoretical, Empirical and Practical Approaches'*, ESPERA 2013. www.sciencedirect.com.

Talavera, M., Miranda, l., Gómez-Mora, J.A. *et al.* 2019. Nematode management in the strawberry fields of Southern Spain. *Agronomy* 9: 252; doi:10.3390/agronomy 9050252.

Taylor, A.G. and Harman, G.E. 1990. Concept and technologies of selected seed treatments. *Annu Rev Phytopathol* 28: 321-339.

Taylor, A.L., Sasser, J.N. and Nelson, L.A. 1982. *Relationship of Climate and Soil Characteristics to Geographical Distribution of* Meloidogyne *Species in Agricultural Soil.* Co-op. Pubn., Dept. of Plant Path., North Carolina State Univ. & U.S. Agency for Intn. Devp., Raleigh, N.C., 65 pp.

Taylor, C.E. 1962. Transmission of raspberry ringspot virus by *Longidorus elongatus* (de Man) (Nematoda: Dorylaimidae). *Virology* 17: 493-494.

Taylor C.E. and Robertson, W.M. 1975. Acquisition, retention and transmission of viruses by nematodes. In F. Lamberti, C.E. Taylor and J.W. Seinhorst (eds.) *Nematode Vectors of Plant Viruses*, pp 253-276. Plenum Press, London, 460 pp.

Taylor, J., Briesbrock, J.A., Hendrix, Jr., F.F. *et al.* 1970. *Peach Tree Decline in Georgia.* Georgia Agri. Expt. Stn. Res. Bulletin 77.

Teotia, R.S., Mandal, S.K. and Sen, S.K. 1992. Can root knot nematodes in mulberry be controlled. *Indian Silk,* 30: 6-9.

Terblanche, J.H., Grech, N., Frean, R. *et al.* 1986. *Good News for Passionfruit Industry.* Citrus and Subtropical Fruit Research Institute Information Bulletin No. 164, pp. 1–2, 4–5.

Tewoldemedhin, Y.T., Mazzola, M., Botha, W.J. *et al.* 2011a. Characterization of fungi (*Fusarium* and *Rhizoctonia*) and oomycetes (*Phytophthora* and *Pythium*) associated with apple orchards in South Africa. *Eur J Plant Pathol.* 130: 215–229.

Tewoldemedhin, Y.T., Mazzola, M., Labuschagne, I. and McLeod, A. 2011b. A multi-phasic approach reveals that apple replant disease is caused by multiple biological agents, with some agents acting synergistically. *Soil Biol Biochem.* 43: 1917–1927.

Tewoldemedhin, Y.T., Mazzola, M., Mostert, L. and McLeod, A. 2011c. *Cylindrocarpon* species associated with apple tree roots in South Africa and their quantifcation using real-time PCR. *Eur J Plant Pathol.* 129: 637–651.

Thirumala Rao, V. 1956. Stray notes on some crop outbreaks of South India. *Indian Journal of Entomology* 18: 123-126.

Tibugari, H., Mombeshora, D., Mandumbu, R. *et al.* 2012. A comparison of the effectiveness of the aqueous extracts of garlic, castor beans and marigold in the biocontrol of root-knot nematode in tomato. *J Agric Tech* 8(2): 479–492.

Tsuchizaki, T. 1975. *Mulberry Ring Spot Virus.* CMI / AAB Descriptions of Plant Viruses, No. 142.

Tsuchizaki, T., Hibino, H. and Saito, Y. 1971. Mulberry ringspot virus isolated from mulberry showing ringspot symptom. *Ann. phytopath. Soc. Japan* 37(4): 266-271.

Umesh, K.C., Krishnappa, K. and Bagyaraj, D.J. 1988. Interaction of burrowing nematode, *Radopholus similis* and VA mycorrhiza, *Glomus fasciculatum* in banana. *Indian Journal of Nematology* 18: 6-11.

Utkhede, R.S. and Smith, E.M. 2000. Impact of chemical, biological and cultural treatments on the growth and yield of apple in replant-disease soil. *Australas Plant Pathol.* 29: 129–136.

Utkhede, R.S., Vrain, T.C. and Orstan, M. 1992. Effect of nematodes, fungi and bacteria on the growth of young apple trees grown in apple replant disease soil. *Plant Soil* 139: 1-6.

Uyemoto, J.K. 1975. A severe outbreak of virus-induced grapevine decline in cascade grapes in New York. *Plant Disease Reporter* 59: 98-101.

Vadivelu, R.S., Rajendran, G., Naganathan, T.G. and Jayaraj, S. 1987. Studies on the nematode pests of banana and their management - A review. *Third Group Discussion on the Nematol. Problems of Plantation Crops*, Sugarcane Breeding Institute, Coimbatore, pp. 32-33.

Van Gundy, S.D. 1957. The first report of a species of *Hemicycliophora* attacking citrus roots. *Plant Disease Reporter* 41: 1016-1018.

Van Gundy, S.D. 1958. The life history of the citrus nematode *Tylenchulus semipenetrans. Nematologica* 3: 283-294.

Van Gundy, S.D. 1959. The life history of *Hemicycliophora arenaria* Raski (Nematoda: Criconematidae). *Proceedings of Helminthological Society of Washington* 26: 67-72.

Van Gundy, S.D. and McElroy, F.D. 1969. Sheath nematode, its biology and control. *Proceedings of Ist International Citrus Symposium, Vol. II*: 985-989.

Vanitha, P., Poornima, K. and Somasundaram, E. 2018. Cover cropping in guava orchards to reduce population of root knot nematode. *Meloidogyne enterolobii. Pest Mgmt. Hort. Ecosys.* 24: 53–57.

Vargas, R., Rodriquez, A. and Acosta, N. 1996. Components of nematode suppressive activity of velvet bean, *Mucuna deeringiana. Nematropica* 26: 323.

Venkitesan, T.S. 1976. *Studies on The Burrowing Nematode* Radopholus similis *(Cobb, 1893). Thorne, 1949 on Pepper* (Piper nigrum *L.) and its Role in Slow Wilt Disease*. Ph.D. thesis, Univ Agri Sci, Bangalore, 122 pp.

Verdejo, S., Calvet, C. and Pinochet, J. 1990. Effect of mycorrhiza on kiwi infected by the nematodes *Meloidogyne hapla* and *M. javanica. Bulletin de Sanidad Vegetal. Plagas* 16: 619–624.

Verdejo-Lucas S. 1992. Seasonal population fluctuations of *Meloidogyne* spp. and the *Pasteuria penetrans* group in kiwi orchards. *Plant Disease* 76: 1275-1279.

Verma, M.K. 2009. Annual Report, Central Institute of Temperate Horticulture, Srinagar

Verma, R.R. 1987. Reaction of some peach cultivars to *Meloidogyne incognita. Indian J. Nematol.* 17: 123-124.

Vidya, K. and Reddy, B.M.R. 1998. Integrated management of *Radopholus similis* infecting banana using plant product, nematicide and biocontrol agents. *Mysore J. Agric. Sci.* 32: 186-190.

Vijaya Kumari, N. and Sujathamma, P. 2014a. Integrated management of root-knot nematode *Meloidogyne incognita* Chitwood in mulberry (*Morus* spp.). *J. Pharmacy Biol. Sci.* 9(1): 30-33.

Vijaya Kumari, N. and Sujathamma, P. 2014b. Root-knot nematode infestation on mulberry (*Morus* spp.). *Intn. Journal of Advances in Agricultural & Environmental Engg.* (IJAAEE) 3(1): 146-149.

Villiers, E.A. and Milne, D.L. 1973. The control of nematodes on passion fruit. *Citrus and Sub-Tropical Fruit Journal* 480: 7-13.

Villota, F.J. and Agudelo, F.V. 1997. Evaluation of guava material (*Psidium guajava* L.) for the damage behaviour of *Meloidogyne incognita. Fitopatologia Colombiana* 21: 31- 38.

Vovlas, N. and Inserra, R.N. 1976. Istopatologia di radici di olivo infestate da *Rotylenchulus macrodoratus* Dasgupta, Raski et Sher. *Nematol. Mediterr.* 4: 223-230.

Vovlas, N., Inserra, R.N. and O'Bannon, J.H. 1989. *The Fig Cyst Nematode*, Heterodera fici. Nematology Circular No. 168. Florida Department Agriculture and Consumer Services, Division of Plant Industry, 3 pp.

Vrain, T., Deyoung, R., Hall, J. and Freyman, S. 1996. Cover crops resistant to root-lesion nematodes in raspberry. *HortScience* 31(7): 1195-1198.

Vrain, T.C. 1993. Restriction fragment length polymorphism separates species of the *Xiphinema americanum* group. *Journal of Nematology* 25: 361-364.

Vrain, T.C. and Copeman, R.J. 1987. Interactions between *Agrobacterium tumefaciens* and *Pratylenchus penetrans* in the roots of two red raspberry cultivars. *Can. J. Plant Pathol.* 9: 236-240.

Vrain, T.C. and Pepin, H.S. 1989. Effect of *Pratylencus penetrans* on root rot of red raspberry caused by *Phvtophthora eythroseptica. Acta Hort.* 262: 231-240.

Walia R.K. and Bajaj H.K. 2003. *Textbook on Introductory Plant Nematology*. Indian Council of Agricultural Research, New Delhi, 227 pp.

Walia, R.K. and Chakraborty, P.K. (eds.). 2018. *Nematodes Problems of Crops in India.* ICAR-All India Coordinated Research Project on Nematodes in Agriculture, New Delhi, 402 pp.

Walker, G.E. 1997. Effects of *Meloidogyne* spp. and *Rhizoctonia solani* on the growth of grapevine rootings. *J. Nematol.* 29: 190-198.

Walker, M.A., Wolpert, J.A. and Weber, E. 1994. Field screening of grape rootstock selections for resistance to fanleaf degeneration. *Plant Dis.* 78: 134-136.

Wallace, H.R. 1963. *The Biology of Plant Parasitic Nematodes*. E. Arnold Ltd., London, 282 pp.

Walters, T.W., Pinkerton, J.N., Riga, E. 2009. Managing plant–parasitic nematodes in established red raspberry fields. *Hortechnology* 19(4): 762-768.

Walunj, A.R. 2013. *Seasonal Incidence and Biological Management of Root-Knot Nematode,* Meloidogyne incognita *(Kofoid and White, 1919) Chitwood, 1949 Infesting Pomegranate,* Punica granatum *L.* Ph. D. dissertation submitted to M.P.K.V., Rahuri-413 722 (MS).

Wang, K.H., Cerruti, R.H. and Antoon, P. 2007. Protecting crops from nematode pests: Using marigold as an alternative to chemical nematicides. *Plant Disease*, UH-CTAHR, PD-35: 1-6.

Wang, K.H. and Sipes, B.S. 2009. *Solarization and Cover Cropping as Alternatives to Soil Fumigants for Nematode Management in Hawaii's Pineapple Fields.* College of Tropical Agriculture and Human Resources, University of Hawaii at Mānoa, Honolulu, Hawaii.

Wang, K.H., Sipes, B.S. and Schmitt, D.P. 2001. Suppression of *Rotylenchulus reniformis* by *Crotalaria juncea, Brassica napus*, and *Tagetes erecta*. *Nematropica* 31: 235–249.

Wang, K.H., Sipes, B.S. and Schmitt D.P. 2002. Management of *Rotylenchulus reniformis* in pineapple, *Ananas comosus*, by intercycle cover crops. *J. Nematol.* 34(2): 106–114.

Warner, F. 2018. *Nematode Problems on Strawberries.* https://pestid.msu.edu/ nematodes-problems-on-strawberries/. Accessed 7 Dec 2018.

Watkins, R. 1976. Cherry, plum, peach, apricot and almond. *Prunus* spp. In Simmonds, N.W. (ed.), *Evolution of Crop Plants*, pp. 242-247. Longman, London, UK.

Webster, A.D., Smith, R.A. and Watkins, R. 1986. Apple rootstock studies. I. Preliminary evaluations of several MM.106 × M.27 hybrids. *J Hortic Sci.* 61: 429–437

Wehunt, E.J., Horton, B.D. and Prince, V.E. 1980. Effect of nematicides, lime and herbicides on peach tree short life in Georgia, U.S.A. *J. Nematol.* 12: 183-189.

Weisz, R., Fleischer, S. and Smilowitz, Z. 1996. Site-specific integrated pest management for high-value crops: impact on potato pest management. *Journal of Economic Entomology* 89: 501–509.

Weller, D.M. 1983. Colonization of wheat roots by a fluorescent pseudomonad suppressive to take all. *Phytopatholgy* 73:1548-1553.

Westerdahl, B.B. 2003. *UC IPM Pest Management Guidelines: Avocado. UCPMG-Publication 3436.* Division of Agriculture and Natural Resources, University of California (ANR Publications), Oakland, California. Available at: www.ipm.ucdavis. edu/PMG/ r261200111.html.

Westerdahl B.B. and Duncan R.A. 2015. Peach nematodes. UC Pest Management Guidelines (Updated 9/15). UCIPM University of California Agriculture and Natural Resources, State-Wide Integrated Pest Management Program. (accessed on 26 December 2022). Available online: http://ipm.ucanr.edu/ MG/ r602200111.html.

Westerdahl, B.B., Giraud, D., Etter, S. *et al.* 2003. Management options for *Pratylenchus penetrans* in Easter lily. *J. Nematol.* 35: 443-449.

Westerdahl, B.B., Westphal, A. and Hasey, J.K. 2017. *Walnut Nematodes. UC Pest Management Guidelines (updated 6/17).* UCIPM University of California Agriculture and Natural Resources, State-wide Integrated Pest Management Program. http://ipm.ucanr.edu/ PMG/ r881200111.html.

Whitehead, A.G. 1968. Taxonomy of *Meloidogyne* (Nematodea: Heteroderidae) with descriptions of four new species. *Transactions of Zoological Society of London* 31: 263-401.

Whiting, M.D., Lang, G. and Ophardt D. 2005. Rootstock and training system affect sweet cherry growth, yield, and fruit quality. *HortScience* 40(3): 582-586.

Wildmer, T.L., Mitkowski, N.A. and Abawi, G.S. 2002. Soil organic matter and management of plant parasitic nematodes. *Journal of Nematology* 34: 289-295.

Willers, P. 1999. Do nematodes play a role in avocados? *Neltropika Bulletin* 306: 7–8.

Willers, P., Daneel, M.S. and De Jager, K. 2001. Banana. In: M.A. Van den Berg, E.A. De Villiers and P.H.

Joubert (eds.) *Pest and Beneficial Arthropods of Tropical and Non-Citrus Subtropical Crops in South Africa*, pp. 34-43. Agricultural Research Council–Institute for Tropical and Subtropical Crops, Nelspruit.

Willett, M., Smith, T.J., Peterson, A.B, *et al.* 1994. Growing profitable apple orchards in replant sites: an interdisciplinary team approach in Washington State. *Horttechnology* 4: 175–181.

Wohlfarter, M., Giliomee, J.H., Venter, E. and Storey, S. 2011. A survey of the arthropod pests and plant parasitic nematodes associated with commercial figs, *Ficus carica* (Moraceae), in South Africa. *African Entomology* 19: 165–172.

Wunsche, J.N. and Lasko, A.N. 2000. The relationship between leaf area and light interception by spur and extension shoot leaves and apple orchard productivity. *HortScience* 35 (7): 1202–1206.

Wyss, U. 1987. Video assessment of root cell response to Dorylaimid and Tylenchid nematodes. In: J.A. Veetch and D.W. Dickson (eds.) *Vistas on Nematology*, pp. 211–220. Society of Nematologists, Inc., Hyattsville, Maryland.

Wyss, U., Lehmann, H. and Jank-Ladwig, R. 1980. Ultrastructure of modified root-tip cells in *Ficus carica*, induced by the ectoparasitic nematode Xiphinema index. *Journal of Cell Science* 41: 193–208.

Xia, Y., Li, S., Liu, X. *et al.* 2019. *Bacillus halotolerans* strain LYSX1-induced systemic resistance against the root-knot nematode *Meloidogyne javanica* in tomato. *Ann Microbiol* 69(12): 1227–1233.

Xu, J., Liu, P., Meng, Q. and Long, H. 2004. Characterization of *Meloidogyne* species from China using isozyme, phenotypes and amplified mitochondrial DNA restriction fragment length polymorphism. *Eur. J. Plant Pathol.* 110: 309-315.

Yagita, H. and Komuro, Y. 1972. Transmission of mulberry ringspot virus by *Longidorus martini* Merny. *Ann. phytopath. Soc. Japan* 38: 275-283.

Yang B, and Eisenback JD. 1983. *Meloidogyne enterolobii* n. sp. (Meloidogynidae), a root-knot nematode parasitizing Pacara Earpod tree in China. *J. Nematol.* 15: 381–391.

Yang, I.L., Deng, T.C. and Chen, M.J. 1986. Sap transmissible viruses associated with grapevine yellow mottle disease in Taiwan. *Journal of Agricultural Research of China* 35: 504-510.

Yao, S., Merwin, I.A., Abawi, G.S. and Thies, J.E. 2006. Soil fumigation and compost amendment alter soil microbial community composition but do not improve tree growth or yield in an apple replant site. *Soil Biol Biochem.* 38: 587–599.

Yassin, A.M. 1974. A note on *Longidorus* and *Xiphinema* species from the Sudan. *Nematol. Medit.* 2: 141-147.

Young, T.W. and Ruehle, G.D. 1955. The role of burrowing and meadow nematodes in avocado decline. *Plant Disease Reporter* 39: 815–817.

Yuksel, H.S. 1981. *Heterodera fici* Kirjanova 1954 in Aaegean Region. *Journal of Turkish Phytopathology* 10: 45–51.

Zasada, I.A., Walters, T.W. and Pinkerton, J.N. 2010. Post-plant nematicides for the control of root lesion nematode in red raspberry. *Hortechnology* 20(5): 856-862.

FSC
www.fsc.org
MIX
Papier | Fördert gute Waldnutzung
FSC® C083411

Kolibri 360 GmbH
Ferdinand-Jühlke-Straße 7
99095 Erfurt, Deutschland
produktsicherheit@kolibri360.de